Routledge Revivals

Britain 1740-1950

Originally published in 1992, this book provides students with a well-illustrated, clearly written text which offers a coherent overview of Britain's development from a pre-modern to a modern economy and society. The key processes that have shaped the geography of modern Britain are rooted in the significant demographic, economic, technological and social transitions of the early eighteenth century, the impact of which was not fully diffused through the nation until the mid-20th Century. This country-wide survey examines the nature of this transformation. The material in the book is accessible because the book is clearly structured into 3 phases: 1740 to the 1830s; the 1830s to the 1890s and the 1890s to 1950. For each period, the principal aspects of change in population, industry, the countryside and urban life are examined, and regional examples given to support the analysis.

Britain 1740-1950

An Historical Geography

Richard Lawton and Colin G. Pooley

Routledge Revivals
Routledge
Taylor & Francis Group

First published in 1992
by Edward Arnold

This edition first published in 2021 by Routledge
2 Park Square, Milton Park, Abingdon, Oxon, OX14 4RN
and by Routledge
605 Third Avenue, New York, NY 10017

Routledge is an imprint of the Taylor & Francis Group, an informa business

Publisher's Note
The publisher has gone to great lengths to ensure the quality of this reprint but points out that some imperfections in the original copies may be apparent.

Disclaimer
The publisher has made every effort to trace copyright holders and welcomes correspondence from those they have been unable to contact.

ISBN 13: 978-1-032-00582-9 (hbk)
ISBN 13: 978-1-003-17478-3 (ebk)
ISBN 13: 978-1-032-00584-3 (pbk)

DOI: 10.4324/9781003174783

BRITAIN 1740–1950

An Historical Geography

Richard Lawton

and

Colin G. Pooley

Edward Arnold
A division of Hodder & Stoughton
LONDON NEW YORK MELBOURNE AUCKLAND

First published in Great Britain 1992

Distributed in the USA by Routledge, Chapman and Hall, Inc.
29 West 35th Street, New York, NY 10001

British Library Cataloguing in Publication Data

Lawton, Richard
Britain 1740–1950: An historical geography.
I. Title II. Pooley, Colin G.
941

ISBN 0-7131-6550-2

Typeset in 10/11pt Palatino by Rowland Phototypesetting Limited, Bury St Edmunds, Suffolk. Printed and bound in Great Britain for Edward Arnold, a division of Hodder and Stoughton Limited, Mill Road, Dunton Green, Sevenoaks, Kent TN13 2YA by Biddles Limited, Guildford and King's Lynn

Contents

List of tables

List of figures

Preface

A brief account of long-term change in a complex economy and society, with considerable regional diversity in environment, landscape and culture, poses considerable problems of organization. Our overall aim has been to provide students on degree courses in geography and related subjects with a balanced overview of the transformation of Britain from a pre-modern to a modern economy and society. Those terms themselves raise fundamental questions about the nature of such transformations, the processes of structural change in British society and its place within an international economy and polity at which we have been able to only hint in specific contexts and in brief overviews of the three distinctive phases of development around which the text is organized (Chapters 2, 7 and 12).

The time period covered, 1740 to 1950, has been adopted in the belief that whilst key processes that shaped the geography of modern Britain are rooted in significant economic, demographic, social and technological changes from the early eighteenth century, their impacts on the scale and location of activity were not fully diffused through the nation and its regions until the mid-twentieth century. Until relatively recently it was customary to encapsulate many of the complex interactions in these changes as a move from a 'pre-industrial' to an 'industrial' society. Nowadays the idea of transforming technological and economic revolutions in agriculture, industry, transport and commerce in the eighteenth and early nineteenth centuries is largely rejected in favour of the concept of more gradual transitions extending over longer time periods and varying from sector to sector and region to region.

Our choice of a starting point is determined principally by an acceleration in the growth of the national economy and related changes in population development from around 1740, and the end point by our belief that many of these economic and demographic transitions did not extend throughout society until after the Second World War. Although manifesting themselves in different ways, such long-run and gradual changes can be seen not only in economic and social development, but also in demographic change, the evolution of landscapes and the emergence of new regional structures. The details of these changes form the substance of this book.

However, the two centuries covered in this volume are not a seamless garment: Britain's garb changed in material, design and mode of manufacture. The three sections into which the text is divided – 1740 to 1830; 1830 to 1890; and 1890 to 1950 – are related to what we see as significant turning points in a number of aspects of economic, social and geographical change. They coincide, broadly, with long cycles in economic development identified by many economic historians (Chapter 1). They represent distinctive, if not always concordant, phases of technological change over a wide spectrum of economic activities. They reflect periods of change in social and political affairs, and in the balance between urban and rural society and regions of growth and decline. They are not neatly arranged around dates of major sources of data, convenient snapshots of Britain, or the sometimes illusory symbols of major political or international events.

Our definition of the subject matter of historical geography is broad: it encompasses the belief that geography is fundamentally concerned with both spatial processes (and resulting patterns) and the relationships between people and the environment in which they lived. Both themes are a strong thread throughout the book. We also believe that historical geography should be firmly rooted in social and economic history, and thus the relationships between economic, social, political, spatial and environmental factors are frequently emphasized.

Inevitably, in a book of this length, we have been highly selective and have chosen to follow four key themes: population; the countryside; industry; and urbanization. Such a systematic framework of analysis has advantages and disadvantages. We hope it will ease the task of a student wishing to obtain an overview of a particular set of activities over the whole timespan and to compare responses of these sectors to changing forces in Britain within the three long cycles of development identified above. However, this framework has been at the cost of an holistic and integrated view of the impact of the totality of change on particular regions. Similarly, while the regional impacts of technological, economic and social changes are indicated in our treatment of individual themes, there is no attempt either to consistently cover these in all regions of Britain or to do so for the same regions in each phase of the country's development.

The choice of themes and regional examples has been guided both by our respective interests and expertise, and by a belief that the characteristics of modern Britain are best reflected in the changing responses of its people to economic, social and political opportunities, internationally and intranationally. In analysing the economic geography of rural, industrial and urban areas it is essential to consider the social consequences for those involved. Although the precise content of chapters dealing with each theme varies from period to period, taking account of the changing emphasis on particular sectors of activity and facets of society, there remains sufficient common ground for comparisons to be made between sections.

Throughout, we have been very aware of continuities as well as changes in the geography of modern Britain. Thus the time boundaries within the three sections of the book have not been rigidly enforced but given flexibility to look forwards and backwards, to cross-link between the themes examined and, in a concluding chapter, to reflect on some of the issues relating to changing international, national and regional awareness. We have both been involved in writing all sections and themes and have contributed to every chapter of the book. We take joint responsibility for any errors and omissions.

In a metric age it has become customary to present statistical data in both traditional and metric measures. However, in a book wholly concerned with a pre-metric era and using a wide range of measures we consider it both impractical – in designing clear statistical tables and in providing an uncluttered text – and unrealistic – in terms of, for example, conversion of wages and prices to present-day metric values – to do so. Apart from providing maps with both metric and non-metric scales, we have used the original units of measurement for all statistics in the text, tables and illustrations. A short table giving metric conversions of the principal measures used is given in an appendix.

Inevitably, a book that ranges widely, if selectively, over a long period must rely substantially on the research of many others. In an attempt to avoid burdening the text with the massive array of material on which we have drawn, we have restricted this to the principal sources consulted: direct references are largely limited to those from which we have quoted or directly drawn data for tables and figures. Rather than giving footnotes and a single bibliography we have arranged all references under further reading placed at the end of each chapter and have mostly limited them to more readily available books and journals. To those quoted (and unquoted) we acknowledge our indebtedness.

Equally we acknowledge the contribution made to our thinking about the subject by numerous colleagues and students in our respective universities of Liverpool (and latterly Loughborough) and Lancaster. Much of the material used in the book evolved from undergraduate courses in historical geography and from responses to projects and tutorials. It has gained immeasurably through collaboration with others with whom we have worked: in particular Paul Laxton and Gerry Kearns at Liverpool; Robin Butlin and Mike Heffernan at Loughborough; Ian Whyte, Marilyn Pooley and many colleagues in the Centre for Social History at Lancaster. It has also benefited from the stimulus as well as the specific researches of many postgraduate students. We are particularly grateful to Robin Butlin who read the final manuscript. Our thanks are extended to the Leverhulme Trust which funded Professor Lawton's research on and writing of the book.

We are equally indebted to those who have helped to translate our ideas and drafts into their finished form. To Sheila Hargreaves, Marlene McNaught and Betty Thomson who typed and word-processed the text. To Claire Jarvis, Mathew Ball and Anne Jackson, who drew the finished maps at Lancaster and to Sandra Mather, Alan Hodgkiss and Paul Smith who drew earlier versions of some of them in Liverpool. Finally, but not least, we are grateful for the patience and care with which Edward Arnold have produced the book: to John Wallace who persuaded us to write it; to Susan Sampford and Laura McKelvie, successive geography editors, who supported us through its writing; and to the admirable sub-editors and graphic designers who have imposed consistency and clarity on the finished product.

Richard Lawton
Marton, North Yorkshire

Colin G. Pooley
Lancaster

March 1991

1

The economic and social transformation of Britain

Introduction

Modern Britain, the 'First Industrial Nation', emerged without precedents. The country was transformed between the early eighteenth century and the First World War by a series of economic, technological, social and demographic transitions which are complex and interrelated. Some have attributed the driving force from the 1780s to economic spin-off from a dominant textiles industry. This is over-simplistic. Economic growth in the late seventeenth and early eighteenth centuries was promoted by improvements in agricultural production that stimulated population increase and, in turn, a range of industries, assisted by growing exports and favourable terms of international trade. Separately insufficient to explain economic take-off and its associated social consequences, these factors are together crucial to understanding.

Britain's basic natural resources – climate, agricultural land and coal, iron and non-ferrous metals – were good (Chapters 4 and 5). A growing and skilled labour force, both adaptable and mobile, generated greater productivity in a more diversified economy. The willingness of entrepreneurs and labour to harness scientific and technological knowledge to technical development were vital. Capital from profits and savings in an expanding market economy was invested in labour-saving forms of production which lowered unit costs, assisted purchasing power at home and increased industrial and commercial price competitiveness in international markets.

Social and political frameworks also favoured change. Population expansion promoted growth through increased demand for food and manufactures and enhanced the supply of labour and human skills (Chapters 3, 8, 13). Social forces – political, scientific, educational and the changing outlook and class structure of a more mobile society – were keys to economic, institutional and societal change. A mainly rural people made the transformation to an urban society and an exploitative wage economy under the stricter discipline of the factory, workshop, warehouse and office.

A narrow political and social hegemony based on the landed gentry gave way to a new, though equally divisive, class structure based on industrial

DOI: 10.4324/9781003174783-1

and commercial wealth in a bourgeois society in which, however, the working class had a distinctive and increasingly important role. Yet a national consensus and a belief in the right-and-might of the British Empire were preserved. Progressive enfranchisement of a male-dominant society, and the response – if belated – of government to social needs in health, housing, and the environment of town and workplace provided long-term political and social stability unparalleled in modern Europe: the result was evolution rather than revolution.

The impacts of change varied considerably in time, space and their effects on the landscape. Thus, while coal-fuel technology made a considerable impact on some processing industries from the seventeenth century the full potential of coal was not realized until it was harnessed to motive power in the late eighteenth century (Chapter 5). Moreover, steam power was not used in many industries until late-Victorian times when it overlapped with electrical power (Chapter 10). Similarly, new crops, rotations, specialist stock-breeding and new forms of tenure and farm organization though common in areas of 'advanced' farming from the mid-eighteenth century (Chapter 4), were often largely absent from backward areas even by late-Victorian times (Chapter 9). The transformation of the economy was a series of transitions – many of them gradual – rather than a single set of revolutions, be they agricultural, industrial, transport or commercial. Extending over a lengthy period from the early eighteenth to the early twentieth century, these require the long-term perspective adopted in this book.

However, over two centuries there were distinctive phases of development related to changes in Britain's international position; to shifts in technology; and to the changing structure of economy and society. These are more fully discussed in the introductions to the three time periods into which the book is divided (Chapters 2, 7, 12). Their legacy on landscape and region differs considerably. While much remains from earlier times in archaeological and landscape survivals in both town and country, there are few periods of history when so much of the British landscape was transformed so quickly. The countryside assumed new patterns of farms, fields, crops and settlements (Chapters 4, 9, 14). Industrial areas bear the marks of mining and massive concentrations of factories (Chapters 5, 10, 15). Enormously expanded towns reflect late-Georgian residential growth but, especially, Victorian impacts on housing, factory and workshop areas, and in their central business districts (Chapters 6, 11, 16). Transport networks of canals and railways were the product of the late eighteenth century and Victorian periods, respectively, though road networks owe much to the turnpike systems of the late eighteenth century. Many of these, in turn, have disappeared under the massive impact of new technologies and changed patterns of production, work and life-style in the twentieth century. Thus from the mid-eighteenth century to the mid-twentieth century the regional geography of Britain was successively reshaped (Chapter 17).

The modernization of Britain

The modernization of Britain, the first such transformation, offers general lessons on the nature and theory of economic and locational change and demographic transition at a variety of scales and from many viewpoints. One aspect is the relationship with Immanuel Wallerstein's 'world system' in which Europe led the way in world-wide economic and political exploitation of resources, trade, peoples and territory. European advances in science, exploration and discovery, and in the settlement of the 'New Worlds' of North America and the southern hemisphere provided the bases of a global economy which Britain dominated in the late eighteenth and nineteenth centuries. That dominance stemmed from the command of raw materials on favourable terms of exchange, abundant inorganic fuel, better constructional materials (particularly iron and steel) and machinery. Commercial, territorial and imperial power gains were a rich return for world-wide investment. In contrast, in the twentieth century power switched from an Atlantic- to a Pacific-centred world economically dominated by the USA and Japan. Britain retains an important commercial role in a truly multinational business structure but since the Second World War, despite continuing political and cultural influence, it has become a second-rank world power.

Rostow (1960) has argued for a 'dynamic theory of production' to explain the economic transition from a 'traditional society' to a 'modern', 'mass-consumption society' in which industrial and commercial skill and enterprise rather than landed wealth generated greater production through more intensive use of physical and human resources. This avowedly capitalist view – Rostow's sub-title is 'A Non-Communist Manifesto' – identifies cycles of economic growth in which investment from savings generated in agriculture and commerce stimulated manufactures and trade.

Rostow's five-stage model is based on such economic indicators as rates of growth in the economy, development of new technology, the relative growth of various sectors and the relationship to patterns of demand (Table 1.1). From a 'traditional society' Britain experienced a 'pre-conditioning', in the early eighteenth century, for 'take-off' with sustained economic growth from the 1780s. In the consequent 'drive to maturity' in mid-Victorian times progressive industrialization changed the structure of the economy and mechanization affected a wider range of industries. That process culminated in an 'age of mass consumption' which, in Britain, did not arrive until the interwar years, a generation behind that in the USA.

While this model takes account of a wide range of economic, technological, historical and social factors, it is essentially focused on phases of economic growth, the rhythms of changing raw material demand and the supply of manufactures in an evolving national and international economy. Economists have long been fascinated by the way in which these manifest themselves in the sequence of growth-boom-recession-slump which characterizes the trade cycle. The Russian economist, Nikolai Kondratieff identified long waves of a cyclical character related to upswings of economic and technological change accompanied in capitalist Europe by substantial investment and territorial expansion which accelerated booms in production,

Table 1.1 British economic cycles of the late eighteenth to the mid-twentieth century

Year	Kondratieff cycles Long waves		Trade cycles			Rostow's 'Stages of economic growth'
1700	I	P –	R -	D –	RI	
						Pre-conditioning
	1787–1842	1787				
						Take-off into sustained growth
1800	'Industrial Revolution' (Cotton, iron, steam power)	1800	1801–13	1814–27	1827–42	
	II 1842–1897	1843–57				
1850	'Bourgeois' (railways, steel, engineering, coal)		1858–69	1870–85	1886–97	Drive to maturity (c. 1850)
	III					
1900	1897–1940 Neo-mercantilist (electricity, oil, chemicals, light engineering, cars)	1898–1911	1912–25	1926–39	1939–45	Age of mass consumption (1930s–)
	IV 1940–(1995(?))	1945–				
1950						

In the Kondratieff cycles: P = Prosperity;
R = Recession;
D = Depression;
RI = Revival

employment and wages. As markets became saturated a downswing would lead to recession with short-lived revivals, then to slump as profits, investment and growth fell. Overlapping with the downturn of old technologies in the declining phase of such long waves, basic and often radically new technical innovations may be introduced which then characterize the industries of the next wave of growth (Table 1.1).

The theory of long waves provides a useful descriptive framework within which to analyse changes in Britain's economic growth and technological development and their consequences for national and regional patterns of production and trade. Elaborating Kondratieff's ideas, the American economist J. A. Schumpeter (1939) showed that successive long waves incorporated shorter-term business waves reflecting fluctuations in trade. He argued that as population and investment grew there were also distinctive phases of innovation in technology and the creation, production and marketing of new products often using new materials. Others have suggested that technological innovation tended to cluster around periods of intense activity, followed by consolidation and then decline in the down-turn of the long wave as the diffusion of new ideas through all sectors of the economy and into all regions slackened. During these phases of decline old plant might be scrapped and, unless reinvigorated, old industries and markets might be abandoned.

In Britain three such long cycles have been identified relating closely to Rostow's phases of 'take-off', 'maturity' and 'mass consumption'. The industrial revolution (the 1780s to 1840s) was characterized by significant innovation in textiles, especially cotton, and iron manufacture. A second long wave to the 1890s saw significant investment and growth through widespread adoption of innovations in the use of steam power, including railways, and steel-making. The third wave, to the Second World War, focused on new industries – chemicals, motor manufacture and many household goods – and such new forms of motive power as electricity and petroleum, as well as the development of new, chemically-derived raw materials. That wave died, haltingly, with a revival of some old staple industries in a wartime boom, but the innovations in electrical engineering and aeronautics which characterize the fourth wave were initiated in the late 1930s and spread to electronics, computers and communication systems from the 1950s and '60s. That phase, it would seem, died in the recessions of the mid 1970s and early 1980s. What a fifth Kondratieff will bring to Britain in the 1990s in a world of deindustrialization, electronic information systems and an economy focused increasingly on tertiary and quaternary activities remains to be seen.

One major force in the evolution from a pre-industrial to an industrial society was the change from an 'organic economy', as E. A. Wrigley (1988) has described it, focused on the land and handicraft industry and using wood as the major source of fuel and construction, to an 'inorganic society' based on coal for fuel then steam power, iron structures and machinery. That transition was gradual, spanning the seventeenth to the late nineteenth centuries in many activities. Before the early nineteenth century an advanced organic economy met accelerating population growth by more intensive utilization of land and natural raw materials, improved yields and lowered unit costs. Rural incomes in particular benefited from a dual economy in which handicraft industry supplemented farming. One crucial element was the substitution of abundant coal as fuel in a growing range of industries for increasingly scarce – and expensive – wood (Chapter 5). Water, then steam power opened up new avenues for growth in a mineral-based inorganic economy from the late eighteenth century. The considerable investment in new technologies diffused only gradually through Britain's industries and regions over the next century, but led to substantial increases in total production and in productivity per unit of labour and as a return on capital. These were reflected in a progressive, if discontinuous rise in real wages and living standards though these were uneven in their spread over time and between classes and regions.

The earlier stages of modern capitalism were characterized by what Franklin Mendels (1972) has called 'proto-industrialization'. The essence of this idea is not new. Domestic handicraft production using family labour in a dual rural economy based on agriculture and industry has long been recognized as important in such marginal farming areas as the south Pennines, parts of eastern Scotland and woodland areas of midland and southern England. Rural out-workers, free of guild restrictions, were employed by capitalist entrepreneurs who put out raw materials to farms and workshops, integrated the various stages of production and marketed the finished product. Many urban workshops were organized on similar principles; the system often focused upon and stimulated commercial growth in organizing centres such as Halifax, Leicester and Northampton.

Cheap surplus labour (general or seasonal) is central to proto-industrialization. Economic diversification was reflected in the growing proportion employed in handicraft and manufacture in the late eighteenth and early nineteenth centuries, and helped sustain more rapid population growth and encourage earlier marriage (Chapter 3). By absorbing women and young children into the workforce it staved off the poverty which such population growth threatened and which Malthus feared. The interdependence of rural producers and urban-focused capital and marketing helped create specialized regional economies which assisted rural-urban migration, both characteristic of the first industrial revolution.

One dimension of this system, the exploitation of surplus labour within a capitalist wage-economy, appeals to Marxist interpretations of modernization: for example the Kriedte *et. al*. (1981) view of 'industrialization before industrialization' as a form of immature capitalism which becomes integrated within the capitalist mode of production through exploitation of surplus and reserve wage-labour. Marxist theory also recognizes the importance of development phases, parallel to Kondratieff's economic cycles, shaped by social and political factors. Rather than smooth waves reflecting shifts in the levels of profit, investment and technical innovation the first long wave reflects the surplus value from enhanced labour-productivity derived from capitalist access to cheap labour and raw materials from agricultural regions. The second phase extended that process to European markets and areas of overseas settlement. The profits drawn from imperialist expansion protected gains by means of trade barriers through a third phase from the late nineteenth century to the Second World War. Since then international corporations have monopolised industrial production and international finance to draw dependent neo-colonial territories into a full world system in which, despite London's important commercial role, Britain's economy – especially manufacturing – has been less successful.

Political and social factors were vital in these transformations. The restructuring of the labour force through the creation of a wage-dependent industrial working class and the decline of the agricultural workforce, was one significant factor. Another was the rise of middle-class industrial entrepreneurs and their conflict with both urban working classes and metropolitan capitalist financiers. Thus, the emergence of capitalism and wage-labour in the domestic system provided, in some regions, a basis for transition to urban and factory labour. Such areas – the textile districts of east Lancashire and West Yorkshire and the Scottish central lowlands – played a key role in a new structure of labour and society. In the case of Lancashire this was manifest in movements towards the reform of the franchise, free trade and Chartism from the 1820s to the 1840s. On the other hand,

London's finance-capital structure had a national and international rather than regional emphasis. Its workforce, scattered among many workshops and casual trades, generally lacked political coherence until the twentieth century.

The changing regional geography of Britain

New social relations were also shaped in different areas and helped to form new regional geographies. In broad terms three major spatial features may be identified. First, the cores of activity and growth and their declining peripheries changed. Up to the mid-eighteenth century London and south eastern England was the most populous and advanced region in agriculture and commerce. In Scotland the eastern lowlands, focused on Edinburgh, was the economic and political centre; the Highlands were remote and politically and demographically unstable. That picture changed dramatically with the shift of industry to water-power, the adoption of steam-power and new forms of proto-industrial then factory production. The economic – and much of the political and cultural – strength of Britain moved to such large provincial cities as Manchester, Birmingham, Leeds, Glasgow, Liverpool and Cardiff, and to the coalfield industrial regions of midland and northern Britain and South Wales. By the late nineteenth century many pioneer staple industries – cotton, coal, iron and steel – were in incipient decline. While in some areas, such as the West Midlands, growth industries replaced older ones, the lion's share of new labour-intensive manufactures went to London and the South East (Chapter 12). The 'regional problem' of contemporary Britain, with its southeast-northwest gradient of employment, social welfare and income reflects a century-old trend and a reversion to the focus of the British economy and polity on a European, rather than Atlantic-orientated, southeast.

A second basic shift has been in the balance between rural and urban areas. In the early phases of industrialization many activities were rural, but the continuing growth of London and the rapid expansion of major provincial cities from the late eighteenth century, then of industrial towns in the railway age produced a predominantly urban population by 1851. Extensive out-movement led, from mid-century, to a decline in rural population (Chapter 8). By the 1890s rapid outward growth of towns was permitted by mass transportation systems and promoted by demand for space for industry and commerce, housing and recreation. Major cities were the foci of a new metropolitan regionalism with a growing awareness of the need for town and country (even regional) planning. Although local and regional loyalties of an earlier age persisted, increasing uniformity of products, lifestyles and many features of the twentieth-century landscape together with the growing dominance of public opinion by London-focused media and advertising tended to weaken the regionalism of the industrial revolution. The response of regional cities to much-changed circumstances varied: great ports such as Glasgow and Liverpool declined rapidly after 1918; Manchester and Birmingham struggled to maintain their nineteenth-century influence. All, including London, suffered progressive problems with a steepening gradient from prosperous outer suburbs to deprived inner city areas in employment and population numbers and economic, social, housing and environmental conditions.

Changing regional economic activity and social life revealed in the following chapters represent varying responses to the main phases of economic change. The cotton textile region of North West England prospered up to the First World War through its role as a major export industry and its related engineering, chemical and commercial sectors, but then declined rapidly. The West Midlands, more diversified and adaptable, engendered growth industries as older ones declined and has only recently been overtaken by a sharp fall in economic strength in the down-wave of the fourth Kondratieff. Clydeside, North East England and South Wales, leaders in the second Kondratieff of coal, steel and engineering, experienced near-collapse of their economies between the wars and are still among the most deprived of Britain's industrial regions despite the development of regional policy for depressed areas from the 1930s (Chapter 15).

Theoretical considerations suggest a number of explanations of such changing inter- and intra-regional levels of development and prosperity. In a free market, labour shortages and rising costs in an area might be met by movement from regions of labour-surplus. Much rural-urban migration and movement of skilled workers from areas of industrial decline to those of growth reflect such trends since the industrial revolution. But imperfect knowledge, family ties, local loyalties, and poor relief have always been an impediment to labour market equilibrium: hence more recent policies of seeking to take work to the workers (Chapters 12 and 15). Moreover, industrial and commercial activity do not invariably, as suggested by classical economic theory, locate at 'least-cost' points of production. Plants may be retained long after access to raw materials, transport networks or cheap and skilled labour cease to be as competitive as those of other areas.

Indeed, some economists argue that as regions or towns diversify (particularly those with large industrial and commercial concentrations) a process of cumulative causation will continue to attract high investment regardless of pressure on labour and space, promoting higher productivity and giving further advantage to leading as opposed to lagging economies. Far from particular industries promoting 'growth poles' (as did the cotton industry in Lancashire during the industrial revolution), much of the industry of the later nineteenth and twentieth centuries was attracted by the size, diversity and flexibility of the labour force and the market of large cities, especially the metropolis, together with strong support and multiplier effects from their wide range of services and social attractions. However, high wages and costs of living, and pressures on land and infra-structure may drive firms to peripheral locations or adjacent regions. Such 'trickle-down' effects are recognized in centre/periphery models of economic development which resemble aspects of regional change in Britain between the wars and at certain stages during the nineteenth century, though they are difficult to identify from the limited regional data available before the First World War.

A market and wage economy brought particular problems of fluctuating labour demand and short- and long-term unemployment. Short-term lay-offs due to bad weather affected not only agricultural labourers but also building workers, dockers and sea-going men. Seasonal fluctuations in demand for coal or, even, certain types of clothing created seasonal unemployment. In a casual labour market with an excess of unskilled workers (Marx's army of the labouring poor) underemployment – as well as unemployment – was a powerful weapon in capitalist control of the workforce to which differences in regional demand (reflected in migration and even emigration) was only a partial answer. Even among skilled workers periodic unemployment produced much mobility, especially through 'tramping', with Union support, in search of work (Chapters 8, 10).

Longer-term adjustments to changes in the structure and location of economic activity also created structural unemployment with varied impacts on different regions. Despite labour migration from areas of decline to areas of growth, the lags in this process were reflected in substantial differences in regional unemployment: for example, the West Midlands' decline in heavy industry led to substantial movements of coal and iron workers to North East England, South Yorkshire and South Wales in the 1870s (Chapter 10). By the early twentieth century its workers were moving from cycle to car manufacture, or from mechanical to electrical engineering, but also from skilled workshop to semi-skilled assembly line jobs often in South East England (Chapter 15). The impact of the trade cycle was always more intensive in areas dominated by a limited range of activities, and unemployment more severe and longer-lasting in areas with a declining major industry. Recovery was quicker and stronger in areas of new growth industries.

Countering economic pushes and the pull of better wages or a job in another trade or another area, there were always powerful social forces – attachment to home area; ties of family and community; knowledge, and the character and purpose to take advantage of other opportunities – which slowed down and for some prevented adjustments to a changing labour market. These probably increased with the spread of social benefits and local authority housing between the wars, inducing further frictional unemployment as labour sought to cling to dying jobs in dying areas.

Sources

Although Britain has some of the best historical data in the world, most statistical time-series and archival sources, whether relating to national and regional trends or to particular sectors of the economy and society, lack continuity and are based on different, often non-comparable, criteria. Unfortunately, from the geographer's point of view, they frequently also lack spatial coherence in their scale and areas of enumeration. Thus many key questions about the social and economic transformation of Britain can only be partially answered. This section briefly outlines some of the main sources on which succeeding chapters of the book are based.

For most purposes the eighteenth and early nineteenth centuries were a pre-statistical era though some of the most reliable and continuous statistics, those on trade, date back to 1696 for England and to 1755 for Scotland.

These, and estimates of national income derived from the Income Tax imposed by Pitt during the French Wars, give some insight into the broad pattern of the national economy. Regional figures of relative wealth and living standards are more difficult to come by and are largely built up from local studies based on the inventories of wills, estimates of earnings in local economies or, by the nineteenth century, government surveys of particular industries (such as coal and cotton). Together with the continuous record of grain prices in regional markets these permit the construction of long-term indices of wages, prices and real earnings.

For England, the general population trends to which these relate are now well understood thanks to the reconstruction of vital trends from parish registers from 1538 to the beginning of civil registration in 1837 (Chapter 3). In Scotland the record is much less complete, and details of regional population change are uncertain in all areas before the first British census of 1801. Effects of mobility are still substantially unresearched, although the considerable social and spatial changes in the population attracted much contemporary comment and are reflected in many documentary sources including the voluminous Poor Law records which shed much light on regional poverty and, from records on settlement of persons applying for relief, on mobility.

Similarly, there are few precise and continuous statistics on particular sectors of the economy. Wartime crises provided a notable series of County Reports to the Board of Agriculture between 1793 and 1817, which give a comparable picture of agriculture and the rural economy to put alongside that derived from eighteenth-century estate records and Enclosure Acts. But statistical information on farming is confined to the Crop Returns of England and Wales gathered by the Home Office in 1801 (Chapter 4). The situation for industry and commerce is even less satisfactory. Information on occupations was collected only in a rudimentary form in censuses before 1841. Occasional Parliamentary or trade enquiries – such as the Iron Masters' deputation to the House of Commons in 1806, the Lords Select Committee on the Coal Trade of 1829, or the Select Committee on the Woollen Manufactures 1820 – provide an overall view, but before the 1830s information on industry has usually to be pieced together from local records (Chapter 5).

There was little direct government intervention in economic and social affairs at this stage, but where matters were regulated by Parliament, as with canal and river navigations and, from the 1830s, the railways, the picture is fuller. The records of turnpike trusts and local advertisements of carrier and stagecoach services give a good basis for studies of local and regional transport networks. As industry, commerce and urban centres grew these were increasingly backed by town, county and trade directories which are invaluable in filling gaps left by the lack of official statistics. Similarly the growing need for towns to control their environments was answered by Improvement Boards and Commissions which provide much information on urban growth and problems in the era prior to government investigations from the 1830s (Chapter 6). These, and the Poor Law records, tell us much about family and community problems in both town and country.

In contrast to the pre-1830s paucity of statistics, the Victorian Age was obsessed with numbers. Problems of urban management, working conditions and the consequences of trade fluctuations for the nation's economy and people led to a growing number of enquiries by Parliament, local government, and trade and welfare organizations. The establishment in 1834 of the Statistical Society of London (later the Royal Statistical Society) epitomizes the Victorians' approach to issues ranging from demographic trends to international trade, and from individual industries to crime and education.

But the main agency of investigation was government. The need for fuller knowledge of population led to the establishment in 1837 of the office of the Registrar General in England and Wales and in Scotland in 1855. Much fuller censuses from 1841 – with a wide range of information on occupations, structure, and mobility – and compulsory civil registration of births, deaths and marriages gave the basis for a much better understanding of both national and regional trends in population and economic activity (Chapter 8). However, the opportunity to gather census data on production units was lost, despite some figures on the size of farms and industrial workplaces in 1851. So too was the chance to systematically investigate cultural and social trends, apart from the never-repeated Censuses of Education and Religious Observance taken in 1851.

The picture of industry remained fragmentary until the first Census of Production in 1907. Concern about basic reserves of coal, iron and other metals was reflected in annual figures of output from 1854 and of the number and size of coal mines from 1864. But the major sources of information on key industries were the enquiries of Select Committees and Royal Commissions often generated by problems in working conditions or fear of decline due to recession or falling resources. Similar concerns over working conditions led to the setting up of the Factory Inspectorate which produced annual figures of employment and power used from 1838, while the threat of chemical pollution added the Alkali Inspectorate to the list of regulatory bodies from the mid-1860s (Chapter 10).

Agriculture was regularly reported upon, notably in a series of essays on the farming of individual counties in the Journal of the Royal Agricultural Society between 1843 and 1878 and in other individual surveys such as James Caird's letters to *The Times* in 1850–51 or Rider Haggard's survey of Rural England in 1902. However, the recording of land use and statistics of crops, stock and production remained incomplete, although detailed farm files and maps were produced in awards made under the Tithe Commutation Act of 1836. Together with the remaining Parliamentary Enclosures (not least under General Acts of 1836 and 1845), these are a good basis for studies of farming and rural landscapes in many areas. But the first systematic statistics of British Agriculture awaited the annual 'June Returns' from 1866. Gathered from farms throughout the country they record parish details of crops and livestock (though not of yields until estimates of crop production were added in 1884) and are an essential adjunct to the frequent Parliamentary Committees on agriculture of the Victorian period, in particular the Royal Commissions of 1881–82 and 1894–97 (Chapter 9).

Rural distress attracted much attention in these, and in evidence to and Reports of the two great Poor Law Commissions of 1834 and 1905–09, but the major focus of social, environmental and health investigations was the towns. Cholera and other epidemics, the scandal of bad sanitation, poor water supply and housing drew government into a wide range of surveys and subsequent legislation (Chapter 11). Such investigations as Chadwick's massive *Report of the Royal Commission on the Sanitary State of Large Towns and Populous Districts* of 1840–45 provide essential information, but the bulk of urban studies is based on the wealth of local records collected by local authorities – including rate books, health reports and corporation minutes – together with census enumerators' books and directories.

Despite progressive government involvement in social problems from the 1830s, implementation needed strong and co-ordinated action at local government level rather than *ad hoc* solutions. Nowhere is this better seen than in studies of incomes, living standards and poverty. Fragmentary information on wages, and working and living conditions of particular groups of workers are available from national sources (e.g. the Poor Law) and local records; from trade associations and unions; and in the many social surveys from early-Victorian times. The latter, promoted in big cities by the National Association for the Promotion of Social Science founded in 1857, led to the great surveys of London by Charles Booth (1890–1903) and Seebohm Rowntree's investigation of poverty in York (1901): but the first systematic government information, on family earnings and budgets by industry and town, awaited the Board of Trade Surveys of 1906.

By that time national concerns for social matters were apparent in the considerable detail of the 1911 Census of Population on housing, fertility and household (Chapter 13). They were also the context for town and regional planning, control of land use, the location of economic activity and local authority housing (Chapter 16). Planning still lacked a national inventory of land-use, despite the long sequence of Ordnance Survey maps and plans from 1801 which made Britain one of the best-mapped countries in the world. That gap was filled by the remarkable Land Utilisation Survey (1930) organized by Professor Dudley Stamp which mapped Britain and published a series of one-inch land-use maps (1931–45) and county reports (1937–44). The gap in data on farm economies was filled partly by the 1941–43 *Farm Survey* and other national censuses such as those on woodlands (1924) and agricultural production (1908, 1925 and 1930–31), giving a much fuller picture of British agriculture from the inter-war years which data on agricultural output (from 1936) and farm incomes (1938 and 1946) helped to extend. The Stamp survey also provided the basis of many of the issues debated in the wartime *Committee on Land Utilisation in Rural Areas* (the Scott Report of 1942), one of the influential reports on which much post-war regional planning was based (Chapter 14).

Parallel surveys of basic resources such as surface mineral and constructional materials helped to extend the basis of annual minerals statistics for planning. These added a new dimension to the Censuses of Production of 1912, 1924, 1930 and 1937 and then annually from 1948 which provide a more complete picture of industrial production in the twentieth century. Comparable data on services, including retail and wholesale trade, were not collected until 1950. Despite the lack of input-output or value-added statistics (such as those in American industrial censuses), knowledge of Britain's national and regional economic and social structures now rests on a much fuller quantitative basis than that of the nineteenth century (Chapter 15).

Regional development is particularly difficult to analyse from available sources. First, basic data on employment, income, prices, population trends and social conditions cannot be integrated into a single spatial framework. Thus the effects of boom and recession on particular regions are difficult to measure, though their consequences may usually be implied from changes in employment, occupational structure and population movement from the early-Victorian period. Secondly, two practical difficulties limit comparable regional studies over a long time period: the breaks in continuity of many data sets; and the lack of a single coherent set of administrative areas on which to base analysis. At local level the parish is most often the best unit of study. At the middle level, historic counties often cut across both distinctive natural regions and the growing urban-centred regional hierarchy of the late nineteenth and twentieth centuries. The latter lacked adequate administrative definition despite successive attempts by the Local Government Board (from 1871), the Local Government Acts of 1888 and 1894, and the Royal Commissions on Local Government of 1923–26 and 1966–69 to impose rationality (Chapters 11, 16). At the broader provincial level not until the adoption in 1966 of Standard Regions for many planning and statistical purposes was there an agreed framework of analysis, though these regions are too large for many purposes.

Conclusion

The development of Britain's national and regional economies from the mid-eighteenth century, the temporal framework of this book, reflects not only long waves of economic development and the classic but generalized interplay of land, labour and capital, but also involves geographical changes in the composition and structure of society and its activities. David Harvey's (1982) Marxist critique underlines a 'spatial configuration' – as well as a social, economic and political one – to the uneven geographical development at local, regional, national and international level which characterized the development of modern capitalism. These are not simply a consequence of differing spatial patterns of production which successively follow major rounds of investment and create, cumulatively, regions of very different character. Location is but one aspect of change. As important are socially-induced changes: the different values and responses of society to new patterns of social activity; and the processes of integration and conflict within new societal structures. As more widely-based studies are developed of Britain's industrial and urban societies in the industrial revolution and after, there is growing recognition of the importance of the regional differences – physical, social and cultural – which induced success in some phases and aspects of change but failure in others.

It can be argued that industrialization and modernization, far from destroying regional identity, created in the eighteenth and nineteenth centuries a strong new regionalism, modified but not destroyed by a metropolitan regionalism focused on major cities. In time, these drew much of their character and purpose from the activities, people and cultural life of their hinterlands. Despite continuing centralization on London of both government and the commanding heights of political, economic and cultural life in the twentieth century, there remains a counterpoise; a distinctive and vigorous provincial life and identity. Its nature and the extent to which it encapsulates forces at work over the past two centuries are discussed in the final section of the book (Chapter 17).

Bibliography and further reading for Chapter 1

For an overview of the social and economic transformation of Britain from the eighteenth century:

CHECKLAND, S. G. *British public policy: an economic, social and political perspective* (Cambridge: Cambridge University Press, 1983)

CRAFTS, N. F. R., LEYBURN, S. J. and MILL, T. C. 'Trends and cycles in British industrial production, 1700–1913' *Journal of the Royal Statistical Society* Series A, **152** (1989), pp. 43–60.

DEANE, P. *The state and the economic system* (Oxford: Oxford University Press, 1989).

DEANE, P. and COLE, W. A. *British economic growth, 1688–1959* (Cambridge: Cambridge University Press, 1967).

DUNFORD, M. and PERRINS, D. *The arena of capital* (London: Macmillan, 1983).

HOBSBAWM, E. J. *Industry and Empire* (London: Weidenfeld and Nicolson, 1968).

LEE, C. H. *The British economy since 1700: a macro-economic approach* (Cambridge: Cambridge University Press, 1986).

MATHIAS, P. *The first industrial nation* (London: Methuen, 1969).

MATHIAS, P. and DAVIS, J. A. *The first industrial revolutions* (Oxford: Blackwell, 1989).

ROYLE, E. *Modern Britain: a social history, 1750–1985* (London: Arnold, 1987).

THOMPSON, A. *The dynamics of the industrial revolution* (London: Arnold, 1973).

THOMPSON, E. P. *The making of the English working class* (Harmondsworth: Penguin, 1968).

THOMPSON, F. M. L. (ed.) *The Cambridge social history of Britain, 1750–1950* (3 vols) (Cambridge: Cambridge University Press, 1990).

WRIGLEY, E. A. *Continuity, chance and change: the character of the industrial revolution in England* (Cambridge: Cambridge University Press, 1988).

For a spatial perspective see:

BAKER, A. R. H., HAMSHERE, J. D. and LANGTON, J. (eds) *Geographical interpretations of historical sources* (Newton Abbot: David and Charles, 1970).

DARBY, H. C. (ed.) *A new historical geography of England after 1600.* (Cambridge: Cambridge University Press, 1976).

DODGSHON, R. A. and BUTLIN, R. A. (eds) *An historical geography of England and Wales* (London: Academic Press, 1978; 2nd edition, 1991).

LANGTON, J. and MORRIS, R. (eds) *Atlas of industrializing Britain* (London: Methuen, 1986).

POPE, R. (ed.) *Atlas of British social and economic history since c.1700* (London: Routledge, 1989).

STAMP, L. D. and BEAVER, S. H. *The British Isles: a geographic and economic survey* (London: Longman, 6th edition, 1971).

SMITH, W. *An economic geography of Great Britain* (London: Methuen, 1959).

TURNOCK, D. *The historical geography of Scotland since 1707* (Cambridge: Cambridge University Press, 1982).

WHITTINGTON, G. and WHYTE, I. (eds) *An historical geography of Scotland* (London: Academic Press, 1983).

A broader perspective on the modernization of economies is given in:
HARVEY, D. *The limits to capital* (Oxford: Blackwell, 1982).
HARVEY, D. *The conditions of postmodernity* (Oxford: Blackwell, 1989).
KRIEDTE, P., MEDICK, H. and SCHLUMBOHM, J. *Industrialization before industrialization* (Cambridge: Cambridge University Press, 1981).
LANDES, D. S. *The Unbound Prometheus. Technological change and industrial development in Western Europe from 1750 to the present* (Cambridge: Cambridge University Press, 1969).
MENDELS, F. 'Protoindustrialization: the first phase of the process of industrialization' *Journal of Economic History* **32** (1972), pp. 241–61.
POSTAN, M. M. and HABAKKUK, H. J. (eds) *The Cambridge economic history of Europe.* Vol. vi: *The industrial revolution and after* (Cambridge: Cambridge University Press, 1965).
ROSTOW, W. W. *British economy of the nineteenth century* (Oxford: Clarendon Press, 1948).
ROSTOW, W. W. *The stages of economic growth* (Cambridge: Cambridge University Press, 1960).
SCHUMPETER, J. A. *Business cycles: a theoretical, historical and statistical analysis of the capitalist process* (2 vols) (New York: McGraw Hill, 1939).
WALLERSTEIN, I. *The capitalist world economy* (Cambridge: Cambridge University Press, 1979).
WRIGLEY, E. A. 'The process of modernization and the industrial revolution in England' *Journal of Interdisciplinary History* **3** (1972), pp. 225–59.

On the changing landscape see:
HOSKINS, W. G. *The making of the English landscape* (London: Hodder and Stoughton, 1955; revised edition, 1988).
Also a series of county volumes in the *Making of the English Landscape* series, edited by W. G. Hoskins, published by David and Charles, Newton Abbot, 1970–73.
LOWENTHAL, D. and PRINCE, H. C. 'The English Landscape' *Geographical Review* **54** (1964), pp. 309–46.
TRINDER, B. *The making of the industrial landscape* (London: Dent, 1982).
WHYTE, I. D. and WHYTE, K. *Exploring Scotland's historic landscapes* (Edinburgh: John Donald, 1987).

For discussions and samples of sources see:
English Historical Documents Vol. xi 1783–1832 (eds A. Aspinall and E. A. Smith, 1959); Vol. xii(1) 1833–1874 (eds G. M. Young and W. D. Hancock, 1956); Vol. xii(2) 1874–1914 (ed. W. D. Hancock, 1977), (London: Eyre and Spottiswood).

On the census:
HIGGS, E. *Making sense of the census: the manuscript returns for England and Wales 1801–1901* (London: HMSO, 1989).
LAWTON, R. (ed.) *The census and social structure* (London: Cass, 1978).
WRIGLEY, E. A. (ed.) *Nineteenth-century society. Essays in the use of quantitative methods for the study of social data* (Cambridge: Cambridge University Press, 1971).

On Government publications:
FORD, P. and FORD G. *Breviates and guides to British Parliamentary Papers* (Oxford University Press, 1953).

On directories:
SHAW, G. and TIPPER, A. *British directories: a bibliography and guide to directories published in England and Wales (1850–1950) and Scotland (1773–1950)* (Leicester: Leicester University Press, 1988).

More generally:

RODGERS, A. *Approaches to local history* (London: Longman, 1972).

For basic national statistics:

MITCHELL, B. R. *British Historical Statistics* (Cambridge: Cambridge University Press, 1988).

SECTION I:

Britain from the 1740s to the 1830s

2

The political, economic and social context 1740–1830

The transformation of the British economy initiated in the early eighteenth century was three-fold (Figure 2.1). First, a mainly rural society became a mainly commercial and industrial nation. Secondly, associated changes in work patterns, mobility of labour and the size and structure of increasingly urbanized communities induced a profound social revolution. Thirdly, associated changes in population behaviour saw the onset of a demographic revolution through the gradual control of mortality and in increased mobility.

In the early eighteenth century indigenous food supplies largely controlled prices and were the major influence on the economy and population. Relatively high mortality and control of family size in a society with a relatively high average age of marriage restricted population growth. That Britain largely avoided demographic crises from the late seventeenth century reflects social as well as economic factors, but it also indicates a varied and generous endowment of natural resources: a climate supporting a range of crops and stock but not given to seasonal or cyclical extremes; varied soils adaptable to the more specialized farming systems of the new agriculture; well-distributed supplies of fuel, particularly abundant coal, and ferrous and non-ferrous metals. In a mainly organic economy based on agriculture and handicraft industries these were sufficient to sustain a relatively small and slow-growing population.

The growth of output in early eighteenth-century England both in agriculture (0.9 per cent per annum between 1700 and 1750) and industry (0.7 per cent p.a.) was not matched in Scotland, but in both countries economic growth exceeded a population growth of some 0.3 per cent p.a. Agricultural production was significantly widened, though not equally in all regions and certainly not on all farms, by innovation in crops and farming methods. Increased productivity in well-farmed areas, both open field and enclosed,

 DOI: 10.4324/9781003174783-2

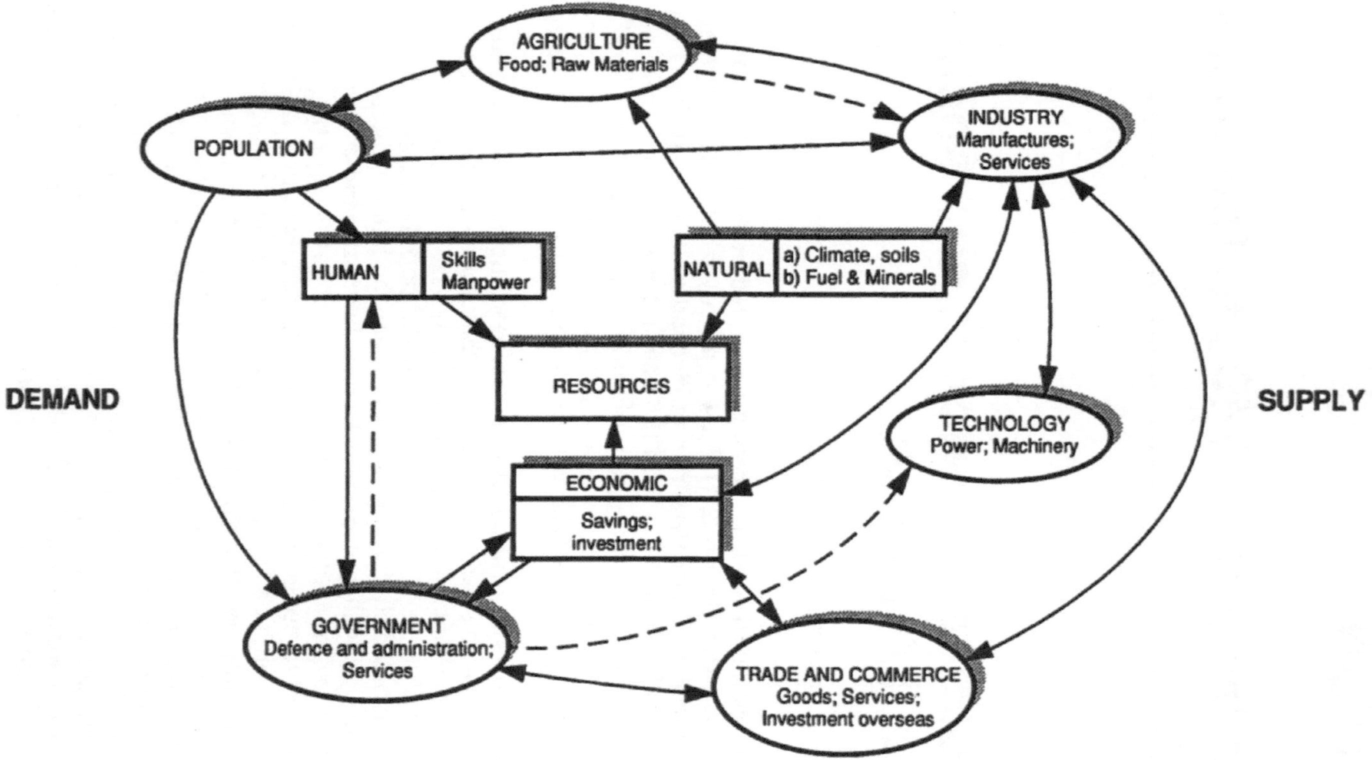

Stronger links between sectors and components are shown in solid lines with arrows indicating the main flows; weaker and/or less direct links are shown by pecked lines

Figure 2.1 A simple model of social, economic and technological relations in the industrial revolution.

and reclamation of wetland and heathland began to transform rural landscapes and settlement patterns in the age of the improver. Substantial growth in traditional industries, especially textiles, was achieved by changes in the organization of production that made fuller use of family labour in many rural economies, not least in some marginal farming areas (Chapter 5). Growing market orientation of a greater range of both agricultural and industrial production promoted and depended upon better transport from the mid-eighteenth century and was reflected in both growing regional specialization and increased trade and commerce.

In 1700 four-fifths of the population was rural and two-thirds agricultural. By mid-century just under half of England's rural population was agricultural and one-third mainly industrial. By 1800 these and the urbanized sectors were roughly equal and, according to the crude census occupational statistics, by 1821 46 per cent of Britain's three million families depended on 'trade, manufacture and handicraft' as against 33 per cent in agriculture. By that time two-fifths of England's population and over one-quarter of Scotland's were town-dwellers, and many countryfolk were industrial workers.

As economic power and social status began to shift from the landed aristocracy to the bourgeoisie, so the political emphasis began to change. Despite continuing dominance of Parliament by the landed interest, commercial and manufacturing interests were influential in both domestic and foreign policy. The loss of the American colonies and the turmoil of European wars, especially against the French, emphasized the role of industry in productivity and overseas trade. Moreover, the need to take account of the urban, trading and manufacturing interests were the subject of necessary moves towards Parliamentary reform and the franchise.

The response of society was vital for economic change. Despite much concern over the consequences of change, direct government intervention was limited. Once the shackles of Mercantilist policies – for example the Navigation Acts and the many excise duties so unpopular with the American colonies – were released, free trade was raised to the status of political dogma from the late eighteenth century. As Britain became more dependent on imported food and raw materials and on growing export markets, so the political power of those representing commercial interests (for example Canning and Huskisson in Liverpool) and the manufacturing districts (Bright, Cobden and the Manchester Free Traders) strengthened. Agricultural production was a source of controversy between farmers and an increasingly industrial population which they could not feed cheaply, especially in poor harvest years and wartime (for example the 1790s). In its approach to social problems, however, government was progressively forced into more utilitarian policies though the full fruits of attempts to promote the greater good of more of the people came only with Victorian reformers.

As Parliament assumed dominance over the Crown from the late eighteenth century, its own representativeness came into question. The dominance of the landed interest, especially the aristocracy, was progressively challenged by the bourgeoisie. The 1832 Reform Act began to switch emphasis from the regions of past dominance, and from the land to the towns and industrial regions though fear of the masses still directed much government action. Brutal and repressive laws, both to felons and civil dissidents, spawned the convict colony of New South Wales in 1787 and an extensive network of Home Office informants kept a close eye on both organized labour and rural and urban protest.

From the mid-eighteenth century resource utilisation shifted significantly in four ways: first, by a growing emphasis on coal, metals and other inorganic raw materials; secondly, through greater dependence on imports of raw materials (notably cotton), timber and foodstuffs (especially grain and tropical products such as sugar); thirdly, by mechanization and restructuring of production methods in such key industries as cotton and iron; fourthly, through much increased capital inputs and increases in productivity, notably in mechanizing industries but also in agriculture. The importance of new technology in the early industrial revolution has sometimes been exaggerated. Many increases in output came from capital and labour inputs (especially in the proto-industrial sector), though coal, water- and steam-power made an impact from the 1760s in the iron and textile industries and in both factory and workshop production. Despite the impact of technological progress on many sectors of the economy by the 1830s, aggregate growth was focused on a few sectors; textiles, iron, transport and shipping.

Human factors, individual and collective, were also crucial to change. Economic stimuli to change depended for success upon innovative individuals, usually practical farmers, craftsmen and industrialists rather than scientists, important though advances in scientific knowledge from the late seventeenth century were for some inventions. Enterprising entrepreneurs and work forces were vital to the adoption and diffusion of new methods of production. Improvements were often through adaptations of existing ideas and patents by industrialists and workmen. Population growth necessitated increased food production and stimulated manufacturing, especially as cheap goods came from better production methods (whether organizational or technological). Together they were to prove the way to avoid the Malthusian trap.

One stimulus came from the rich and upper-middle class demand for luxury goods (clothing, furniture and household goods) and services, including travel to and recreation at spas and seaside resorts or in 'town' at the theatre or assembly. Such opportunities did not begin to reach the working classes until the onset of late-Victorian factory production of consumer goods and not fully until the age of mass consumption. But there was, progressively, some money for second-hand clothing, cheaper soap, a few items of furniture or an iron cooking pot (arguably the biggest single contribution made by the Darbys of Coalbrookdale to social well-being): such contributions helped to improve living standards.

Enhanced production in one sector of industry created demand in others: machine-spun yarn for more weavers; textile machinery for more and better wrought iron products; supply of raw materials, food and manufactures to the market for better transport. Such demands could be met only by workable ideas on the speeding up of production, thereby lowering unit costs, and, where expensive plant and machinery was involved, the raising of capital. Both often required re-organization of labour and manufacturing processes. Yet there were many areas of production in agriculture as well as industry where new methods and machines did not demonstrably lead to cheaper production, or where the returns on new capital investment were not attractive, with consequent time-lags in the adoption of inventions in almost all sectors of activity and within and between industries. Textile manufacture was transformed relatively quickly (though more gradually in weaving than spinning) to provide much cheaper and, on the whole, better-quality cloth. But fine goods (linen, lace, some silks) and clothing for the home market came from a multiplicity of outlets (often direct to the customer from tailors and dressmakers) and remained one of the smallest and least mechanized modes of production until late-Victorian times.

The turmoil in the lives of many people caught up in agrarian and industrial 'improvements' from the late eighteenth century produced tensions and often considerable personal deprivation as their livelihood was destroyed or earnings seriously lowered. Families and communities were disrupted as people sought work in the new centres of activity. For example, there was severe unemployment among handloom weavers in the 1820s and '30s; nail-makers were under growing pressure from the machine-made product from the early nineteenth century; agricultural labourers feared loss of work through enclosure of pasture or the adoption of threshing machines. Thus there was considerable poverty in some areas of decline and periodic outbreaks of unrest.

Such protest was born more of despair than of revolt in communities caught up in the social upheaval of rapid change. The Poor Law, which until 1834 was still based on the principle of relief of the parish poor, ameliorated rural poverty in many parts of southern Britain and, more patchily, in Scotland. Indeed outdoor relief was held by some to be overgenerous, allegedly promoting improvident marriage and over-large families. But Parish relief did little to prevent rural protest such as food riots (for example in 1794–96) or machine-breaking and rick-burning in the impoverished countryside (the Captain Swing disturbances of 1830). Chartism, a more explicitly political movement attempting to secure political recognition through a workers' Charter presented to Parliament in 1839, was mainly focused on the industrial regions of the textile north and central Scotland. But its urban forebears were also protests against the instability of work and wages in the rapidly-changing industrial society which underlay Luddite machine-breaking (in the handloom and stocking-frame areas) as well as Manchester's notorious Peterloo massacre of 1819.

Economic growth also levied a social price, particularly in the rapidly growing towns. The inadequacy of political representation in large cities and industrial areas, as compared with the continuing scandal of rotten boroughs, was gradually tackled after the 1832 reforms. Environment and health, a much greater threat to well-being, reflected the inadequacy of the framework of local government and the limited effectiveness of *ad hoc* improvement commissions, watch and police committees and the like in dealing with problems of housing, sanitation and water supply, disease and crime. By the 1830s debate on the urban problem came to a head stimulated by the first cholera epidemic of 1831–32. The Municipal Corporations Act of 1835 was the first step towards local government reform, though significant intervention in the urban environment did not occur until Victorian times.

The challenge of poverty was also tackled from the 1830s. The Old Poor Law's system of parish relief and settlement laws was increasingly at odds with a more mobile, urbanized society. Moreover, the open-ended demands of outdoor relief on parish rates was disliked and avoided by many big landlords. The solution of the New Poor Law Act of 1834 was to impose a national system, with the principle of providing work (rather than assistance) within a framework of Poor Law Unions and workhouses in England and Wales, though Scotland retained a more localized system. Many such changes were inspired by the radical philosophy and utilitarian principles of Jeremy Bentham and his followers; but the Poor Law's ethos was harsh, invoking a continuing debate on conditional relief of poverty and need as against the right to social benefits.

The contrast between growing laissez-faire in economic affairs and progressive involvement in social policy, and between centralization of administration and devolution to local communities are aspects of the political consequences of change. Another was the extent to which modernization destroyed regional identity by increasing involvement in a national and international economy and shifts in population from country to town and between regions. However, old regional loyalties died hard: for most people London or Edinburgh remained remote until the age of cheap transport, mass literacy and growing political and media control from the capitals – Highlanders remained Highlanders; Yorkshiremen, Yorkshiremen; and Lancastrians or Devonians retained independence and distinctiveness of outlook. Whilst economic and social changes in this period initiated a new phase in the regional geography of Britain and made new impacts on settlement, population distribution and on the landscape, they were only gradually reflected in the political and administrative map. London's dominance of government, finance and culture remained strong but there were powerful regional counterpoises.

Bibliography and further reading for Chapter 2

The following selected texts provide background to the period:

BAGWELL, P. S. *The transport revolution from 1770* (London: Batsford, 1974).

BARKER, T. C. and SAVAGE, C. *An economic history of transport in Britain* (London: Hutchinson, 1974).

BRIGGS, A. *The age of improvement 1783–1867* (London: Longman, 1959).

CRAFTS, N. F. R. *British economic growth during the industrial revolution* (Oxford: Oxford University Press, 1985).

DEANE, P. *The first industrial revolution* (Cambridge: Cambridge University Press, 1969).

DERRY, T. K. and WILLIAMS, T. I. *A short history of technology from the earliest times to A.D. 1900* (Oxford: Oxford University Press, 1960).

EVANS, E. J. *The forging of the modern state. Early industrial Britain 1783–1870* (London: Longman, 1983).

FLINN, M. W. *Origins of the industrial revolution* (London: Longman, 1966).

FLOUD, R. and MCCLOSKEY, D. (eds) *Economic history of Britain since 1700.* Vol. 1 *1700–1860* (Cambridge: Cambridge University Press, 1981).

GILBERT, A. D. *Religion and society in industrial England. Church, chapel and social change 1740–1914* (London: Longman, 1976).

HADFIELD, C. *British canals* (Newton Abbot: David and Charles, 1950).
HOBSBAWM, E. J. *The age of revolution, 1789–1848* (London: Weidenfeld and Nicolson, 1969).
HOUSTON, R. A. and WHYTE, I. D. (eds) *Scottish society, 1500–1800* (Cambridge: Cambridge University Press, 1989).
MANTOUX, P. *The industrial revolution in the eighteenth century* (1928; revised edition, London: Cape, 1961).
MARSHALL, D. *Eighteenth century England* (London: Longman, 1962).
MARSHALL, D. *Industrial England, 1776–1851* (London: Routledge, 1973).
PAWSON, E. *The early industrial revolution: Britain in the eighteenth century* (London: Batsford, 1979).
POPE, R. (ed.) *Atlas of British social and economic history since 1700* (London: Routledge, 1989).
PORTER, R. *English society in the eighteenth century* (Harmondsworth: Penguin, 1982).
SMOUT, T. C. *A history of the Scottish people, 1560–1830* (London: Collins, 1969).
THOMPSON, E. P. *The making of the English working class* (Harmondsworth: Penguin, 1968).
TURNOCK, D. *The historical geography of Scotland since 1707* (Cambridge: Cambridge University Press, 1982).

3

Demographic Change from the 1740s to the 1830s

Introduction

From the early eighteenth century the demographic setbacks of the previous 50 years were gradually overcome. Despite continuing fluctuations, for example the high mortality of 1727–31 and 1740–42 in England and of the 1740s in Scotland, population growth accelerated from the 1740s and increased rapidly from around 1780 to 1820 when it reached rates never achieved before or since in Britain. New patterns of regional growth and distribution were established as urban and industrial areas, many of them in hitherto thinly populated regions, grew rapidly and attracted large numbers of migrants. Thus from the 1740s to the 1820s there were significant changes in demographic trends – the first stages of the so-called demographic (or vital) transition – and in the geography of population.

The precise reasons for changes in mortality and fertility are still poorly understood and their impact at regional level is difficult to estimate in the absence of accurate and regular population counts prior to the first census of 1801. They involve economic, social, environmental, and medical factors. Economic forces and growing demand for labour in both agriculture and industry stimulated movement to areas and sectors of innovation and expansion; improved living standards encouraged earlier marriage and bigger families especially at a time when lack of regulation and wide-spread domestic industry enabled children to contribute to family income from an early age. Social responses to perceived opportunities relaxed restraints on early marriage from the 1780s to the 1820s. On the debit side, from the late eighteenth century concentration of population exposed more people to risks of infectious disease and imposed an urban penalty which held back other improvements in life expectancy and, from the 1820s, checked the general fall in mortality. To what extent improvements from the mid-eighteenth century were due to medical factors is uncertain. Plague, one of the most virulent epidemic diseases of the past, had disappeared from Britain after 1665. Smallpox, one of the scourges of the eighteenth century, was substantially controlled by inoculation and, particularly, Jenner's method of vaccination (1798). But as towns grew, poor environmental

DOI: 10.4324/9781003174783-3

conditions and inadequate water supplies and sanitation created more adverse conditions for a growing proportion of the population. Medical practice had few effective answers to the infectious diseases which flourished in these conditions until the bio-medical revolution from the late nineteenth century (Chapters 6, 8, 11).

Given these changes, it is not surprising that there was considerable contemporary controversy over eighteenth-century population trends. Some late eighteenth-century commentators, focusing on rural areas of little change, believed that Britain's population was at best slow-growing. Others alerted by rapid increases in towns and in industrial and innovative farming areas, feared a population explosion and, like the Reverend Thomas Malthus, urged restraint on growth. In his *Essay on the Principle of Population* (1798, revised and modified in successive editions to 1830) Malthus saw the critical problem as the relationship between population and food supplies. Over-rapid population growth would exceed the rate at which agricultural production could expand, forcing up prices and reducing real wages leading to increased mortality (his 'positive check'). Unless a reduction in birthrate could be achieved via the 'preventive checks' of abstinence ('moral restraint') and deferred marriage, demographic disaster would ensue.

These ideas reflect a population/resource system in which the critical relationships in pre-industrial times were between population and food prices. A poor harvest could trigger off high crisis mortality. This would be likely to lead to deferred marriage and reduced numbers of births. Either or both would reduce population growth and ease pressure on food supply and, hence, prices, a situation common up to the late seventeenth century (Figure 3.1a). By the late eighteenth century, however, the system was more flexible (Figure 3.1b). Positive feedback in the non-agricultural sector played a fuller role in employment and wage levels. Improved purchasing power and better transport gave access to food supplies from outside the locality or through imports. Reduced dependence of population on local food supply allowed fertility to rise and provided wider outlets for the additional labour.

Mortality trends

In England and Wales the crude death rate fell from 32 per thousand in the early eighteenth century to 22–26 per thousand in the early nineteenth century (Figure 3.2) with similar falls in Scotland, although the poor quality of its parish registers makes it impossible to give precise figures. English life expectancy at birth improved from 33–35 to 36–37 years as famine ceased to be an absolute threat to life: the last major famine in Britain struck Scotland in the 1690s. A century later the failure of the Highlands' harvest was met by bringing in food on the new roads and canals and demographic disaster was avoided. While it is difficult to provide firm evidence of the direct link between food supply, a better diet and improving life expectancy, increased home food production and better access to markets through improved communications and the ability to import when necessary was reflected in greater availability of foodstuffs at, generally, a lower real cost. This must surely be a key factor in a steadily falling death rate during this period.

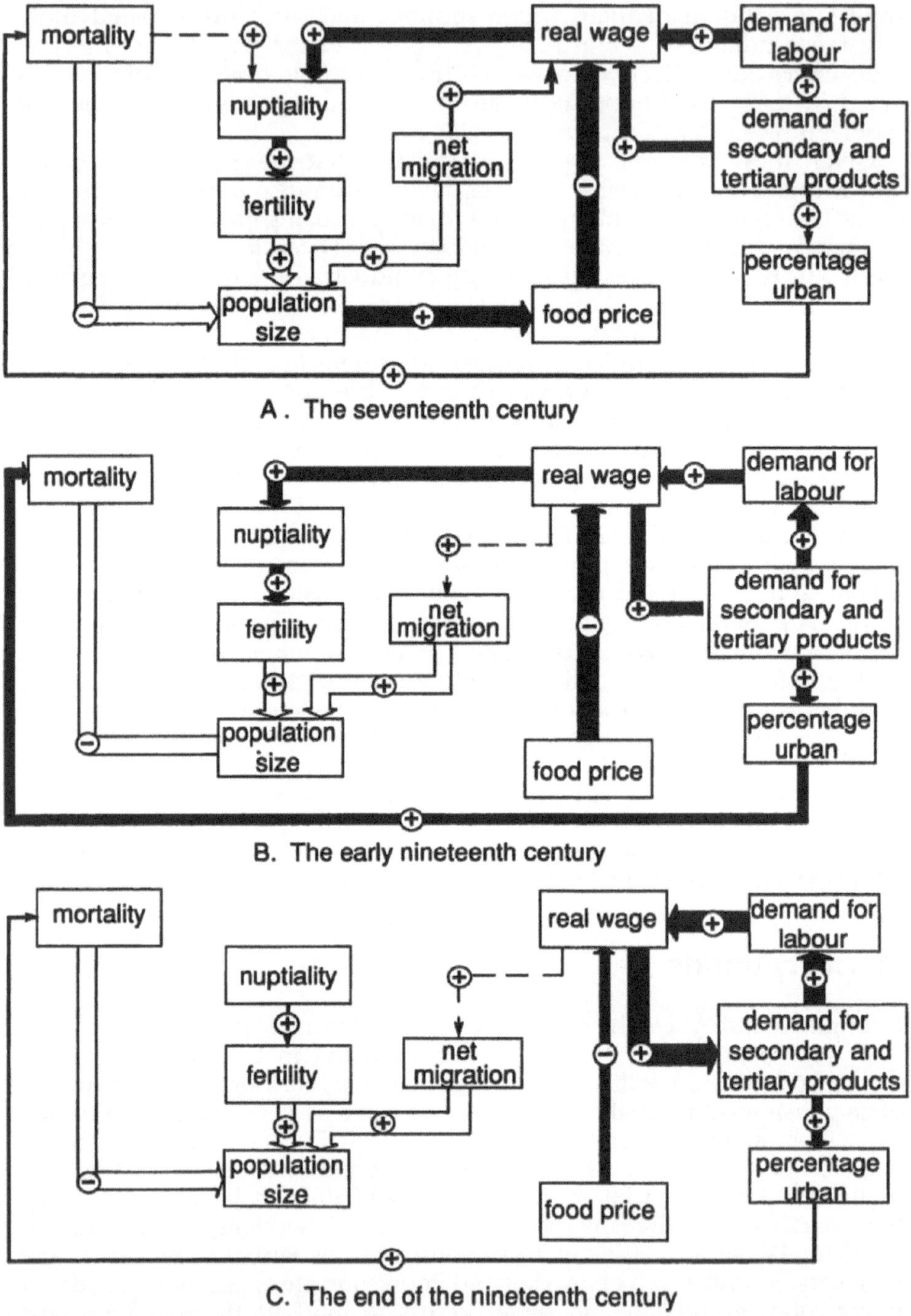

Figure 3.1 A system of demographic/economic relations in British population development. Mortality, nuptiality, fertility and population size are a natural growth system influenced by various exogenous influences and feedbacks which check (negative) or promote (positive) population trends. Based on Wrigley and Schofield (1981) pp. 470, 474 and 477.

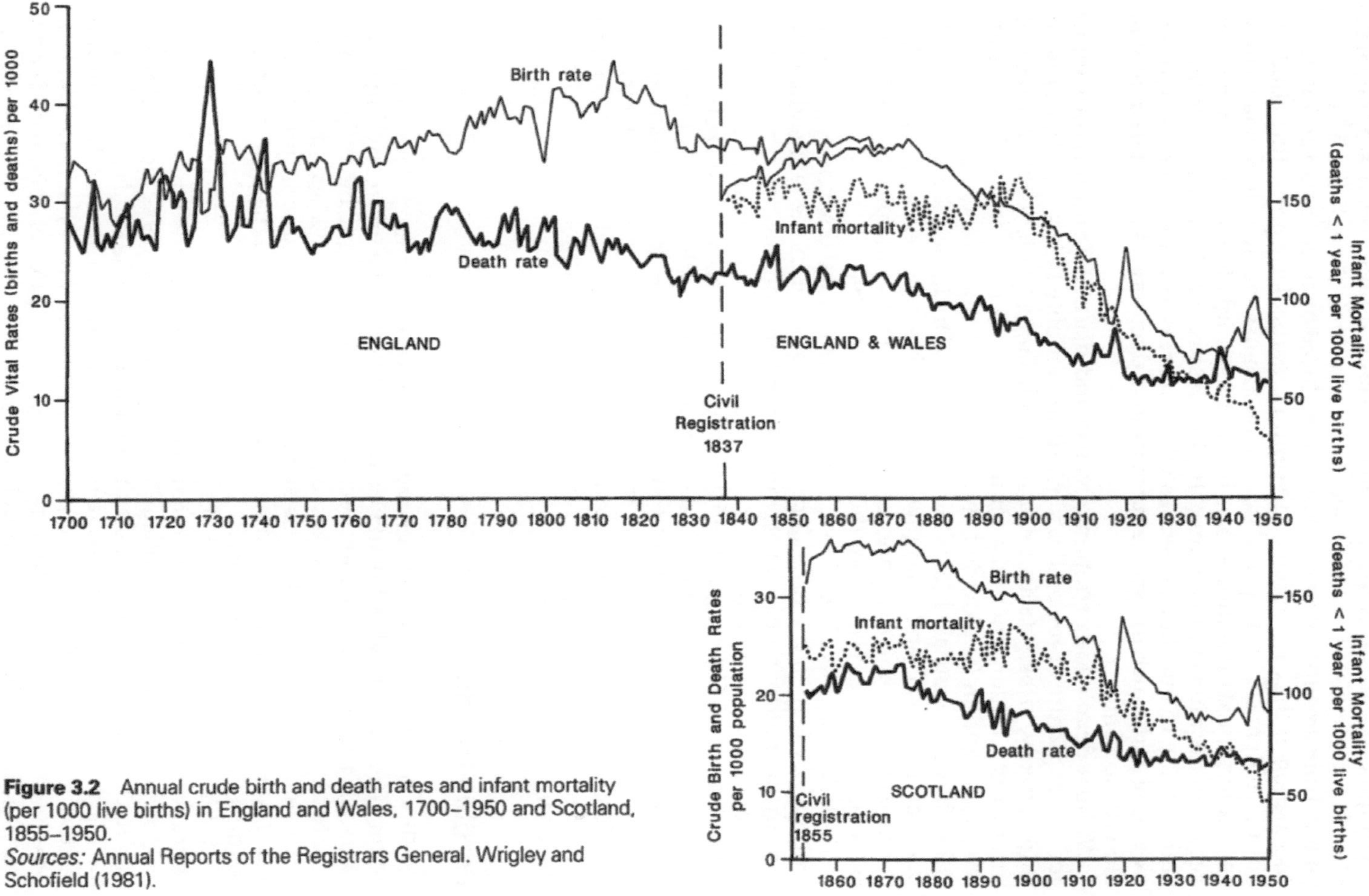

Figure 3.2 Annual crude birth and death rates and infant mortality (per 1000 live births) in England and Wales, 1700–1950 and Scotland, 1855–1950.
Sources: Annual Reports of the Registrars General. Wrigley and Schofield (1981).

On the other hand, control of disease showed little advance in general. Most epidemic diseases continued to defy control, though the 15 per cent fatality from smallpox was substantially reduced by widespread adoption of vaccination in the early nineteenth century, particularly protecting children during epidemics. Widespread protection against smallpox in Scotland may have contributed one-third to one-half of the population increase in the first decade of the nineteenth century. In cities, however, there was widespread vulnerability to infectious diseases which increased with progressive overcrowding and which the greater provision of doctors and hospitals in towns did little to combat.

Without detailed registration of cause and place of death it is difficult to get a clear picture of regional variations in mortality before Victorian times. Undoubtedly, urban death rates were substantially higher than those of rural areas. London suffered a continuous excess of deaths over births until the early- (and in its crowded central areas the mid-) nineteenth century. Other large industrial cities – Birmingham, Manchester, Leeds and such ports as Liverpool and Glasgow – shared this experience. Even smaller towns were vulnerable to periodic set-backs due to epidemics when deaths exceeded births, as seen from their bills of mortality.

Mortality may have been lower in purely agricultural districts of southern and eastern England, where deprivation was offset in the late eighteenth century by 'outdoor' poor relief (the so-called Speenhamland system), than in more varied proto-industrial economies of parts of the rural midlands and south Pennines. In Scotland the Highlands seem to have lagged behind the central and eastern lowlands, despite a considerable broadening of its economy in the late eighteenth century. Already the northwest-southeast, and inner city-suburb gradients from higher to lower mortality which are still features of Britain's demographic map may be discerned, though considerable local and annual variations makes it difficult to detect clear regional patterns in mortality.

Fertility trends

Fertility trends are now considered to be much more important in population increase in England than was thought before the impressive analysis of parish registers and family reconstitution carried out by Wrigley and Schofield (1981). Social responses to economic factors had always kept births in pre-industrial Britain at levels substantially below those in present-day less-developed countries. Marriage rates and age at marriage were crucial factors in determining fertility.

Marriage was not universal in England and Wales where perhaps 15 to 20 per cent of women remained unmarried in the early eighteenth century; the proportion was higher in Scotland though cohabitation was more common as reflected in higher levels of illegitimacy in many of its rural areas. During the eighteenth century the average age of marriage fell steadily from the relative high of 28 years for men and 26.8 for women of the mid- and late seventeenth century (reflected in the lower birth rates of this period) and marriage rates rose sharply, stimulated by rising real wages in an accelerating economy. Scotland undoubtedly shared in this stimulus, especially in the Highlands and Islands. The sparse evidence suggests earlier marriage, with the average age for women falling from 25.3 to 23.4 years between the first and third quarters of the eighteenth century, though with a lower marriage rate and substantial male emigration from mid-century this did not necessarily produce larger families than in England.

Despite a fall in real wages at the end of the eighteenth and beginning of the nineteenth centuries, early and more universal marriage was encouraged in many parts of Britain by the rapid growth of industry and the employment afforded to women and children, especially in handicraft industry. In the early nineteenth century the mean age of marriage for women in England was 23.4 and only one-tenth never married. During the years 1783–1828 the crude birth rate in England increased from around 35–36 per thousand to a peak of over 44 in 1815 before returning to the previous level in the 1830s (Table 3.1). In Scotland, with less evidence to go on, there seems to have been an increase in family size in the industrial areas between mid- and late century. This is consistent with a lower average age of marriage for women in urban rather than in rural areas. Therefore, in most parts of Britain a combination of earlier marriage – accounting for perhaps half the increase in gross reproduction rates – and somewhat bigger families contributed, along with declining mortality, to accelerating natural increase and faster population growth from the 1780s to the 1820s. Moreover, it gave a youthful population structure predisposed to continuing growth as suggested by information on ages from the 1821 census (Table 3.2).

Differences in the fertility of the population of town and country affected regional rates of increase. Increased natural growth in the developing textile areas of West Yorkshire and the North West of England and in the East and

Table 3.1 Summary of mortality and fertility trends in Great Britain, 1740s–1950s

Decadal average	England (to 1830s) and Wales (1830s–)			Scotland (1800s–)		
	Average crude rates per 1000					
	Birth	Death	Infant mortality	Birth	Death	Infant mortality
1740s	33.02	29.32	—	—	—	—
1770s	36.20	26.61	—	—	—	—
1800s	39.22	25.86	—	35.12[1]	24.12[2]	163.8
1830s	35.70	22.18	—	—	26.84	—
1860s	35.15	22.50	154.1	35.03	21.97	121.0
1890s	29.87	18.21	153.5	30.25	18.47	127.7
1920s	18.33	10.91	71.8	21.48	13.76	88.8
1950s	15.70	11.63	25.7	18.39	12.14	31.7

Sources: England, 1740–1830s: Wrigley and Schofield (1981), Table A3.3
Scotland, Flinn *et al* (1977)
1860s–1950s: Annual Reports of the Registrars General for England and Wales and for Scotland
The crude rates are per 1000 population (birth and death) and per 1000 live births (infant mortality) for the decade indicated.
[1]Flinn, Table 4.4.5
[2]A weighted rate estimated for the four principal cities from Flinn, Table 5.5.4.

Table 3.2 Age structure in Great Britain, 1821–1951

Census Year	Percentage of population aged			
	0–14	15–44	45–64	65+
1821	48	29	16	7
1851	35	46	14	5
1871	36	45	15	5
1891	35	46	14	5
1911	31	48	16	5
1931	24	47	22	7
1951	22	43	24	11

Source: Censuses of Population

West Midlands was, according to Deane and Cole (1967), due mainly to increased fertility. However, the large towns depended on migration to make up natural deficiencies in population though by the early nineteenth century high suburban birth rates in provincial and industrial cities were reflected in natural growth. London's fertility levels remained low with a continuing deficiency in natural growth due to high mortality. In many rural areas higher birth rates contributed to above-average natural increase, notably in eastern England and in southern and south-western counties. Despite somewhat bigger families in the late eighteenth and early nineteenth centuries there is no clear evidence that relatively generous outdoor poor relief promoted higher fertility in such mainly agricultural areas. It may have helped more children to survive and, given demand for labour in agriculture and local handicraft industry, led to bigger households.

Fragmentary Scottish evidence suggests faster growth with rather higher fertility in the eastern lowlands (especially the Lothians). There were considerable contrasts between north-east Scotland where farm servants lived on the farm, marriage was later and marriage rates lower but illegitimacy higher, and the southeastern and western areas of the country where earlier and higher rates of marriage were reflected in somewhat higher birth rates. In the Highlands and north of Scotland natural growth was relatively rapid in the late eighteenth century, not least because of increasing birth rates and relatively early marriage: but by the 1820s relative decline in the non-agricultural sectors of the economy were a prelude to falling marriage and fertility rates and, especially from the 1840s, increased emigration.

Migration

Despite such differences in vital rates, migration was increasingly the key to differential regional growth from the mid-eighteenth century. Urban growth was particularly dependent on recruitment of labour from surrounding rural areas and, notably in London and the big cities, from other regions. Preindustrial society was much more mobile than once thought: large numbers of young people entered apprenticeships and service in London from all parts; county and market towns recruited from up to 30 to 40 miles around. With the onset of larger-scale industrial activity in areas of proto-industrialization and, from the late eighteenth century, factory industry, the degree and range of mobility increased. Much of this movement was from rural areas to nearby towns, although London and the bigger cities drew upon regional and, for certain types of skilled labour, national hinterlands. These are the hallmarks of the second phase of the early transitional society in Wilbur Zelinsky's five-stage mobility transition (Zelinsky, 1971).

Migration is difficult to measure in the absence of population registers. Calculation of the net gain or loss by migration must be speculative in view of the lack of population censuses before 1801 and of vital registration before 1837 in England and 1855 in Scotland. Nevertheless, comparisons of parish listings and names in parish registers suggest a high turnover throughout rural Britain where annual hiring of farm labour continued to promote considerable local movement. Poor Law records also point to substantial migration of craftsmen, some over considerable distances.

The impact of external migration on both general and regional population is equally hard to assess. On balance, England's net loss was probably modest – less than 0.1 per cent per annum in the early nineteenth century – though the total numbers of emigrants were substantial between 1801–31 and increased sharply in the depressed 1820s, with a net loss estimated at 125,000. Precarious estimates for the eighteenth century suggest that England may have had a net loss overseas of half a million, though substantial movements to the North American colonies up to the American war of Independence (1776–1783) were off-set by growing Irish immigration to Britain. In the late eighteenth and early nineteenth centuries there were considerable numbers of Irish in London, Liverpool, Glasgow and in many of the industrial areas of North West England and western Scotland. Emigrants came from all parts, and from urban and industrial as well as rural backgrounds. In the 1740s much of the substantial Scottish migration went to the Appalachian frontier and after the American War to Canada. The greater proportion was from the western Highlands which also provided a large part of the renewed flood of emigration of the 1820s and early 1830s.

Low natural increase or, as in London and large provincial cities, loss by excess mortality meant that most late eighteenth-century towns depended on migration to sustain their growth. So too for a generation or two did recently-developed spas and residential towns as well as the new industrial centres. London's population increased from 675,000 to 750,000 between 1750 and 1780 with an estimated migrational gain of over 400,000 in the metropolitan counties. Many cities doubled in size during that period: Manchester to 42,821 in 1788; Birmingham to around 50,000; Glasgow from 25,000 to over 36,000 between 1755 and 1785; Leeds to 17,117 in 1775; and Liverpool from maybe 20,000 in 1750 to 54,000 in 1790. All experienced substantial in-movement. Moreover, the pace and level of migration to towns accelerated between 1780 and the 1820s, the period of most rapid increase for many large British cities. Over two-thirds of Liverpool's growth in the 1790s was from migration and included many Irish as well as Scots and overseas-born. More than half of the increase in London and the metropolitan counties between 1780 and 1830 came from an estimated net in-migration of three-quarters of a million. Many of the new industrial towns had even faster relative growth between 1780 and 1830: ten-fold increases in such textile towns as Bury, Bolton, Bradford and Huddersfield; seven-fold increases in Manchester and Leeds to 182,000 and 123,000 respectively; substantial growth in metal-working towns such as Wolverhampton and Sheffield; rapid growth in ports, with a five-fold increase in both Liverpool and Glasgow (each to 202,000) and similar rates in the naval ports of Plymouth and Portsmouth and such coastal resorts as Brighton.

In contrast older regional centres and the county and market towns of non-industrializing areas grew relatively modestly, mainly through natural increase or short-range movement from adjacent rural areas. Exeter's population of 28,000 in 1831 was little more than twice that of the late seventeenth century; Norwich, England's second city before the industrial revolution, had only 62,000 people in 1831 as compared with 40,000 in 1786; Edinburgh, despite its capital status and primacy in financial and professional services, grew more slowly than Glasgow though it doubled its population to 162,000 between 1801 and 1831.

Despite substantial migration from the countryside, rural population growth accelerated to peak increase between 1811 and 1821 in most areas of Britain. Demand for labour increased, especially in industrial villages and good farming areas, though the proportion of farmworkers in the population fell. There is little evidence of loss of population due directly to enclosure and improvements: indeed these usually stimulated growth except where land was converted to grazing (Chapter 4). However, in parts of midland and southern Britain large landowners often sought to limit the numbers of cottages in 'closed' parishes to avoid high Poor Rates. In such areas as the East Midlands there was considerable concentration on 'open parishes', often the source of cheap labour for handicrafts. In this and other rural manufacturing areas in the Pennines, the West Country and South West England, the linen areas of eastern Scotland, and in many metal-working districts diverse sources of income in a dual manufacturing and farming economy absorbed much of their high natural increase.

Nevertheless, urban growth stimulated substantial migration from the countryside and produced complex currents of movement towards cities and a cumulative shift of emphasis in population growth towards midland and northern England and South Wales, although London continued to dominate both south-eastern England and the national system. In Scotland these flows were principally towards central Scotland. Many people tended to move initially within their locality, then in a pattern of chain migration from smaller to larger towns. Often they did not sever links with their home area since much migrant labour (in both agriculture and industry) was seasonal. However, as industrial work, in particular, focused more on continuous production processes so, stage by stage, people moved up the settlement hierarchy to larger settlements.

Population distribution

By the early nineteenth century a new pattern of population distribution had been achieved in Britain (Figure 8.1). The percentage of the population

Table 3.3 Regional population trends in Great Britain, 1751–1951

Region	Total (000s)					Percentage increase over				Percentage share of GB population				
	1751	1801	1851	1901	1951	1751–1801	1801–51	1851–1901	1901–51	1751	1801	1851	1901	1951
ENGLAND	5691	8305	16765	30509	41159	33	102	82	35	76.8	79.1	80.5	82.5	84.3
South East	1690	2503	5121	10508	15127	52	104	106	44	22.8	23.8	24.6	28.4	31.0
West Midlands	558	858	1707	2997	4423	56	100	74	48	7.5	8.2	8.2	8.1	9.1
East Midlands	472	640	1149	1998	2893	47	80	75	43	6.4	6.1	5.5	5.4	5.9
East Anglia	484	626	1041	1147	1382	30	68	8	22	6.5	6.0	5.0	3.1	2.8
South West	1109	1344	2248	2553	3229	26	67	15	26	15.0	12.8	10.8	6.9	6.6
Yorkshire–Humberside	461	817	1811	3515	4521	82	121	94	29	6.2	7.8	8.7	9.5	9.3
North West	425	882	2519	5291	6447	110	186	109	22	5.7	8.4	12.1	14.3	13.2
North	494	634	1166	2516	3137	33	83	115	26	6.7	6.0	5.6	6.8	6.4
WALES (and Monmouth)	449	588	1165	2019	2599	35	98	73	29	6.1	5.6	5.6	5.4	5.3
SCOTLAND	1100	1625	2889	4472	5096	48	100	55	14	17.1	14.4	13.9	12.1	10.3
GREAT BRITAIN	8230	10518	20816	37000	48854	28	98	78	32	100	100	100	100	100

The regions are the 1971 Standard Regions

Sources: Adapted from Lawton (1977) using data from Deane and Cole (1962), Flinn *et al* (1977) and Censuses of Population 1801–1951

living in the mainly agricultural and in 'mixed' counties of England and Wales, which had remained little changed (at around 31 per cent in each group) up to 1781, fell to 26 and 19 per cent, respectively, by 1831. In contrast rapid acceleration in growth in the industrial counties gave them 45 per cent of the population of England and Wales by 1831 as compared with 38 per cent in 1781 and 36.7 in 1751. It is impossible to make similar eighteenth-century estimates for Scotland but the mainly rural areas of the northern Highlands, north east and Borders had 44.3 per cent of Scotland's population in 1801 compared with 54.7 in the western and eastern Lowlands: in 1831 the figures were 40 and 60 per cent respectively.

Of the major regions South East England increased its share of Britain's population (Table 3.3) due essentially to London's large, continuous growth at the expense of most other parts of the region and considerable migration from all other areas of Britain, Ireland and overseas. Of the other regions only the West Midlands, Yorkshire and, most notably, North West England increased their share of the national total; Scotland's proportion declined but had an increasing percentage in the central lowlands by 1831: all were areas of early industrialization, relatively high levels of urbanization and substantial net migrational gain. Wales as a whole lost ground a little, though many of its rural migrants went to the South Wales coal and iron districts or to nearby areas of North West England and the West Midlands. In the most marked relative decline of population from the mid-eighteenth century South West England lost ground due to a relative decline in rural industry and commerce and a slower growth in labour demand in an increasingly specialist stock-farming area. Indeed from the 1830s, population migration from the British countryside was increasing in contrast to rapid growth in industrial and urban Britain (Chapter 8).

The impact of population change

In a period of relatively high fertility, high but declining mortality and high rates of mostly short-distance mobility there were progressive impacts on families, households and communities. These impacts were felt most clearly in large towns and new industrial areas.

The most obvious structural characteristic of the population from 1740 to 1830 is the high proportion of children and the consistent but small excess of women. The proportion of 25–59 year-olds in the population of England and Wales fell from over 40 per cent in 1740 to approximately 35 per cent in the 1830s (Table 3.2). All age groups under 25 increased significantly, especially those aged 0–4, which rose from around 12 to 15 per cent of the population. Scotland's age structure was much more stable until the surge in natural growth from the late eighteenth century. However, the increase in the juvenile dependency ratio was less disadvantageous than might be thought since many children contributed to productive processes in the early stages of industrialization.

Considerable male emigration from Scotland from the 1740s produced a lower male: female ratio than in England, although both were consistently in the range 85–98 in the first half of the nineteenth century. This excess of women was accentuated as young women moved to work in nearby towns where availability of eligible women led to higher marriage rates and earlier age of marriage. Urban work opportunities made it easier to set up an independent home and contributed, along with the industrialization of the countryside, to a falling mean age of first marriage: for example mean age at first marriage for brides fell dramatically from 27.4 years (1745–54) to 23.5 years (1820–29) at Stratford upon Avon.

Despite the falling age at marriage, there was little change in mean household size during this period (or indeed between the seventeenth and the early twentieth centuries). The nuclear family of man, wife and children was the norm with a fairly stable mean household size of between 4.3 and 4.9 for England and Wales as a whole, but varying between regions and social classes. Household size appears to have declined in such rural counties as Gloucester, Worcester, Norfolk, Suffolk and Cumberland, and to have increased most rapidly in fast-growing industrial settlements of North West England reflecting the pressures on housing in rapidly-growing towns where children may have continued to live with parents after marriage. In both rural and urban areas, however, 'houseful' size was swelled by the presence of migrant kin, of living-in apprentices and, in some cases, of servants, placing extra pressure on living space in often overcrowded housing.

The impact of late eighteenth-century population change on everyday life often depended on age. In the first half of the eighteenth century those born in rural areas often lived in the same or an adjacent parish all their lives. They were little affected by the considerable demographic and social changes taking place around them. Men and women born in the countryside in the second half of the eighteenth century would have been much more likely to feel the impact of change. In northern England the growth of new water-powered factories might attract young men and women from domestic to factory textile work in a growing industrial village. Alternatively a craftsman might learn completely new skills, for example in metal and machine manufacture, leading eventually to migration to industrial areas. Such movements from home increased opportunities for early marriage and fuelled population growth.

However, a good deal of labour migration was forced by economic decline (particularly in domestic weaving and spinning) or by down-turns in trade, especially during and after the Napoleonic wars. Although much of it was over relatively short distances, and undertaken almost entirely on foot, migration had a big impact on people's lives. In many cases, a move of some 50 miles was enough to cut migrants off from their community and restrict opportunities for visits home. Although contact with more distant kin may have been maintained in times of crisis (especially death or illness) for most migrants life would be played out within their new community.

Acute illness and poverty could, however, lead to enforced return migration, especially where settlement restrictions under the Poor Law were strictly enforced. Destitute families could be forcibly removed to their parish of 'settlement', however distant, before qualifying for poor relief. This could lead to the break-up of the family home and increased the risk of destitution and early death. Such forced migration could have serious demographic impacts.

One important factor which structured the pattern of migration within particular regions, and which lessened its impact on individuals, was the strength of family ties. Brothers and sisters, although perhaps moving individually, often ended up in the same town and subsequently helped one another get settled in, to find work or housing, and provided support in times of personal crisis. Individual migrants adopted many strategies for providing stability within their everyday lives.

Many of these are illustrated in the family history written in 1826 by Benjamin Shaw who lived variously in north Yorkshire, south Cumbria and north Lancashire. Benjamin's father, Joseph was born in Garsdale near Sedbergh, North Yorkshire in 1748 and was apprenticed for nine years to a weaver in Westmorland. Marrying in 1771, he settled near Dent in the next valley south from Garsdale where Benjamin was born. As well as weaving, Joseph Shaw acquired self-taught skills in clock repairing and metal working and, in 1791, moved to work in the new worsted mill at Dolphinholme (Lancashire) some 35 miles from Dent, a direct result of labour recruitment in the Dent area by the mill owner. Joseph was laid off from the mill after only two years and returned to the Sedbergh area, but Benjamin continued at the mill until the end of his apprenticeship and in 1793 married a girl from Lancaster who had also lived for a short time in Preston. In 1795 he too was laid off and they reluctantly left Dolphinholme for Preston (Lancashire), some 20 miles distant, where they were to live for the rest of their lives. Later the same year they were joined there by Joseph and other members of the family: Preston was to be the main focus of the Shaw family for the next thirty years.

Such migration was generally stimulated by a combination of economic necessity and family ties. There is no reason to believe that the Shaws were atypical of other families of the time and their history gives valuable insights into the nature, causes and impacts of the extensive local and rural-urban migration of the later eighteenth and early nineteenth centuries.

Bibliography and further reading for Chapter 3

The standard work on demographic change is:

WRIGLEY, E. A. and SCHOFIELD, R. S. *The population history of England 1541–1871: a reconstruction* (London: Arnold, 1981).

See also:

BONFIELD, L., SMITH, R. and WRIGHTSON, K. (eds) *The World we have gained: Histories of population and social structure: essays presented to Peter Laslett on his seventieth birthday* (Oxford: Blackwell, 1986).

DUPAQUIER, T. (ed.) *Malthus past and present* (London: Academic Press, 1983).

FLINN, M. W. *British population growth 1700–1850* (Macmillan, 1970).

FLINN, M. W. *et al. Scottish population history from the seventeenth century to the 1930s* (Cambridge: Cambridge University Press, 1977).

GLASS, D. V. and EVERSLEY, D. E. C. *Population in history* Part 2, Great Britain. (London: Arnold, 1965)

HABAKKUK, J. *Population growth and economic development since 1750* (Leicester: Leicester University Press, 1972).

LASLETT, P. *The world we have lost: further explored* (London: Methuen, 3rd edition, 1983).
MALTHUS, T. R. *An essay on the principle of population* (Penguin edition with introduction, 1798, Harmondsworth, 1970).
TRANTER, N. *Population and Society, 1750–1940* (London: Longman, 1985).

Specifically on migration see:

CLARK, P. and SOUDEN, D. (eds) *Migration and society in early-modern England* (London: Hutchinson, 1987).
HANDLEY, J. E. *The Irish in Scotland, 1789–1845* (Cork: Irish University Press, 1943).
HOLDERNESS, R. A. 'Personal mobility in some rural parishes of Yorkshire, 1777–1822; *Yorkshire Archaeological Journal* **42** (1970), pp. 444–54.
NICHOLAS, S. and SHERGOLD, P. R. 'Internal migration in England 1818–39'. *Journal of Historical Geography* **13** (1987) pp. 155–68.
REDFORD, R. *Labour migration in England 1800–1850* (Manchester: Manchester University Press, 1926).
SOUTHALL, R. 'The tramping artisan revisits: the spatial structure of early trade union organization' in Withers, C. (ed.) *Geography of population and mobility in nineteenth century Britain* (Cheltenham: HGRG, 1986).
WITHERS, C. 'Highland migration to Dundee, Perth and Sterling, 1753–1891' *Journal of Historical Geography* **11** (1985) pp. 395–418.
ZELINSKY, W. 'The hypothesis of the mobility transition' *Geographical Review* **61** (1971) pp. 219–49.

4

The countryside from the 1740s to the 1830s

Introduction

The achievements of Britain's rural areas under an innovating inorganic economy were considerable. Increases in agricultural output grew from 0.6 per cent per annum in the early eighteenth to 0.8 by the late eighteenth century and to 1.2 per cent per annum 1801–31, largely meeting the food and many raw material requirements of a substantially larger population. Secondly, with two-thirds of capital stock and one-third of national income in 1801, agriculture continued to attract substantial investment and to promote increases in farmland, buildings and produce, though this slackened substantially in the post-1815 depression. Thirdly, considerable growth in rural industry strengthened the basis of the economy of many areas and promoted further population growth (Chapter 5). Of the rural two-thirds of England's population in 1801, half were involved in manufactures, crafts and services, though outside the main areas of rural industry most country folk, an increasing proportion of whom were landless labourers, depended on agriculture for a living. Nevertheless, the relative importance of employment on the land was falling. By 1841 only 28 per cent of Britain's population was supported mainly by agriculture, although the proportion varied from three-fifths in the Scottish Highlands and the most rural counties of England and Wales to under one-fifth in central Scotland, South Wales, the West Midlands, and North East England and only one-tenth in Lancashire.

In the late eighteenth century a number of forces were changing the nature and balance of the rural economy. Periodic dependence on imported grain in years of dearth added to the growing range of exotic food imports, particularly tea and sugar (Table 5.1). Despite the protection of home producers by the Corn Laws, imports of food and formerly home-produced natural raw materials – such as wool, flax and hides – gradually increased during the post-1815 depression, though early-Victorian farmers still largely fed a population nearly three times that of the early eighteenth century. The structure of landownership and farming, still heavily dominated by large landowners particularly the landed aristocracy, reflected both improvements in farming and the impact of enclosures. The landless peasantry and

DOI: 10.4324/9781003174783-4

many smallholders – ousted by more substantial tenant and, in areas of fragmented ownership, family farmers – swelled the ranks of agricultural labourers or moved to the towns. Increased accessibility through improved turnpike roads, canals and transportation systems (whether local carters and carriers, or regional and inter-regional coach and narrow boat services) increasingly brought most rural areas into regular contact with growing urban/industrial markets and were a stimulus to specialization and improvement in all branches of farming.

Three sets of forces shaped the nature of change: the structure of the economy and international trade; agricultural methods, including land management and crop and stock improvements; and the tenurial and institutional frameworks within which farming operated. In none of these can it be truly said that there was an agricultural revolution in the eighteenth century: rather a quickening of the pace of improvement, reflected in rising yields and productivity of both land and labour, and a move to more fully market-orientated farming systems with increased regional specialization adapted to physical, tenurial and market conditions.

Agrarian change

Most changes in crops were associated with the wider introduction of 'rest crops' into arable rotations. Under traditional open-field grain-growing systems, fallowing to rest and cultivate the soil was essential while grazing of fallows and commons was crucial for stock. New crops introduced from the early seventeenth century enabled farmers to cultivate continuously, to increase animal fodder and (through greater stock capacity) to provide more animal manures. Clovers and other leguminous plants such as lucerne and sainfoin, added substantially to soil nitrogen. Together they enhanced soil fertility and the yield of grain crops. By the early nineteenth century average wheat yields in England were 20–21 bushels per acre, double those of 1650 and one-third greater than a century previously. Turnips were eaten off the field by sheep flocks that were systematically folded over the arable, dunging and treading – a sort of walking 'muck machine' – to the benefit of both soil texture and fertility.

Turnip husbandry was best suited to light soils and, in the balanced Norfolk four-course system, achieved spectacular results on improved arable and reclaimed heath and down throughout eastern Britain. Other roots such as the swede and the mangold, introduced from the 1760s and 1780s respectively, were better suited to heavier soils and more often fed to cattle along with hay from both traditional meadows and the 'new' rotation grasses and legumes. The emphasis was on fatstock housed in bullock yards from which manure was collected to be spread when fully rotted. Pastures in both arable and upland areas benefited all stock: milk cattle for butter and cheese; sheep for the growing eighteenth-century market in wool and mutton. Although the effect of improved fodder supply on livestock is difficult to estimate, sheep flocks possibly doubled between the early eighteenth and early nineteenth centuries. Cattle numbers may have fallen, partly due to breeding of improved dual-purpose animals for milk and meat, partly to their replacement as draught animals by horses, a change largely completed by the early nineteenth century.

Indeed the growth in stock, including horses in all forms of transportation, was a factor in the continuing dominance of grain in most types of British arable farming. Wheat, the preferred bread grain, accounted for nearly half the cereal acreage by 1836: having ousted rye by the end of this period, it was the main crop throughout southern Britain wherever climatic and soil conditions permitted. Barley, principally malted for beer and, in Scotland, whisky, remained an important spring-sown rotation crop on some 30 per cent of England's cereal acreage in the 1830s and rather more in Scotland. Oats, a more tolerant and versatile crop for damp western and upland conditions, maintained its role as a food cereal in upland areas, particularly of Scotland, but was increasingly used as fodder giving it second place to wheat (about one-third of the cereal acreage) by the nineteenth century.

Though much of its produce was consumed locally, British – especially English – agriculture had long been a market rather than subsistence economy. Growing demand and better bulk transport along expanding canal systems from the 1760s promoted farming as a business responsive to changing market opportunities and competition. Environmental factors favoured adaptability to new technical and market considerations in a period of considerable change. Climatic extremes and disasters seldom affected the whole national harvest, and violent price fluctuations because of glut or dearth were constrained under the Corn Laws by permitting imports once prices rose above a certain level. Soils are adaptable to a variety of crops and farming systems and the variety of British conditions offered farmers a range of crops, rotations and stock. Such flexibility and convertibility within a mixed farming system provided opportunities to balance production of crop and stock products and to adapt to factors of changing demand (the market) and supply (yields and costs of production).

The responses to innovation were diverse. Progress was first made where 'improvers', often large aristocratic landowners, enclosed open fields, reclaimed land and introduced new crops and rotations, improved stock and more intensive methods of cultivation using such machinery as the horse-hoe and iron ploughs, harrows and rollers. There was also more liberal use of marl and natural manures. On such estates customary forms of tenure (including copyhold) were generally replaced by tenancies controlled by leases specifying standards of cultivation that enhanced and conserved fertility. Especially in the arable areas of eastern Britain these effected substantial change in farming from the 1740s. In the 'good sands' region of north-west Norfolk four-course husbandry was general on such estates as the Cokes of Holkham, 'turnip' Townsend at Raynham and the Walpoles at Houghton: balancing food and fodder crops, autumn-sown wheat was followed by turnips then spring barley undersown with clover and seed grasses to give temporary pasture in the fourth year.

This rotation and its many variants was adopted in widespread reclamation and cultivation of light downland soils from the 1770s in Wessex, the Lincoln Heath and Edge, and on the Yorkshire Wolds where the improvements and landscaping of the Sykes estate at Sledmere provides a classic example. Parallel developments were slower to take root in Scotland. But on both landlord and tenant land in the Lothians and in Ayrshire the new husbandry was practised by the late eighteenth century, while on the Gordon estates of the Buchan plateau of Aberdeenshire four-course agriculture left an abiding mark on the farming landscape.

Innovation was not confined to large-scale farming. Despite their complex web of common and grazing rights, open fields were progressively modified in many parts of England especially where soil conditions favoured more intensive cropping. In Oxfordshire fallows were reduced and open-field rotations doubled into six courses: elsewhere reclamation of additional fields produced four-course and longer rotations. Yet alongside such flexibility and innovation many farmers changed little until the mid-nineteenth century, especially on heavy clayland where soil and drainage conditions limited the use of new crops and new methods of grain husbandry. Specialization on heavy land, for example in the Oxford and Liassic Clay belts of the midlands, was more often associated with enclosure and conversion to grass. Early developments in stock breeding associated with such pioneers as Robert Bakewell of Dishley, Leicester, used selective inbreeding to enhance particular qualities: Bakewell's New Leicester sheep put on weight well and added to carcase value though at the expense of the quality of wool and, some maintain, meat. Similar methods were also used to develop cattle breeds (such as the Colling brothers' Durham shorthorn) with better milking or fattening qualities.

While bullock yards for fatstock were as much a feature of the arable stock systems of eastern Britain as the sheepfold, in the 'grassy shires' of midland England and most parts of western and upland Britain grazings were the key to successful stock farming. In a 'grazing kingdom', as Gilbert White of Selborne called it in 1788, permanent grassland exceeded arable until the Second World War. Although the plough-up of lowland and upland grazings increased the proportion of arable during the Napoleonic Wars, most of the increased crop acreage of the eighteenth and early nineteenth centuries came from elimination of fallows and reclamation of the waste. Grassland provided around three-fifths (up to two-thirds including rotation grass) of the cultivated area of England and Wales in the early nineteenth century: the proportions in Scotland may well have been as great.

Despite the initial success of Fenland drainage in the seventeenth century much seasonally inundated land remained as grazings. Other wetlands – the Somerset levels, the Humber warplands and the Lancashire mosses – notwithstanding Acts for enclosing and drainage were still mainly cattle grazings, though Romney Marsh continued to graze its fine-woolled sheep.

The quality of many permanent grasslands left much to be desired. Although traditional methods of ploughing up ridges and new undersoil hollow drains (filled with stone, brushwood or tile) assisted drainage, grassland management lagged behind improvements on the best arable farms. However liming and dressing with natural manures enhanced the quality and carrying capacity of grass in many areas. There were distinct differences in emphasis between districts fattening stock in lowland Britain, such as the Vales of Aylesbury and Belvoir and much of Northamptonshire, and those specializing in cheese production, for example Cheshire, the lower Severn and parts of Somerset, or butter-making (for example Thomas Hardy's 'vale of little dairies' in south Wiltshire). Near big towns, especially London, meadow hay was a valued crop for both the horse population and the town dairies which, before the railway age, supplied most of the urban milk market.

Upland pastures were another matter. Although cultivation reached considerable elevations, up to 1700 feet in Wales during the drive for more grain of the Napoleonic Wars, the successful upper limit of cultivation was around 800 feet in western Britain, though the 40 inch isoheit was a more critical divide. Historically most of Britain's uplands were cattle-raising country. From the Middle Ages, however, England and Wales substituted moorland breeds of sheep adjusted to heather or grass while downland and heath breeds remained important for wool. In Scotland sheep farming in the Border abbeys spread through the Southern Uplands from the fifteenth century. Cattle dominated the Highland economy until after the '45 rebellion and, despite the efforts of large landowners such as the Dukes of Sutherland to change the farming system and replace joint farms by massive, cheaply operated sheep runs, it was only with the collapse of the Highland economy based on farming, some fishing and kelp (for soda ash for the chemical industry) after the Napoleonic Wars that the emphasis on the store cattle trade, with its annual drovings to the cattle 'tryst' at Falkirk, weakened. Indeed, despite substantial emigration from the mid-eighteenth century (Chapter 3), the population of the Highlands increased and, particularly in the small communities of the western glens and the islands thrived with the adoption of potato cultivation.

While key innovations helped to accelerate agricultural improvement from the early eighteenth century, the diversity of reaction to changing economic, technical and tenurial circumstances emphasizes the importance of looking at their spread (or diffusion), effectiveness and impact on different regions: rather than a period of rapid general improvement, it was a transition marked by improvement in yields, increasingly market-orientated farm economies and considerable flexibility in the use of physical, capital, and human resources. Whereas in some areas there was much progress in adopting new crops and new methods of farming in efficient, well-managed farms by the 1830s, in others innovation and improved production came slowly from the mid-nineteenth century (Chapter 9).

Reclamation

Much of the growth in output in this period came from extension of cultivation, one-third of enclosures by Act being of common and waste. Most woodland had already disappeared, though remnants remained in forests such as Delamere, enclosed in 1812 and subsequently reafforested. Until the late eighteenth century charcoal burners continued to take timber for iron smelting from the Weald, the Forest of Dean and at new furnaces in South Wales, North West England and western Scotland (Chapter 5). In an organic economy small woodlands continued to provide villages with fuel and building and constructional material though major users, such as shipbuilders, increasingly relied on imported timber.

Wetlands, from seasonally inundated valleys and regularly flooded estuaries to extensive fenland and lowland bogs, offered potentially valuable extensions of farmland. Marshland reclamation depended on embanking and regulating natural water courses and cutting additional drains. Reclamation in the lower Trent Valley and the southern Fenland by Cornelius Vermuyden in the mid-seventeenth century was the model for considerable eighteenth-century activity in the Fens, the Hull valley (from the 1760s) and the Somerset levels (from the last decade of the century).

Problems of peat shrinkage following cultivation, maintenance of banks and drains, and the difficulties of raising water into drainage channels in the days before steam pumps were adopted from the 1820s, led to reversion of cropland to waste, especially in parts of the southern Fenland, and impeded progress in the Lincolnshire fens. Though the incentive of substantial urban markets for vegetable and dairy products led to small-scale reclamation, larger peat mosses and meres of south Lancashire (e.g. Chat Moss), Cheshire and north Shropshire went unreclaimed until the railway age. Elsewhere, as around the Humber estuary, warping – impounding flood water and allowing alluvium to settle – extended traditional coastal and estuarine grazings and cash-crop arable farming. But despite attempts at under-drainage, heavy soils from the estuarine carse lands of Scotland to the clay vales of midland and south-eastern England suffered from high costs of working and limited cultivability because of excess ground water, a problem not adequately met until the widespread adoption of pipe drains from the 1840s (Chapter 9).

The most significant extensions to cropland in the eighteenth and early nineteenth centuries came from reclamation of light soils. Former sheep-walks on the chalk and Jurassic limestone and lowland sandy heaths of south-eastern England, the Triassic sandstones of the midland and Lancashire-Cheshire plains, and grazing on coastal and fluvio-glacial deposits could all be cultivated using the new husbandry. The introduction of fodder crops with arable rotations developed the potential for grain growing. Considerable investment in marling (spreading calcareous loam from the many diggings that pit such areas), building up good-quality Southdown and New Leicester flocks and, in some areas, fattening bullocks provided manure to maintain the land in good heart. Although grain yields were not as great as on loam and clay soils, greater ease of working, horse-hoe and drill cultivation and more regular harrowing made for cheaper, more profitable production. Intensive investment and careful control of leases on substantial tenant farms saw grain production and mixed farming systems successfully switch the emphasis in British arable farming to the lighter loams of eastern areas. Even so, many very light, hungry soils (Norfolk's Breckland, the heaths on the Bagshot Beds of Surrey and the Bunter Pebble Beds of Cannock Chase and mid-Cheshire) awaited the early-Victorian improvement boom and many were planted with conifers, to William Cobbet's disgust in the 1820s.

Most of the estimated area of waste, one-fifth of England and Wales in 1800, was moorland and heath. Throughout the eighteenth century patient small-scale extensions of cultivation continued as stone was cleared and small enclosures added to the patchwork of fields in the Lake District, the south Pennines, parts of South West England and mid-Wales. Hill grazings were chiefly for sheep, though the black cattle trade continued to thrive in Wales and Scotland and, where pastures could be improved close to urban markets, dairying flourished as in the Craven district of Yorkshire. Frequent liming, vital to improvement of acid upland soils, led to great increases in lime burning in Derbyshire and parts of the northern Pennines producing better grazings and, where rainfall and soil conditions permitted, fodder crops. The stimulus of high grain prices which rose in the French wars to 119–126 shillings per quarter (as compared with 35–55 shillings pre-1790), and easing and cheapening of enclosure under a General Act of 1801, saw much marginal land reclaimed for cultivation at altitudes not farmed since the early Middle Ages.

This slow process encountered frequent setbacks. For example, 20,000 acres of the Royal Forest of Exmoor were enclosed by Act in 1815, but only through substantial investment of capital and energy by the Knights, a family of Black Country ironmakers, were its farming and landscape transformed in a costly enterprise which was not completed until the 1860s. In Scotland, under different legal and tenurial conditions, many large landowners sought profitability on their Highland estates through large-scale sheep rearing though tenant opposition as well as problems of acquiring skilled Border shepherds held back these 'improvements'. Even after the collapse of cattle and kelp prices in the 1820s and the potato famines of the late 1840s destroyed many small communities, there was much opposition to the clearances in the islands and glens of the western Highlands (Chapter 9).

Enclosure

Enclosure was the mechanism of change in tenure and farm structure which made a marked and lasting impact on the rural landscape of much of Britain. New farming techniques could be and were adopted in open fields but their scattered holdings were difficult and costly to work and common rights, although beneficial to the landless and small farmers, often inhibited good stock management and extension of cultivation. By the early eighteenth century one-third of England's farmland and most of the tillage in Wales were already enclosed. A further 10 million acres (30 per cent of their area) were enclosed after 1730, two-thirds by Act of Parliament, mostly in midland open-field areas (Figure 4.1). The pace of enclosure increased in the 1760s and '70s as improvement quickened; a second surge promoted by the wartime boom in agriculture peaked in 1810–14, accounting for half of English enclosure by Act before 1830, before the pace of reclamation slackened in the depressed 1820s (Table 4.1).

In Scotland, later improvement generally delayed replacement of runrig farming until the 1760s. Consolidation of infield arable strips and large-scale waste improvement, especially on big estates, was facilitated under an enclosure Act of 1661 (which encouraged planting and enclosed grazing)

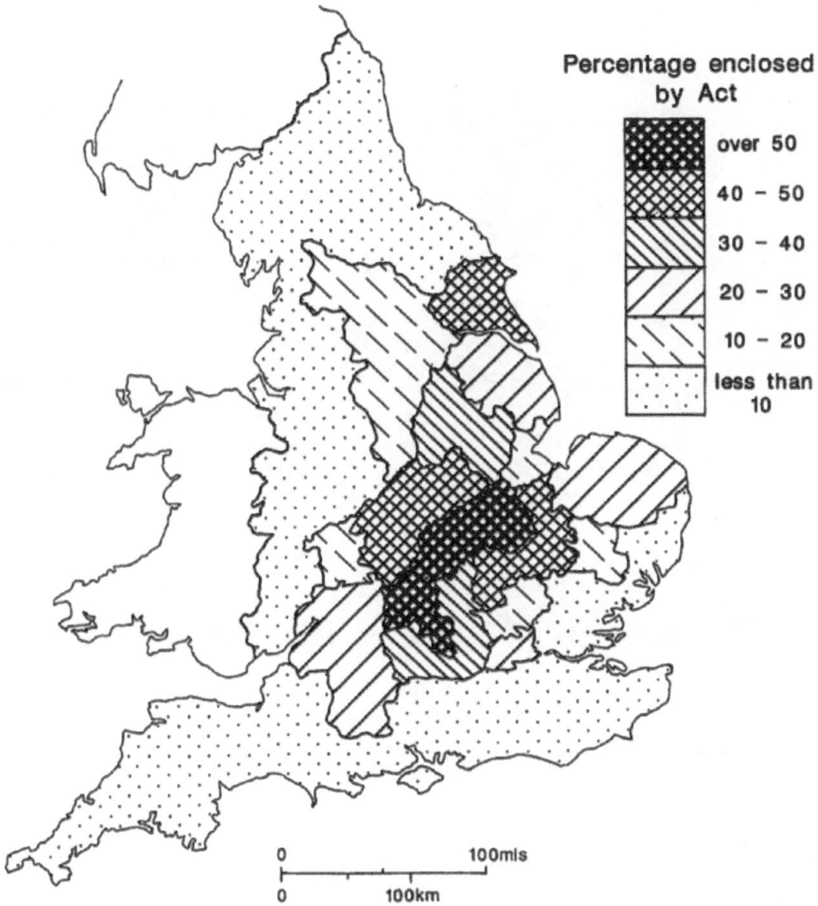

Figure 4.1 Enclosure of open fields in England, 1750–1870. Based on Turner (1980) p. 59.

Table 4.1 Parliamentary enclosure in England, 1730–1830

	Enclosure by Act					
	Open field arable		Common and waste		Total	
Number of acts	3093		2172		5265	
Acres (million)		4.5		2.3		6.8
Percentage of total area		13.8		7.1		20.9
Period of Enclosure	Total (million acres)	Percentage of England	Total	Percentage	Total	Percentage
pre–1793	1.9	5.7	0.7	2.2	2.6	7.9
1793–1815	2.0	6.1	0.9	2.8	2.9	8.9
1816–29	0.2	0.7	0.1	0.4	0.4	1.2
All pre–1830	4.1	12.6	1.8	5.4	5.8	18.0

Source: M. E. Turner, (1980) pp. 62 and 71

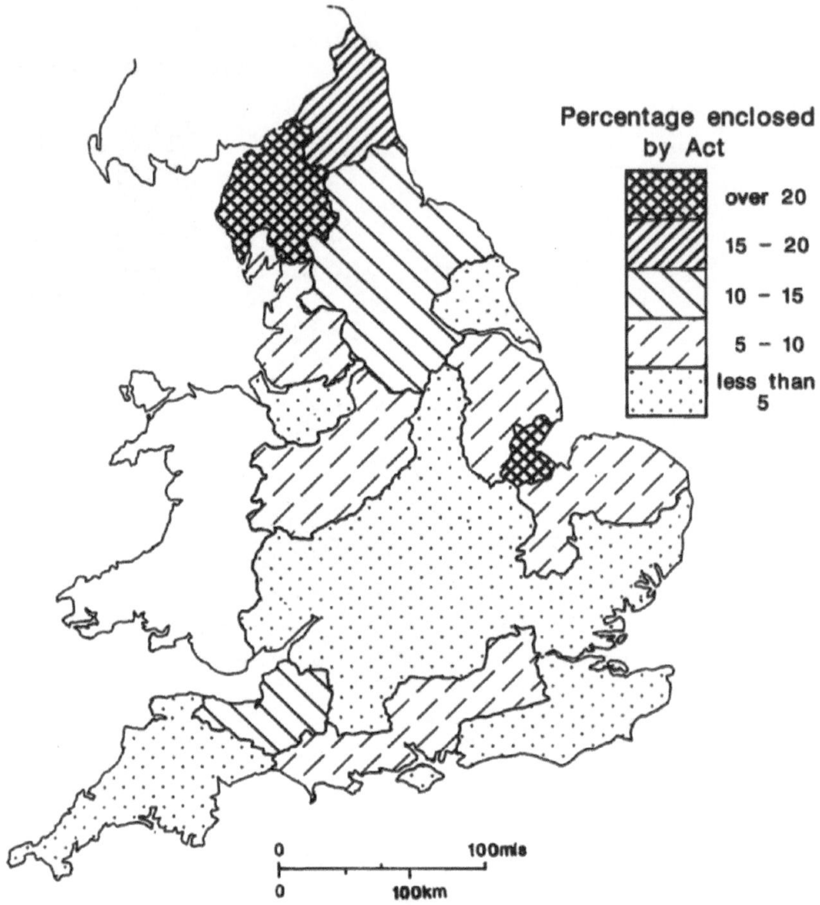

Figure 4.2 Enclosure of commons and wastes in England, 1750–1870. Based on Turner (1980) p. 61.

and general Acts of 1695 'Anent Lands Lying Runrig' and relating to 'the Division of the Commonties'. From areas of early innovation in Ayrshire, the Lothians and the lower Tweed Valley enclosures on separate farms spread throughout lowland Scotland and, through improvement and enclosure of waste, into the eastern Grampian and Moray Firth areas. But full reorganization of the Highland economy and communities awaited mid- and late-Victorian times (Chapter 9).

The pattern of Parliamentary enclosure reflects economic, tenurial and agrarian forces. Upland, wetland and heathland reclamation accounts for high proportions of enclosure of common waste in Cumbrian, Pennine and some eastern counties (Figure 4.2). In Wales half an estimated 385,000 acres of enclosure occurred between 1790 and 1815, one-eighth of the Principality's common and waste land. In midland and eastern England local variations in farming, tenure and soils affected both the time over which enclosure took place and its impacts (Figure 4.1). For example, most of east Durham's open fields were enclosed before the mid-eighteenth century and

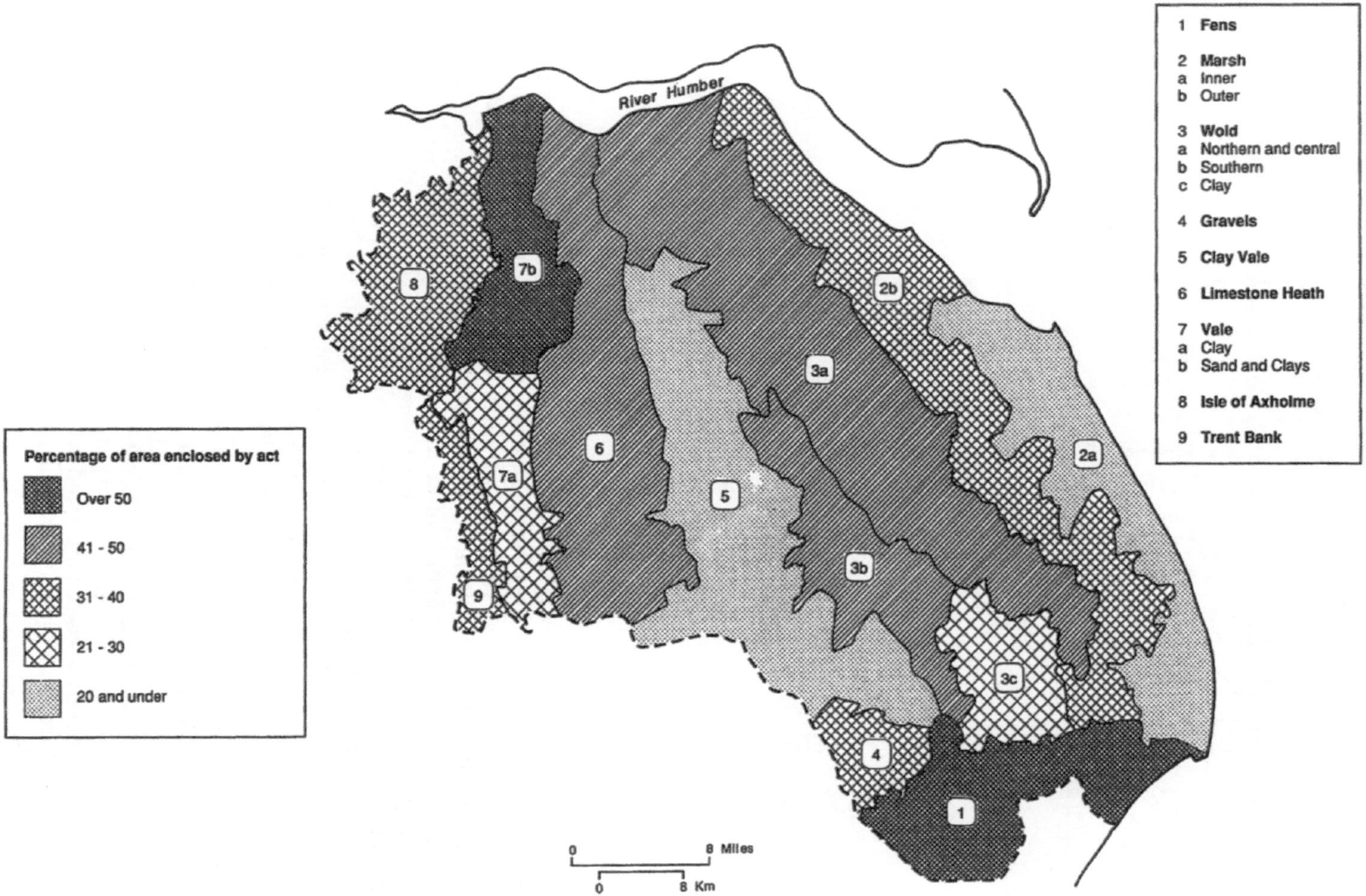

Figure 4.3 Parliamentary enclosure in Lindsey (Lincolnshire). Based on S. A. Johnson, 'Enclosure and the changing agricultural landscape of Lindsey from the sixteenth to the nineteenth century.' Unpublished MA thesis, University of Liverpool (1957).

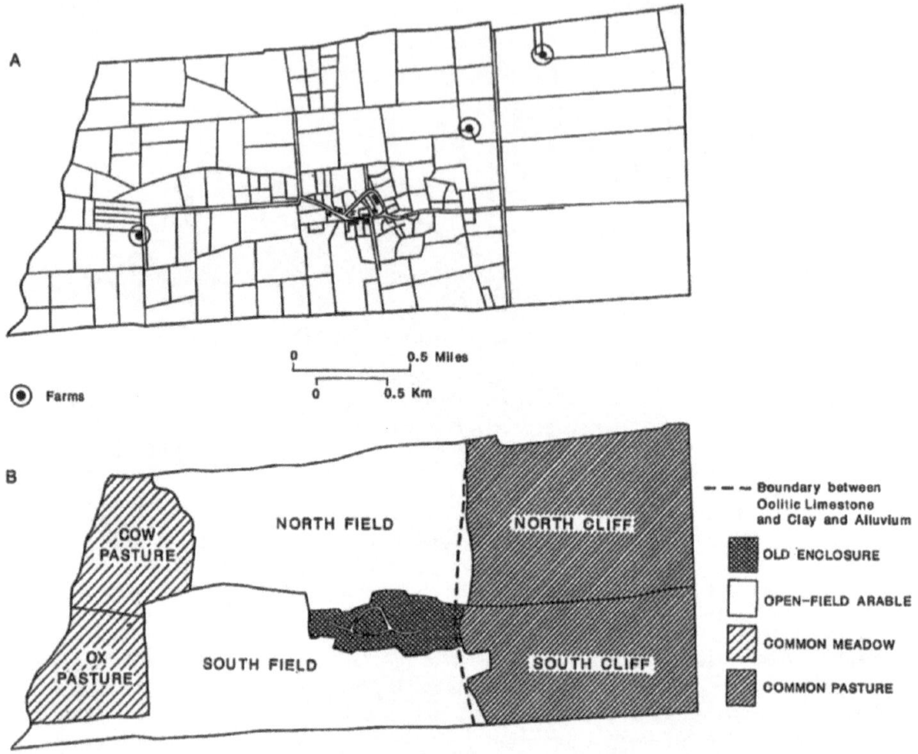

Figure 4.4 Field patterns and enclosure in Ingham (Lincolnshire). A. Field patterns of the 1950s; B. Landuse at the time of the enclosure award of 1770 (Act of 1769). Based on Johnson (1957).

nearly all its 107,000 acres of Parliamentary enclosure was of Pennine wastes and commons. Lincolnshire's varying terrain of marsh and fen, limestone and chalk uplands and clay vales had varied levels of enclosure (Figure 4.3). Though an open-field area, it had considerable 'ancient enclosure': first, on light land converted to pasture after medieval depopulation; secondly, through piecemeal reclamation of waste especially wetland; thirdly, by partial engrossing of open fields by agreement. In parts of the south Wolds 40 to 60 per cent of the land was already enclosed before the Enclosure Act of 1773. Distinctive small, irregular fields contrast with the straight-sided boundaries set out under the enclosure awards providing clues to landscape evolution until they were enlarged and rationalised for modern mechanized farming (Figure 4.4).

The impacts of enclosure on landscape, settlement and community were considerable. Within the boundaries laid out by the Commissioners, field sizes and shapes reflect farm economy and land use: early enclosures near to villages were for small grazing paddocks, orchards and gardens; larger fields on later enclosures were for arable; those used for grazing, whether on former common meadows or converted open field or upland grazings (such as the Derbyshire Peak District), were smaller with substantial stockproof hedges or walls.

Many farm buildings gradually moved into new holdings after enclosure, creating a secondary dispersal of settlement in areas previously dominated by compact villages. In Scotland older 'fermtouns' and Highland 'clachans' were replaced by scattered farms or, on some large estates, by new planned villages. While such rebuilding was beneficial to farm tenants, the landless labourer often fared badly. Many big farms in Scotland and northern England only provided communal 'bothies' for their male labour force. In English parishes, dominated by large landowners, tied cottages were mostly for key workers (e.g. ploughmen and stockmen); farm labourers had to manage with hovels in overcrowded villages and a long walk to their daily toil.

Farming regions

Such changes in agriculture were spread over many years and the full reshaping of the countryside was not achieved until Victorian times (Chapter 9). However, by the end of the eighteenth century distinctive patterns of regional farming were emerging in response to market forces and the relative costs of production on different types of soil. These are explicit in the numerous accounts of such writers as Arthur Young and William Marshall and in the various county reports to the Board of Agriculture (1793–1817) and Sir John Sinclair's Statistical Account of Scotland in the 1790s, as well as in statistics of the 1801 Crop Returns and the surveys of land use, ownership and settlement under the Tithe Commutation Act, 1836 (Chapter 1). One of the major characteristics of improved agriculture was that the balance between crop and stock could shift in response to changes in demand. Climatic limitations on grain crops in western and upland regions and the new opportunities for wheat and barley cultivation on lighter soils made for a basic division between an arable-dominant eastern and stock-dominant western Britain. But within each there was considerable diversity related to soils, market opportunities and the degree of innovation.

In western areas the reclaimed mosslands and light soils of south west Lancashire were already providing potatoes and other cash crops for the regional market in a generally pastoral region, while some Cheshire cheese-makers began to see new opportunities in milk production near the towns. Within the English midlands further conversion of traditional grain-growing areas to fattening pastures was reflected in Leicestershire's verdant ten-acre fields, high stocking densities, careful grassland management and early prowess in stock breeding. In the arable east cheaper costs of working and good, if not outstanding, yields under the Norfolk husbandry and its many variants, saw it progressively dominate grain production and contribute much to lowland arable stockfeeding. Yet the two systems were flexible: when grain prices were high in the 1790s and 1800s cultivation returned to some heavy soils in the midlands or pushed the margins of cultivation upwards from the valleys in the pastoral uplands; reversion to grass came as wheat prices fell to 40–70 shillings per quarter in the post-1816 depression.

Economic margins of cultivation were increasingly determined by prices and access to market. Perishable commodities – vegetables, fruit and milk – had to be grown near their markets while bulky crops such as hay withstood transport costs less well than grain. The relationship of Middlesex farming to the London market as described by John Middleton in 1798 and Rev. Henry Hunter's fuller analysis of 1811 illustrates the point. The built-up area was ringed by brick pits which when worked out were temporarily cultivated before being built on. Beyond were stock grazings for the capital's milking herds, with market gardens, vegetable and fruit farms on the good loams of the Thames terraces. Hay for the city's horse population dominated the next belt along the river meadows and on the clay soils beyond grain dominated.

At a time when many farming regions and systems were defined by or related to soil and physical characteristics, the emphasis here was on the market, anticipating J. von Thünen's 1826 analysis of land use in terms of '. . . the laws which govern the prices of agricultural products and . . . by which price variations are translated into land use'. In his model of an 'isolated state' production costs (labour), inputs (seed, fertilizer, etc.), physical attributes of soil and weather, and varying transport costs are dynamic variables in a land use – rent – profit equation.

Life in the countryside

Although London's influence extended over much of Britain, and the nature and role of many provincial towns was changing, British economic activity was still focused principally on a series of regional economies. The character of rural Britain was thus diverse, with distinctive landscapes and economic systems, populations with distinct regional dialects and cultures, and varied experiences of rural life. National and international trends certainly affected most regions and as labour markets extended some people migrated over long distances. But for the majority of the rural population mobility and horizons were limited.

The diversity of housing was reflected in the varied local building materials used: limestone and sandstone were frequently used in the north; limestone and timber in much of the midlands; a mixture of flints, chalk, clay and sandstone in East Anglia and southern England; and mainly timber in the Weald. During the eighteenth century brick was increasingly used for new building where stone was not available and slate (especially from Wales) began to replace thatch, stone and tiles for roofing.

These varied building materials were also related to different vernacular styles, and in turn affected the durability and comfort of rural housing. Stone-built houses, although reasonably solid and weatherproof, were usually very small with two tiny rooms providing accommodation for a family of perhaps eight or ten. The timber frame and dried mud cottages of clay regions of southern England, were not only small but, if poorly maintained, could rapidly become very uncomfortable to live in. In the stock districts of northern England and in Wales the single-storey long house, built of a variety of materials, was common. Typically, both living quarters (perhaps a single large room) and accommodation for livestock would be under a single long roof with only a flimsy partition between. In Scotland rural housing was generally worse than in most of England. Typically, cottages had unlined walls, no proper floor or ceiling, unsound foundations and no drainage. The traditional 'black house' of the western Highlands and Islands was constructed of stones gathered from the fields, roughly fitted together and roofed with turf resting on wooden timbers. Built like an elongated beehive it accommodated both people and cattle in an area some 30 feet long and 15 feet wide.

Housing quality also varied with the economic prosperity and range of activities of different regions; and with the individual's status within rural society. More prosperous farmers, craftsmen and skilled workers (especially in arable areas) could afford decent brick or stone houses: rural labourers, smallholders and squatters on the margins of cultivation might live in a mud and timber hovel. In Lincolnshire in 1813 it might cost £30–£40 to build a two-room cottage of mud but £60 for the equivalent cottage built of brick and tile. The low wages of most agricultural workers necessitated correspondingly low rents, perhaps 2s per week which might be some 20 per cent of cash wages. As most rural dwellers rented their home, they were restricted to cheap, poor quality housing and were dependent on the landlord for maintenance of the property.

Some farmworkers managed to save enough to buy their own cottage or build their own home on waste land, but their numbers were relatively small and their housing quality and living space were not necessarily any better than those of tenants. In 1797 an estimated 10 per cent of rural families owned their own home or lived rent free: the majority of households surveyed probably lived in tied cottages. The rapid turnover of rural labourers, the short periods for which most men were hired, and the variability in hiring practices and wages from one area to another all militated against homeownership. Those that lived in tied cottages had their wages reduced proportionately, but seasonal variability of family income (with most work available in harvest time) meant that a family could not commit itself to a high regular outlay on rent.

Rural housing was thus a mixture of poor quality decaying older property, poorly built newer houses and a minority of decent stone or brick-built cottages for the more prosperous. From the mid-eighteenth century substantial rural population growth put housing under great pressure (Chapter 3). Yet, despite progressive overcrowding, the additional space available around rural housing and especially the availability of gardens and vegetable plots gave a better living environment than for most town-dwellers.

Many large landowners responded to population pressure and the poor quality of older housing by building new cottages. Agricultural improvement, enclosure and diversification of many rural economies in areas of craft industry provided the means of investment in building, not least on the estates of 'improving' farmers, and rising real incomes enabled yeoman and tenant farmers, and some smallholders, to rent better cottages. For instance, in Norfolk in 1813 two-room cottages of flint with pantile roofs were being built with a wooden lean-to providing two more small rooms. Some of the most active house construction in rural areas was aimed at industrial workers and not agricultural labourers. In the handloom textile and knitwear districts the long 'weavers' windows' of often substantial stone or brick cottages testify to the impact of domestic industry.

In addition the development of water-powered factory industry led to the growth of factory villages as manufacturers provided housing for workers attracted in from the surrounding countryside. In England this was often in the form of a row of terrace cottages built on to an existing village but in Scotland, where there were few nuclear villages, wholly new rural settlements were founded at remote water-power sites. For example, David Dale's first industrial village at Catrine in Ayrshire consisted of a crescent of two-storey stone cottages around a village green, but in 1792 he turned his attention to the much larger project of a new mill and industrial village at New Lanark. With 200 stone and slate-roofed houses built in rows two or three storeys high, his Highland workforce found it difficult to settle either to factory work or to life in high-density tenements. He was forced to rely mainly on pauper children to work in the mill until lowland adult labour became available. After 1800 when Robert Owen took over management of New Lanark, it became one of the most celebrated examples of a planned industrial village in Britain.

The crisis in rural housing of the later eighteenth century had a number of consequences. Many families were permanently overcrowded; individual privacy was difficult and much of life, especially the development of friendships and courtship, was lived outside the home in lanes, woods and fields. Marriage was often delayed due to the lack of opportunity to set up a home. Epidemic diseases such as smallpox or typhus fever spread rapidly in overcrowded insanitary conditions and living conditions could be as unhealthy as in many towns.

The constant presence of parents and siblings was a spur to migration for many young people. However, by no means all migrants went to towns: many moved to another parish to seek work or were placed by their parents as farm servants (field or domestic). The practice of some large landowners of maintaining 'closed' villages – that is with limited accommodation designed to keep down the number of residents who might require support from the Parish – exacerbated the housing problem. Moreover, operation of the Laws of Settlement, under which the destitute could be removed to their parish of settlement, limited such movement in many rural areas. The effectiveness of closed villages and the Settlement Laws in limiting migration is, however, the subject of considerable debate (Chapter 3).

The nature and adequacy of rural housing was also closely tied to work. Living space was more important for the domestic weaver or knitter who spent much time indoors, than for the farm worker who toiled for 12 hours a day in the fields. In a domestic economy (Chapter 5), in which several members of a family may have been employed at home, a larger outlay on housing or an adjoining workshop to provide sufficient space for a loom or knitting frame became an essential part of the household budget. In contrast, the single migrant who left home to seek employment in another parish might have been hired at a hiring fair and either given accommodation as a lodger in the master's house (most common in the north and west of England) or, alternatively, housed and fed in sheds or outhouses along with other hired hands as in the 'bothies' of eastern Scotland and the arable counties of England in the late eighteenth century. Housing thus considerably influenced other aspects of everyday life. The domestic weaver of Yorkshire or Lancashire might commit part of his accommodation to work, but he had a level of independence not available to the farm worker in a tied cottage or, especially, to the young hired hand living in his master's house. The Scots working at New Lanark found that most of their life, both inside and outside the mill, was influenced by Robert Owen's ideas. Good housing and assured work was secured at the cost of a degree of social control that would have been anathema to the domestic weaver or the Highland crofter.

Although rural life in the late eighteenth and early nineteenth centuries was undoubtedly hard, it was not without pleasure. The calendar was peppered with market days, fairs, festivals, feast days and other religious holidays which provided opportunities for rest, relaxation and social interaction. The precise nature of these varied from one part of the country to another, but cumulatively they helped to bind a community together whilst at the same time maintaining rural tradition and social hierarchy.

Although some aspects of rural life were predictable, others changed unexpectedly and could cause great hardship and disruption. Both rural industry and agriculture were affected by fluctuations in employment (quite different from normal seasonal fluctuations), while enclosure had a major impact in many parts of the country. The extent to which enclosure significantly affected rural living standards and the availability of housing is a matter for debate. By the mid-eighteenth century many areas were already largely enclosed and Parliamentary Enclosure was focused mainly on the core of open-field farming in midland England and on unenclosed wastes. Though the latter were relatively sparsely populated, the livelihoods of smallholders, cottagers and squatters who depended on access to common land to graze their animals were severely threatened by enclosure. Many such families were forced to migrate to growing industrial towns.

Although the rural world of the eighteenth and early nineteenth century was more circumscribed and isolated than in the late nineteenth century when railways had penetrated most parts of Britain, most rural dwellers had a good knowledge of their local market town and rural to urban migration need not have been unduly difficult or traumatic. Contact between countryside and town was frequent as farmers, tradesmen and craftsmen sold their goods or services in local or regional markets and hired labour at annual hiring fairs. Most ordinary folk would visit a local town at some time, perhaps to seek work, visit a doctor or attend a fair. Although most journeys were on foot, for the relatively affluent, or in an emergency, the expanding network of turnpikes, coaching and carrier services eased movement from countryside to town and between towns.

Within particular regions there was often close economic interaction between countryside and town. In much of Lancashire, Yorkshire and the East Midlands domestic textile industries were the mainstay of the economy and depended heavily on a putting-out system (see Chapter 5). The East Midland's domestic knitters collected factory-spun yarn, increasingly produced in the towns, and travelled to market centres to sell finished goods. On a slightly different scale, in eighteenth-century Cumbria, families who had accumulated wealth largely from rural estates invested in the industries of west Cumbria. This transformed the Cumberland coast from an isolated and backward rural area to a thriving industrial region. For example, the Lowther family, in addition to owning large estates in the county, built the new harbour of Whitehaven to facilitate exports from the Cumberland coalfield: elsewhere along the coast the ports of Workington and Maryport were developed by the Curwen and Stenhouse families.

In the debate on the extent to which social, economic and technological change in the eighteenth century transformed Britain from a series of localized regional economies into an integrated economic system it is usually argued that strong regional economies either persisted or changed to a more unified economic system. However it is probable that both trends were operating at the same time (Chapter 17). Whilst improved transport and communications, faster circulation of information and ideas, and the development of larger and more uniform economic enterprises stimulated the creation of a single economic system, the horizons of most country-dwellers were limited to a restricted range of villages and towns and they maintained staunch local loyalties. Despite many kinds of change in eighteenth-century rural life, in many respects continuity with the past was maintained until late in the nineteenth century (Chapter 9).

Bibliography and further reading for Chapter 4

On agricultural change see:

BECKETT, J. V. *The agricultural revolution* (Oxford: Blackwell, 1990).

ERNLE, Lord. *English farming past and present* London: Heinemann, (6th edition with introduction, 1961).

FUSSELL, G. E. *The farmers' tools 1500–1900* (London: Andrew Melrose, 1952).

GONNER, E. C. K. *Common land and inclosure* (London: Cass, 2nd edition, 1966).

HALLAM, H. E. (ed.) *The agrarian history of England*. Vol vi: *1750–1850* (Cambridge: Cambridge University Press, 1988).

HANDLEY, J. E. *The agricultural revolution in Scotland* (Glasgow: Burns, 1963).

HARVEY, N. *The industrial archaeology of farming in England and Wales* (London: Batsford, 1980).

JONES, E. L (ed.) *Agriculture and economic growth in England, 1650–1815* (London: Methuen, 1967).

JONES, E. L. *Agriculture and the industrial revolution* (Oxford: Blackwell, 1974).

KERRIDGE, E. *The agricultural revolution* (London: Allen and Unwin, 1967).

MINGAY, G. E. *The agrarian history of England*. Vol ix: *1750–1850* (Cambridge: Cambridge University Press, 1989).

MINGAY, G. E. *Enclosure and the small farmer in the age of the industrial revolution* (London: Macmillan, 1968).

OVERTON, M. 'Agricultural revolution? Development of the agrarian economy in early-modern England' in Baker, A.R.H. and Gregrory, D. (eds) *Explorations in Historical Geography* (Cambridge: Cambridge University Press, 1984), pp. 118–39.

PARKER, R. A. C. *Coke of Norfolk. A financial and agricultural survey* (Oxford: Oxford University Press, 1975).

TROW-SMITH, R. *A history of British livestock husbandry 1700–1900* (London: Routledge, 1959).

TURNER, M. *Enclosures in Britain 1750–1830* (London: Macmillan, 1984).

TURNER, M. *English Parliamentary enclosure* (Folkestone: Dawson, 1980).
YELLING, J. A. *Common field and enclosure in England, 1450–1850* (London: Macmillan, 1977).

Some studies of regional change include:

DARBY, H. C. *The changing Fenland* (Cambridge: Cambridge University Press, 1983).
GRAY, M. *The highland economy, 1750–1850* (Westport, Conn: Greenwood Press, 1976).
GRIGG, D. B. *The agricultural revolution in south Lincolnshire* (Cambridge: Cambridge University Press, 1964).
HARRIS, A. *The rural landscape of the East Riding of Yorkshire, 1780–1850* (Wakefield: S R Publishers, 1969).
HUNTER, J. *The making of the crofting community, 1746–1930* (Edinburgh: John Donald, 1976).
ORWIN, C. S. and SELLICK, R. J. *The reclamation of Exmoor Forest* (Newton Abbot: David and Charles, 2nd edition, 1975).
PARRY, M. and SLATER, T (eds) *The making of the Scottish countryside* (London: Croom Helm, 1980).
RICHARDS, E. *A history of the highland clearances. Agrarian transformation and the evictions, 1746–1886* (London: Croom Helm, 1982).
THIRSK, J. *Agricultural regions and agrarian history in England, 1500–1750* (London: Macmillan, 1987).
THOMAS, D. *Agriculture in Wales during the Napoleonic wars* (Cardiff: University of Wales Press, 1963).
WILLIAMS, M. *The draining of the Somerset levels* (Cambridge: Cambridge University Press, 1976).
YOUNGSON, A. J. *After the 'forty-five* (Edinburgh: Edinburgh University Press, 1973).

On housing and rural life see:

ARMSTRONG, A. *Farmworkers: a social and economic history, 1770–1980* (London: Batsford, 1988).
CHALKIN, C. and HAVINDEN, M (eds) *Rural change and urban growth 1500–1800* (London: Longman, 1974).
CHAMBERS, J. D. and MINGAY, G. E. *The agricultural revolution 1750–1880* (London: Batsford, 1966).
CHARLESWORTH, A (ed.) *An Atlas of rural protest in Britain* (London: Croom Helm, 1982).
COBBETT, W. *Rural Rides* (London: Dent, Everyman Library Edition, 1912).
GAULDIE, E. *Cruel habitations: a history of working-class housing 1780–1918* (London: Unwin, 1974).
HAMMOND, J. L. and B. *The village labourer 1760–1832* (London: Longman, 1911).
HARVEY, N. *A history of farm buildings in England and Wales* (Newton Abbot: David and Charles, 1970).
HORN, P. *The rural world 1780–1850: social change in the English countryside* (London: Hutchinson, 1980).
HORN, P. *Life and labour in rural England 1760–1850* (London: Macmillan, 1987).
HOUSTON, R. A. and WHYTE, I. D (eds) *Scottish Society, 1500–1800* (Cambridge: Cambridge University Press, 1989).
KUSSMAUL, A. *Servants in husbandry in early modern England* (Cambridge: Cambridge University Press, 1981).
KUSSMAUL, A. *A general view of the rural economy of England, 1538–1840* (Cambridge: Cambridge University Press, 1990).
MARSHALL, J. D. *The old poor law 1795–1834* (London: Macmillan, 1968).
MINGAY, G. E. *English landed society in the eighteenth century* (London: Routledge, 1963).

MINGAY, G. E. 'The rural slum' in Gaskell S. M. (ed.) *Slums* (Leicester: Leicester University Press, 1990).

PORTER, R. *English society in the eighteenth century* (Harmondsworth: Penguin, 1982).

SNELL, D. M. *Annals of the labouring poor: social change in agrarian England 1640–1900* (Cambridge: Cambridge University Press, 1985).

5

Industry and industrialization from the 1740s to the 1830s

The structure and growth of the national economy

Despite a diversified industrial structure and a significant woollen cloth industry, early eighteenth-century Britain remained a predominantly pre-industrial economy with two-fifths of its national product from agriculture as against one-fifth from mining, manufacturing and building. Two-thirds of England's and four-fifths of Scotland's population was rural though many were handicraft workers. Nevertheless, the foundations for growth had already been laid (Chapter 1): fuller utilisation of basic raw materials (wool, leather and other natural products), fuel and power (from wood, coal and water) and mineral ores partly reflected the demands of a growing population. It also had other essential pre-requisites: the ability to harness and deploy capital; labour, entrepreneurial and technical skills; social and political adaptations promoting and responding to the consequences of an accelerating economy; and increasing prosperity, the effects of which were felt in due course throughout industry.

Around 1770 agriculture had 48 per cent, industry and commerce 24 per cent and professions and services 28 per cent of the national income. By the early nineteenth century those figures were 24 per cent for industry, 17 per cent trade and transport, 21 per cent professional, government and personal services and only 33 per cent for agriculture. By the 1830s, over one-third of national income came from industry as against 23 per cent from agriculture, with trade and transport remaining at about one-sixth. Investment in industry and trade increased nearly six-fold by the 1820s and '30s to over one-third of the investment in the nation's fixed capital as against 22 per cent in the 1760s; whilst agricultural fixed capital formation had doubled, its share declined from 33 to 13 per cent.

Much of the increase in industrial output was for the home market, in which consumption grew by a factor of 2.5, 1740–1800 and by as much again by the early 1830s. England's international trade, valued at £9.7 million per annum at the beginning of the eighteenth century, increased to £13.8 million in the 1740s, Britain's to £25.8 million by the 1780s and the United Kingdom's to £128 million by the 1830s (Table 5.1). Export of manufactures

DOI: 10.4324/9781003174783-5

Table 5.1 British imports, 1740–1950

	Major categories of goods as a percentage of all imports									
	(England and Wales)	(Great Britain)		(UK)						
	1740	1772	1807	1840	1860	1880	1900	1910	1930	1950
Food, drink and tobacco	59.8	65.2	54.6	39.7	38.1	44.1	42.1	38.0	45.5	39.5
Raw materials and semi-manufactures	29.2	24.4	42.7	56.6	56.5	38.6	32.9	38.5	24.0	38.2
Manufactured and miscellaneous goods	11.0	10.4	7.5	3.7	5.5	17.3	25.0	23.5	29.4	22.3
TOTAL VALUE (in £000s)	6704	13305	53800	91200	210500	411200	523100	683700	1044000	2608200

Approximate percentages for major commodities only: total values are 'official values' for 1740–1867, 'computed values' for 1840 and 'current values' for 1860–1950

Source: Mitchell and Deane (1962), pp.279–84

Table 5.2 British commodity exports, 1740–1950

	Principal exports as a percentage of total domestic exports of the United Kingdom									
	1740†	1772†	1800	1830	1850	1870	1890	1910	1930	1950
Cotton yarn and manufactures	0.3	2.2	24.1	50.8	39.6	35.8	28.2	24.4	15.3	7.3
Woollen yarn and manufactures	59.8	41.6	28.4	12.7	14.1	13.4	9.8	8.7	6.5	6.5
Linen yarn and manufactures	1.6	5.7	3.3	5.4	6.8	4.8	2.5	—	—	0.9
Silk and art silk	1.4	1.9	1.2	1.4	1.5	0.7	1.0	0.5	0.3	2.3
Apparel	—	—	—	2.0	1.3	1.1	1.9	2.9	3.5	1.6
Iron and steel manufactures	5.2	8.0	6.6	10.2	12.3	14.2	14.5	11.4	10.3	9.5
Machinery	—	—	—	0.5	0.8	1.5	3.0	6.8	8.2	14.3
Coal, coke, etc	3.2	3.0	2.1	0.5	1.8	2.8	7.2	8.7	8.6	5.3
Earthenware and glass	—	—	—	2.2	1.7	1.3	1.3	1.0	2.1	2.5
Vehicles*	—	—	—	—	—	1.1	3.5	3.8	9.0	18.6
Chemicals	—	—	—	—	0.5	0.6	2.2	4.3	3.8	5.0
Electrical apparatus	—	—	—	—	—	—	—	—	2.1	3.9
Total value (£000s)	5111	10974	24304	38300	71400	199600	.263500	430400	570800	2171300

Values at current prices: not all exports are included so percentages do not total 100.
†England and Wales.
*Carriages, wagons, ships, cars, cycles, aircraft.

Sources: Mitchell and Deane (1962) pp.280–81 and pp.293–95 (for 1740–1800) and Dean and Cole (1969), Table 9, p.31 (for 1830–1950)

greatly increased in quantity and range. In the late seventeenth century woollen cloth accounted for 56 per cent of exports (Table 5.2). By the early 1800s it still provided one-quarter of exports but it was rivalled by cotton. By 1830 cotton yarn and manufactures provided just over one-half of Britain's exports, woollens one-eighth and metal manufactures and minerals a little over one-eighth.

While capital formation in agriculture kept pace with industry and transport up to the late eighteenth century, from the 1790s it was overhauled then surpassed by industry which grew at 1.7 per cent per annum as compared with agriculture's 0.6 per cent in the 1820s and '30s. Much of the capital growth was in machines and buildings, but increasingly in machinery. Moreover, over half the economic growth, 1780–1860 was in modernized industries: productivity in cotton manufacture grew by 2.6 per cent per annum; in worsted by 1.8 per cent; in woollens and iron by 0.9 per cent; and in transport and shipping by 1.3 and 2.6 per cent, respectively.

In England 3.14 million people depended on each of the agricultural and non-agricultural sectors in 1801, substantially more than the 2.8 million urban dwellers. By 1831 perhaps one-tenth of the male labour force worked in factories and large workshops as compared with 32 per cent in retail trade and handicraft, 16.6 per cent as non-agricultural labourers and 32.6 per cent in agriculture, and fuller occupational detail of the 1841 census indicates that those employed in traditional craft and retail activities considerably exceeded those in modern industry. Of the three staple industries discussed in detail below – mining, metal manufacture and textiles – none exceeded employment in agriculture (1.52 million) in 1841; textiles (0.88 million) employed fewer than domestic service (1.24 million), metal manufacture (0.41 million) fewer than clothing (0.56 million), and mining (0.23 million) fewer than building (0.38 million) or food, drink and tobacco (0.31 million). Indeed in the 1830s the leading retail and handicraft trades employed 5 people for every 3 in manufactures: many lived in rural areas and worked in small workshops or family units rather than factories.

Nevertheless, the leading sectors which supplied much of the early stimulus to growth experienced considerable technical innovation and increased productivity. Three aspects are significant: advances in fuel technology, notably through the wider use of coal; the spread of motive power, both water and steam; and the adoption of these to wider-ranging and increasingly higher-powered machines in industry, transport and, though slowly, agriculture.

The gap between innovation and widespread adoption of many new technologies was, however, often considerable. Table 5.3. dating a wide range of inventions and patents, represents only the beginnings of progressive improvement in key processes and their harnessing to the needs of particular industries. Since new technologies frequently demanded new plant, new labour skills and new locations, there was often a substantial time lag in their adoption. Access to knowledge, to capital, the trade-off between costs under established and new methods of production, and potential gains in the volume and quality of output were all considerations for entrepreneurs in deciding whether to adopt new techniques, in new premises on new sites and to recruit or retrain workers. Such practical

Table 5.3 Dates of major technological innovations in the British economy, 1700–1950s

Period	Coal, Energy and Iron		Engineering, Transport and Communications		Textiles and Chemicals		Agriculture and Food	
			Mersey and Irwell Navigations	(1690–1720)				
1700			Savery 'fire engine'	(1698)				
			Turnpike Trusts	(1706–)				
	Darby coke-iron process	(1709–17)	Newcomen engine	(1709)				
					Lombe's silk mill, Derby	(1717)		
1720							Parliamentary enclosure	(1720s)
							Rotherham mouldboard plough	(1730)
							Tull's horsehoeing	(1731)
					Kay's 'flying shuttle'	(1733)	husbandry and drill sowing	(1730s)
1740	Huntsman crucible steel	(1740–42)			Wool carding machine	(1740s)		
							Bakewell's livestock experiments	(1745–)
			Sankey Canal	(1757)				
1760					Roebuck 'lead-chamber' sulphuric acid process	(c. 1760)	Elkington's spring-drains	(1760)
					Hargreaves' spinning jenny	(1764)		
			Watt's steam engine	(1769)	Arkwright's 'frame'	(1769)		
			Ramsden screw lathe	(1770)				
			Boulton and Watt engine	(1775)				
	Underground rail transport	(1777)						
	Improved ventilation							
	First cast iron bridge, Coalbrookdale	(1779)			Crompton's 'mule'	(1779)		
1780			Macadam road surfacing	(1780s)			Colling brothers' Shorthorns	(1780s)
			B and W rotary engine	(1781)				
							Improved seed drill	(1782)
	Cort puddling and rolling process (wrought iron)	(1784)			First steam-powered cotton mill	(1784)		
							Meickle's threshing machine	(1786)
					Leblanc soda process (France)	(1785)		
					(GB)	(1787)		
					Cartwright's loom	(1787)		
							Ransome's iron plough	(1789)

Table 5.3 _Continued_

Period	Coal, Energy and Iron	Engineering, Transport and Communications	Textiles and Chemicals	Agriculture and Food
			Whitney cotton gin (USA) (1793)	
		Maudsley's lathe (1797)		
			Tennant bleaching process (1799)	
1800		Trevithick high pressure engine (1800)		
			Jacquard loom (1801)	
	Gas lighting (1806)			
		Maudsley table engine (1807)		
		London–Holyhead Post Road (Telford) (1810–)		
		'The Comet' steamship (1812)		Common's reaper (1812)
	Davy safety lamp (1816)			
1820	Improved shaft sinking and deep mining (c. 1820)			
			Roberts' power loom (1822)	
		Stockton and Darlington Railway (1825)		
				Bell reaper (1826)
	Neilson's hot blast method (1828)			
		Liverpool–Manchester Railway (1830)	Roberts' self-acting mule (1830)	
	Faraday electro-magnetic induction (1831)	Whitworth machine tools (1833)	Phillips contact process (SO4) (1831)	
				Peruvian guano fertilizer (1835)
		Great Western (Atlantic steam crossing) (1838)		
		Nasmyth steam hammer (1839)		
1840				Liebig's 'On Soil Chemistry' (1840)
				Ransome's steam thresher (1842)
				Scragg's pipe-tiles (1842)
	Underground rope haulage (1844)			Lawes superphosphates (1843)
		McNaught compound engine(1845)		Deare steel plough (USA) (1846)
				Fowler's draining plough (1848)
			Lister machine-comb (1851)	
			Singer sewing machine (USA) (1851)	
	Perkins coal-tar derivatives (1856)		Perkins synthetic dyestuffs (1856)	
	Bessemer steel process (1856)			

1860			Turret lathes (USA 1855)	(1860s)	Bandsaw (cloth-cutting)	(1860s)		
			London Underground		Solvay soda process	(1861)		
	Mechanical coal cutters	(1863–)	(Metropolitan)	(1863)				
	Siemens-Martin open hearth							
	steel	(1866)						
	Mushet steel alloys	(1868)						
							First British cheese factory	(1870)
							Mass imports of New World	
							grain	(1870s)
							Compressed air refrigeration	(1871)
					Brunner-Mond ammonia-			
					soda process	(1872)		
			Bell telephone (USA)	(1876)				
			Otto gas engine (Austria)	(1876)				
	Swan and Edison lamps	(1878–9)						
	Gilchrist-Thomas steel							
	process	(1879)						
1880							Basic slag (phosphoric	
							fertilizer)	(1880s)
					Button-holing machine	(1881)		
			Parsons' steam turbine	(1884)				
	Welsbach gas mantle	(1885)	Benz motor car	(1885)				
			Rover safety bicycle	(1885)				
	Ferranti Deptford power							
	station	(1889)						
					Chance-Claus sulphur		Sterile food canning	(1890s)
					process	(1890s)		
			Electric trams	(1891)	Castner-Kellner alkali			
					process	(1894)		
					Northop automatic loom	(1895)		
			Diesel oil-engine	(1897)	Synthetic indigo dyes	(1897)		
1900							Milk pasteurization	(c. 1900)
							Improved margarine	(1900)
			Domestic vacuum cleaner	(1901)				
	Blackett conveyor belt	(1902)	Wright aircraft	(1903)				
	Electric furnaces for steel		First radio valves	(1904)				
	(USA)	(1900s)						
					Courtauld's viscose rayon	(1907)		
					Bakelite (phenol resin)	(1907)		

Table 5.3 *Continued*

Period	Coal, Energy and Iron		Engineering, Transport and Communications		Textiles and Chemicals		Agriculture and Food	
			Gas turbine engines	(1908)				
			Mass produced Ford model 'T' car (USA)	(1908)				
1920							Adoption of first chemical pesticides	(1920s)
			BBC radio (Nov)	(1922)				
					Synthetic nitrates	(1924)		
	Houdry catalytic process (oil)	(1925–30)						
	National electricity grid	(1926–33)			Cellulose acetate yarn	(1926)		
	Continuous strip mills (in USA)	(1926)						
	(in GB)	(1938)						
	Automatic coal cutters and loaders	(1930s)						
							Birdseye quick-freeze process	(1929)
							Freeze-drying	(1930s)
							Combine harvesters	(1930s)
			BBC TV (Baird)	(1936)				
			Jet aero engine	(1937)				
					PVC (polymer)	(1939)		
					Nylon (British Nylon Spinners)	(1939)		
1940							(DDT) pesticides	(1942)
			IBM electronic calculator	(1944)			Organophosphorous pesticides	(1944)
	Gas trunk lines	(1950s)			Centrifugal (pot) spinning	(1950s)	Organophosphorous	
	Oxygen steel	(1950s)			insecticides	(1950)		
	Nuclear power reactor Calder Hall	(1956)						

Sources: Various; principally Derry and Williams (1960) and Williams (1982)
Note: Specific dates are of British inventions and/or patents unless otherwise stated

considerations – affecting investment: profit ratios, competitiveness and efficiency – as well as sheer industrial inertia, influenced the speed at which new methods spread. Hence the long drawn-out transition from handicraft workshop to factory in many branches of the textiles industry (for example, hosiery and knitwear or quality lace-making) and the still incomplete move to mass-production in industries such as clothing, furniture-making and building (Chapters 10, 15).

Thus, although the industrial revolution is often portrayed as an age of steam, steam-powered machinery and large-scale factory production were far from general at the end of its first phase, and were largely absent from many parts of Britain. Such pioneers of machine textile yarn production as Thomas Lombe of Derby (1717), Jedediah Strutt of Belper (Derbyshire, 1793), Arkwright at Cromford (1771), the Gregs at Styal (north Cheshire, 1784) David Dale (from 1792) and then Robert Owen (from 1800) at New Lanark (Ayrshire) all used water power. The first Factory Inspectorate Report of 1838 showed that water still supplied 23 per cent of the power used in the cotton industry, despite the use of steam from 1784, and 43 per cent of the power used in woollen mills and most of that in the West Country and the Scottish borders, for example, came from water power.

Coal

While the age of steam was some way into the future for most activities, the use of coal fuel widened considerably from the late seventeenth century (Figure 5.1): in mining areas it was often cheaper than and preferred to wood. By the early eighteenth century coal fuelled a number of types of processing: traditional boiling industries such as salt pans, dyers vats, or breweries; refining of commodities such as sugar and beer; and also a growing range of non-ferrous smelting, iron working (though not yet smelting) and glass-making. It was also increasingly used in early steam engines, principally draining mines. Moreover, it supplied a growing domestic market notably London, Edinburgh (justly called 'Auld Reekie') and Dublin.

Over the next century, the rapidly growing populations of the major cities and coalfield industrial areas increased coal's domestic markets. Accessibility at the coal face and transport to markets were crucial for coal's cost-competitiveness. Carriage of such a heavy, bulky commodity doubled its cost 15 to 20 miles from the pithead. With the economic range of land-sale coal as little as seven or eight miles, the main early eighteenth-century suppliers were coastal coalfields and those with direct access to river navigations. Fifeshire and Mid-Lothian supplied Scotland's east coast and contributed to English and North Sea coastal trade; Ayrshire supplied Glasgow and shared in the Irish Sea trade with Cumberland which, through Whitehaven and Maryport, dominated Dublin's market of 200,000 tons of coal per annum in the 1790s.

From Elizabethan times the major producer of coal was North East England. Linked to keel transport along the rivers by a substantial network of wagonways, the 'sea-sale' districts of the Tyne and Wear were commer-

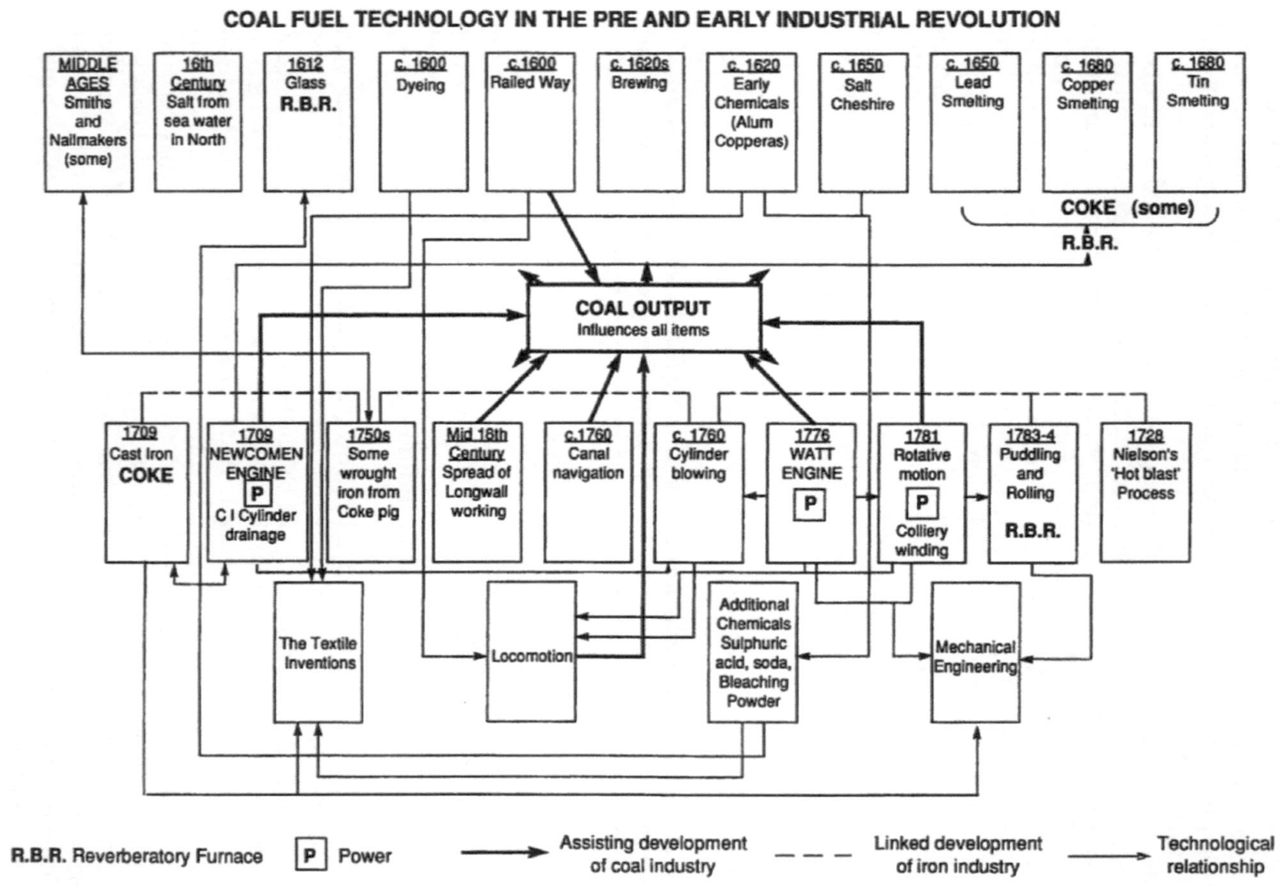

Figure 5.1 Coal fuel technology in the pre- and early industrial revolution.

cially and technologically the most advanced eighteenth and early nineteenth-century mining areas. There was a hierarchy of producers from such major royalty owners as the Dean and Chapter of Durham and wealthy land owners – for example the Lambtons, Lumleys, Ridleys and Ravensworths – to smaller coal owners who operated and leased collieries, the 'fitters' and shipowners who transported coal to the coal factors, and retailers who supplied the major markets, particularly London. London took around two-fifths of the North East's sea-sale trade and was the principal market for both domestic and industrial coal. The Tyne and Wear also supplied east and south coast towns from Whitby to Falmouth and, through river transport, their immediate hinterlands. They also exported to northern France and parts of the Rhine-Scheldt and Baltic areas – in particular Amsterdam, Rotterdam and Hamburg – thereby contributing a not-insignificant share of Britain's export trade (Table 5.2). Financed principally from within the region (though London also played a key role from the late eighteenth century) coal was a driving force in its development and a substantial influence in the growth of its settlement pattern, population and both rural and urban economies. Land-sale coal assisted the North East's industry through its use in salt manufacture, at glassworks (greatly extended in the eighteenth century) and in a wide range of operations in iron-working and ship-building, though that industrial role was not to be fully realized until the mid-nineteenth century (Chapter 10).

Transport was a key factor in the production and use of coal both regionally and from individual collieries. Fluctuating demand, especially for domestic coal, made mining a seasonal occupation in many areas. Competitive access to good markets was vital for investment in both capital and labour which, together with the quality and accessibility (above and below ground) of the coal, determined the conditions under which the decision to produce coal was taken (Figure 5.2).

In West Lancashire, an inland coalfield, the stimulus of large industrial and domestic markets in Liverpool and Manchester promoted investment in collieries by such entrepreneurs as the Duke of Bridgewater and, later, cotton factory owners, as well as local landowners such as the Earls of Crawford and Balcarres. These interests did much to promote improved transport: river navigations, such as the Mersey and Irwell (1694–1721); turnpike roads, for example from Liverpool to Prescot (1726) and Warrington (1752) linking with the trans-Pennine (1732–35) and Cheshire systems (1752–65). Coal transport also played a major role in canal promotion. West Lancashire coal was a major factor in the construction of the Sankey (St Helens) canal (1757) which ran to the Mersey estuary and for which Liverpool businessmen raised much of the capital. The Worsley canal (1761), promoted by the Duke of Bridgewater, carried coal direct from his collieries to Manchester, reducing its price to one-tenth of the previous level. Its extension, the Bridgewater Canal (1776), linked Manchester with the Mersey estuary and the Leeds and Liverpool canal (1774–1811) linked the Wigan coal mines with Liverpool and the emerging cotton textile areas to the benefit of both the coalfield, its industrial areas and commercial outlets (Figure 6.3).)

Accessibility to coal also depended on geological conditions and mining

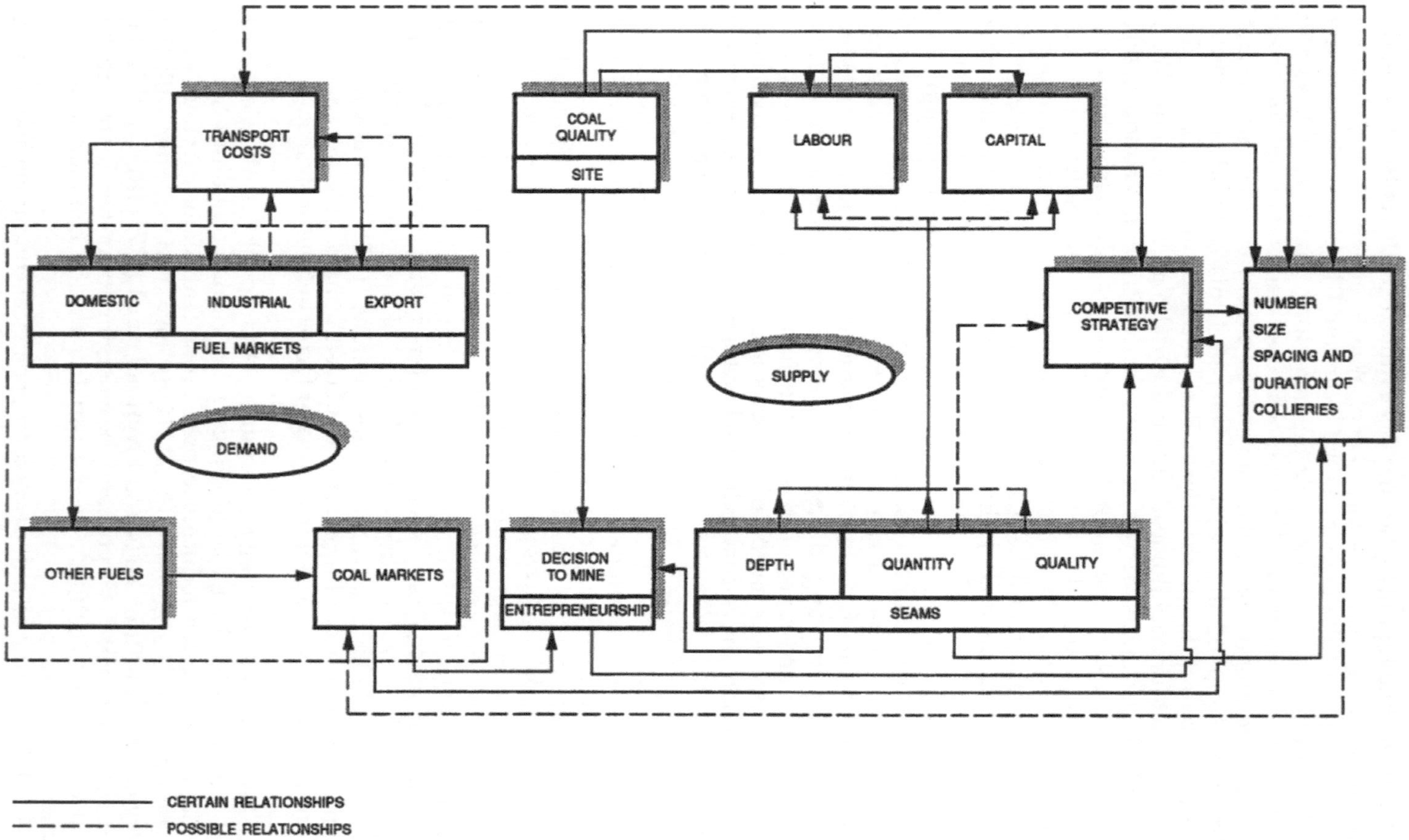

Figure 5.2 Relationships between variables in a mining system. Based on Langton (1979) p. 27.

Table 5.4 Coal output (by Region) in Great Britain, 1750–1950

TOTAL (percentage)	North East	Yorkshire	Midlands	Lancashire	Staffordshire	South Wales	Scotland	Cumberland	North Wales	Salop	Rest	TOTAL (m.tons)
[1]1750–55 T(%)	1.6(35.5)	0.5(11.1)	0.2(4.4)	0.2(4.4)	0.3(6.7)	0.1(2.2)	0.9(20.0)	0.1(2.2)	0.1(2.2)	0.3(6.8)	0.2(4.4)	4.4
1771–75 T(%)	2.0(30.3)	0.8(12.1)	0.3(4.5)	0.3(4.5)	0.5(7.6)	0.2(3.0)	1.3(19.7)	0.3(4.5)	0.2(3.0)	0.4(6.1)	0.3(4.5)	6.6
1791–95 T(%)	2.8(27.5)	1.0 (9.8)	0.5(4.9)	0.8(7.8)	0.9(8.8)	0.8(7.8)	1.7(16.7)	0.5(4.9)	0.3(2.9)	0.4(4.0)	0.5(4.9)	10.2
1811–15 T(%)	3.7(21.3)	1.8(10.3)	1.2(6.9)	1.6(9.2)	2.4(13.8)	1.8(10.3)	2.5(14.4)	0.6(3.4)	0.4(2.3)	0.7(4.0)	0.7(4.0)	17.4
1831–35 T(%)	6.1(19.3)	3.6(11.4)	2.2(7.0)	4.5(13.9)	4.0(12.6)	4.0(12.6)	4.0(12.6)	0.7(2.2)	0.7(2.2)	0.8(2.5)	1.0(3.7)	31.6
[2]1854 T(%)	15.4(23.8)	7.3(11.3)	3.9(6.0)	9.9(15.3)	7.5(11.6)	8.5(13.1)	7.4(11.4)	–	see 'Rest'	–	4.7(7.3)	64.7
1875 T(%)	32.3(24.2)	15.9(11.9)	12.4(9.3)	21.0(15.8)	14.5(10.9)	14.2(10.7)	18.6(14.0)	–	–	–	4.4(3.3)	133.3
1913 T(%)	56.4(19.6)	43.7(15.2)	38.8(13.5)	28.1(9.8)	14.9(5.2)	56.8(19.8)	42.5(14.8)	–	–	–	6.2(2.2)	287.4
1938 T(%)	44.7(19.7)	42.4(18.7)	37.8(16.7)	14.3(6.3)	13.4(5.9)	35.3(15.5)	30.3(13.3)	–	–	–	8.8(3.9)	227.0
N.C.B. Regions	Northern & Durham	North-Eastern	East Midlands	North Western	West Midlands	South Western	Scottish	–	–	South-Eastern	Open-cast	
[3]1950 T(%)	39.5(18.3)	42.6(19.7)	40.1(18.5)	14.7(6.8)	17.6(8.1)	24.3(11.2)	23.3(10.8)	–	–	1.7(0.8)	12.2(5.6)	216.3

Sources: [1]S. Pollard, 'A new estimate of British coal production 1750–1850' *Economic History Review* **33** (1980), Table 14
[2]B. R. Mitchell and P. Deane, *Abstract of British Historical Statistics* (1962), Chap. IV, Table 3
[3]B. R. Mitchell and H. G. Jones, *Second Abstract of British Historical Statistics* (1971), Chapter IV, Table 1
T = Total regional production in millions of tons
Figures given in brackets represent the percentage of British production for each region

techniques. In the early eighteenth century the depth of mines was limited by shaft sinking and drainage problems. Steam-powered atmospheric engines, especially the Newcomen engine (1709), allowed water to be removed from progressively greater depths. The adoption of the Boulton and Watt engine (1776) for ventilation replaced old-fashioned fire-draught systems in the ventilation shafts. Underground tramways and better roofing of workings enabled extraction to extend further from the base of the pit shaft. As a consequence the depth at which coal could be mined increased from three or four hundred feet before the mid-eighteenth century to 800 or 900 feet by the 1800s. With the sinking of the first shafts through the Magnesian Limestone of the west Durham plateau in the 1820s, the 'concealed' coalfields began to be exploited (Figure 10.1). In the larger, more efficient, mines of some areas improved methods of long-wall working replaced wasteful 'room and pillar' methods of extraction which left considerable amounts of coal as supports. Nevertheless, most pits still had fewer than 100 employees in the 1830s and many of the older workings were crude, small-scale and sporadic in operation.

As the market for coal grew sharply production more than doubled between the 1750s and the end of the century, and tripled again by the 1830s (Table 5.4). All coalfields benefited but not in proportion to their reserves. Access to water transport and a strong domestic market or to local coal-fuelled industry (such as in the West Midlands) were crucial. Small coalfields such as Cumberland (with its strong Irish market) and Shropshire (with its early coke-iron industry) benefited alongside major producers for the new industries (such as Lancashire and South Wales). The dominant producers of the early nineteenth century – North East England and Scotland – had both strong local industrial demand and a domestic sea-sale trade. However, by the 1830s regional production was gradually coming more in line with resources and levels of regional demand and the leading coalfields were those which were to dominate production for the next century: North East England, Lancashire, Staffordshire, Scotland, South Wales, Yorkshire and the Midlands (Derby and Nottinghamshire), not necessarily in that order (Chapters 10 and 15).

Iron

One of the major coalfield industries of the early nineteenth century was iron. Yet before 1760 little progress had been made to release iron manufacture from traditional charcoal smelting in the Weald, the Forest of Dean and newer furnaces – often in isolated rural settings or along waterways, such as the Severn – where ore and charcoal could be assembled. Despite experimental smelting of coal measure iron ores with coal the product was unsatisfactory. Hence, notwithstanding substantial iron exports in the eighteenth century, rising demand from iron manufacturers had to be met, up to the 1780s, by imported high-quality bar iron, chiefly from Sweden.

The technical problems were two-fold. First, impurities in most coals, especially sulphur, produced pig iron that was too brittle for good castings and was costly to work. The experimentation of the Darbys of Coalbrookdale successfully overcame this, using the local sulphur-free 'sweet clod' coals to develop pioneer blast furnaces and iron works at a good water-power site near the Severn gorge. Secondly, substantially better blast than that in traditional furnaces was needed for efficient coke-iron smelting. This was met by reciprocating beam engines to power furnace bellows using the Watt steam engine (1769), then by more efficient cylinder engines using new methods of 'boring' metal pioneered by the Wilkinson brothers at Broseley in 1774 and driven, from 1781, by Boulton and Watt rotary engines. These bigger, more efficient blast furnaces greatly enhanced extraction rates of iron from ore. Hence, the output of pig iron accelerated from the 1760s after a lengthy period of transition from charcoal iron smelting; by 1788 production

Figure 5.3 The distribution of iron ores in Great Britain.

Table 5.5 Regional pig iron output in Great Britain, 1788–1950

	North East	West Yorkshire Notts and Derby	North West England North Wales	West Midlands	Shropshire	East Midlands	South Wales	Scotland	TOTAL
				(Thousands of tons)					
1788	–	10	4	32		–	13	7	68
(%)		(14.7)	(5.9)	(47.0)			(19.1)	(10.3)	
1806	2	37	6	50	55	–	71	23	244
(%)	(0.8)	(15.2)	(2.5)	(20.0)	(22.5)		(29.1)	(9.4)	
1839	13	87	34	364	81	–	454	197	1249
(%)	(1.0)	(7.0)	(2.7)	(29.1)	(6.5)		(36.3)	(15.8)	
1855	299	208	48	856	122	–	840	823	3218
(%)	(9.3)	(6.5)	(1.5)	(26.6)	(3.8)		(26.1)	(25.5)	
1870	1440	294	721	1004		65	979	1206	5963
(%)	(24.2)	(4.9)	(12.1)	(16.9)		(1.1)	(16.4)	(20.2)	
1880	2416	674	1599	698		387	890	1049	7749
(%)	(30.2)	(8.7)	(20.4)	(9.0)		(5.1)	(11.5)	(13.5)	
1890	3110	713	1635	626	–	633	825	737	7904
(%)	(35.9)	(9.0)	(20.7)	(7.9)		(8.0)	(10.4)	(9.3)	
1900	3110	853	1586	710	–	637	908	1157	8960
(%)	(34.7)	(9.5)	(17.7)	(7.9)		(7.1)	(10.1)	(12.9)	
1913	3869	1002	1364	850	–	917	889	1369	10260
(%)	(37.7)	(9.7)	(13.3)	(8.3)		(8.9)	(8.7)	(13.3)	
		(Yorkshire and Lancashire)				(including Derby, Notts)			
1938	1833	349	722	317		2466	655	409	6761
(%)	(27.1)	(5.2)	(10.7)	(4.7)		(36.5)	(9.7)	(6.1)	
1950	2402	437	824	551	–	3448	1232	739	9633
(%)	(24.9)	(4.5)	(8.6)	(5.7)		(35.8)	(12.8)	(7.7)	

Source: Mitchell and Deane (1962) pp. 131–33 and (for 1950) Mitchell and Jones (1971) p. 78.

was nearly three times that in 1720. In a more highly-capitalized industry, with more costly plant, a new breed of entrepreneurs played a key role in remodelling and relocating iron manufacturing.

Whereas charcoal iron manufacture focused on easily accessible, if often limited, ores and abundant timber and was close to major markets (the Weald with its proximity to London is the classic case), coke-iron blast furnaces were overwhelmingly on coalfield sites using medium grade coal measure ores with extraction rates of 30 to 40 per cent, though some of Britain's limited high-grade haematite ores (in the Forest of Dean, for example) were also exploited (Figure 5.3). The pioneering areas were not necessarily those with the greatest reserves of either ore or coal. In the late eighteenth century Shropshire, an area of limited resources but considerable pioneering technological and entrepreneurial skills such as those of the Darbys and Wilkinsons, was the leading producer (Table 5.5). This area and the West Midlands, another region of early iron-working from clayband iron ores and good coking coal, produced over two-fifths of Britain's pig iron. South Wales, which initially smelted clayband ores with charcoal, benefited from coal smelting along the north-eastern rim of the coalfield. Local ironmasters, the Homfrays, were joined in 1760 by others from Shropshire (the Guests) and Yorkshire (the Crawshays) to produce an archetypal specialist manufacturing region and Wales's first industrial town (Merthyr Tydfil) in a boom which raised its iron production to three-tenths of the national output by the 1800s. Similarly John Roebuck's technical expertise and William Caddell's entrepreneurial skills promoted the first phase of Scotland's coke-iron industry at the Carron works near Falkirk from 1759. Based on clayband coal measure ores, it quickly achieved one-tenth of the national production. South Yorkshire's early lead in, and continuing dominance of, specialist steel production for the edge-tool and cutlery trade was enhanced by the improved Huntsman crucible method of manufacture from the 1740s yet, in the absence of substantial local ores, its iron production, one-seventh of the national total, was modest.

New smelting techniques were one reason for the increasing competitiveness of iron as a structural material. Cast iron beams and girders replaced timber in many forms of construction: in northern factories and, for example in Liverpool, for churches as well as fire-resistant dock buildings; in bridge-building the first cast-iron bridge appropriately spanned the Severn at Ironbridge, near the Darby works, in 1779. Machine parts increasingly used iron in their construction and with progressive use of iron in fitting and later building ships, as well as railways and eventually locomotives, the use of iron in engineering came of age in the early nineteenth century.

Wider use of iron also depended on improving wrought iron for the forge. In order that pig iron could be shaped into bars and plates from which to produce nails, horseshoes, ploughs, locks and machine parts excess carbon (which made it brittle and unmanageable) had to be drawn off. Traditionally, wrought iron was produced in the finery and chafery and was shaped into the finished product in thousands of smithies. It was a slow, relatively costly process. Coal was used in much of this work by specialist producers, such as in the Birmingham region, and in general industrial markets. In 1784 Henry Cort of Fareham (Hampshire), a specialist in naval work, pioneered his process of puddling – stirring iron in a heated furnace – to draw off excess carbon then rolling to produce wrought-iron bars and plates. Parallel work in other new iron smelting areas – by the Cranage brothers in Shropshire and Peter Onions in South Wales – transformed wrought iron production and focused it on coalfield sites, often achieving maximum economies of fuel in integrated plants alongside the new blast furnaces.

Similar efficiency in the use of fuel marked the final phase of the early industrial revolution in iron manufacture. Many blackband ores were unsuited to the early coke iron industry. But, by using J. B. Neilson's hot blast method (1828) that increased the oxygen supply by passing heated air over the charge, more powerful blast furnaces produced purer iron. Combined with David Mushet's discovery that Scottish blackband ores and splint coal could be fed direct into the furnace without additional fuel, it reduced the cost of fuel to one-third, eliminated coking costs and greatly increased fuel efficiency. As compared with seven tons of coal to produce one ton of cold blast pig, the hot blast method used only two tons. This led to a sharp increase (to one-seventh of the national total) in pig iron from Scottish blast furnaces, much of which went for export to North East England and the West Midlands for finishing and for plate for Clydeside shipbuilding and engineering.

The Baird iron-works at Gartsherrie (Coatbridge) was second only to Dowlais (Glamorgan) in size in the late 1830s. However, South Wales, using both blackband and clayband ores, by then dominated production of rails and plate (the latter finished in the Swansea black-and tinplate industry) and provided two-fifths of British iron manufacture (Table 5.5). It surpassed even the Black Country whose archetypal industrial landscape became synonymous with iron. But, with a long tradition of handicraft metal-working, the West Midlands continued to dominate wrought-iron manufactures – ranging from plates, nails, chains and holloware to specialist engineering products – and reinforced its specialist regional economy focused on coal and heavy industry.

Textiles

For many contemporaries and latter-day analysts, textiles symbolise the early industrial revolution. A consumer rather than a capital goods industry, its response to a sharp increase in demand from a growing population with increasing disposable income arguably provided the initial stimulus for rapid expansion in exports (Chapter 1). Rather than a cause of general economic growth, technological innovation in textile manufacturing was a response to it.

In the mid-eighteenth century woollen manufactures were among the most widespread of British industries and dominated overseas trade (Table 5.2). In parts of northern England and, especially, eastern Scotland flax spinning for coarse linens was important. Much of the fine quality market in textiles, for example silk and mixed cloths, was focused on London and East Anglia. But experimentation with a wide range of mixed yarn fabrics was spreading in many areas, not least in Lancashire's linen/worsted fustians and later its linen/cotton mixtures.

Woollen manufacturing still reflected historical factors – old-established skills and quality; water for fulling and shearing mills; and entrepreneurial linkages – such as those found in broadcloth production in the Cotswolds and the fine mixed-fabric cloths from the Norwich area. Nevertheless, spinning and weaving remained one of the most ubiquitous of domestic activities and provided a growing workforce for newer regional specialization. During the eighteenth century West Yorkshire seized a growing proportion of woollen and worsted markets, especially for cheaper cloths. The finest quality broadcloths (heavier fulled cloths made from short-staple wools) were still principally produced in the Stroudwater Valley, but most of the considerable expansion in woollens was met by Yorkshire raising its production three-fold, to 60 per cent of England's total by 1800. Its price advantage reflected three main factors: skilful entrepreneurial exploitation of cheap domestic labour in small, dispersed production units; water supplies for washing, dyeing and, in due course, power; and coal used in the finishing processes.

Yorkshire's dominance of worsted production was more rapid and complete. From the early eighteenth century it ousted the Suffolk and West Country serges and, from the 1770s, overtook the worsted production of the Norwich area. By then three-quarters of West Yorkshire's textiles were worsted and, quickly adopting innovations pioneered in the cotton industry, it soon dominated British yarn and eventually cloth manufacture.

Handicraft manufacture of knitwear and stockings had provided one of the earliest innovations, Thomas Lee's knitting frame of 1589. Taken up in the seventeenth century principally in London and the East Midlands, it promoted a classic workshop industry. Organized from major regional centres such as Nottingham (principally for lace), Leicester (for knitwear), Derby (for silk) and smaller towns such as Belper, Loughborough and Hinckley, frame-knitting used rural outworkers in surrounding villages. By the early nineteenth century 85 per cent of the much-expanded total of frames were in the East Midlands.

Despite widespread textiles production, Scotland did not immediately share in their growth. Political Union in 1707 adversely affected its significant linen industry until innovations in bleaching and adoption of water-powered spinning led to a three-fold expansion between 1767–68 and 1821–22. In the 1820s nine-tenths of Scotland's linen was made in Fife, Forfar and Perth as against one per cent in West Central Scotland, compared with 63.6 and 22.6 per cent respectively in the 1760s.

Commercial enterprise and highly competitive capitalist organization of a cheap but experienced domestic labour force initiated rapid regional growth and increasing specialization of cotton manufacture in the first phase of industrialization. John Kay's 'flying shuttle' (1733), the first technological innovation in handloom weaving greatly increased weavers' productivity. Their demand for more, and better quality, cotton yarn – still much inferior to Indian production – saw much innovation over the next 50 years. James Hargreaves's jenny (patented in 1770) enabled hand spinners to operate up to a couple of dozen spindles. Improved strength of warp threads, essential

Table 5.6 Cotton spinning in Great Britain 1811

Area	Number of spindles			Number of firms (spindles)				
	Mule	Water frame	Jenny	<5000	5–10000	10–25000	25–50000	>50000
Manchester (and Stockport)	1415644	51252	96072	79	56	14	7	2
Bolton (Bury, Ramsbottom and Heywood)	421392	57107	882	35	17	10	4	–
East Lancashire (Rochdale and Oldham)	242488	13040	–	23	16	6	–	–
Preston (and Chorley)	337264	9840	–	2	6	10	–	1
Cheshire (New Mills, Glossop, Ashton-u-Lyne, and Macclesfield)	684948	51696	8440	77	34	19	1	–
South Lancashire (Wigan and Warrington)	72856	25312	40386	25	3	2	–	–
West Yorkshire (Halifax and Todmorden)	191734	81850	6300	108	10	1	–	–
Sundry places (rest of England and Wales)	43244	20420	3800	N.A.	–	–	–	–
Scotland (in 1808)	800000	–	–	N.A.	–	–	–	–
Total (GB)	4209570	310516	155880					

Source: Samuel Crompton's Census of the cotton industry, 1811 (from G. W. Daniels, *Economic History Review* **2** (1930), pp.107–10)

for fine cottons, was achieved through Richard Arkwright's spinning frame (1769). The ability to spin fine strong 'counts' was finally provided by Samuel Crompton's mule developed in the 1770s and patented in 1779. The way was now open to successful competition in all markets for fine, as well as coarse, cloths since a skilled Lancashire mule-spinner could out-produce an Indian muslin handspinner 25-fold.

The enormously rapid growth of the market for cottons, reflected in exports (Table 5.2), outstripped traditional sources of raw cotton supply from the Levant and India. The problem was met through experimentation with new areas of production in the Caribbean and, then, the southern United States. Moreover, costs were much reduced through the mechanization of combing the cotton lint by Eli Whitney's cotton gin (1793). The vast increase in cotton production in the US from 1.5 to 85 million lbs between 1790 and 1810 was accompanied by the reorientation of Britain's cotton markets from London to Manchester, Liverpool and Glasgow.

The response to late eighteenth- and early nineteenth-century change involving both motive power and scale of organization, produced a basic and lasting relocation of the textile industries. In the East Midlands water power, essential to successful roller spinning, produced a string of mills along the Derwent, including Arkwright's Cromford mills (1771–79) and Jedediah Strutt's mills at Belper and Milford. The factory industry grew out of silk throwing at Thomas Cochett's water-powered factory in Derby (1702), further developed by Thomas Lombe from 1717. Nottingham yarn spinners also took up water power and the area had over 90 cotton mills in the first half of the nineteenth century. In Scotland, cotton manufacture replaced that of linen in and around Glasgow, which was able to harness steam power relatively early to Lancashire technology in large mills such as New Lanark which employed 1700 hands by the early nineteenth century.

The cotton-producing area of east Lancashire and the adjacent Pennine valleys quickly became the focus of mule-spinning in small water-powered factories in an archetypal specialist industrial region. With the successful adaptation of the Boulton and Watt engine (patented in 1775) in 1795 steam-powered factories came to dominate mule-spinning. By 1835 over four-fifths of the area's cotton industry was in steam mills, as against only one-quarter in the midlands (Tables 5.6 and 5.7).

As cheap yarn supplies multiplied, handloom weaving expanded rapidly in the late eighteenth and early nineteenth centuries. Edmund Cartwright's experimental power loom (patented 1775) was harnessed to water then, in

Table 5.7 Water and steam power in the British cotton industry in 1835

Region	Number of mills	Horse power (percentage)	
		Steam	Water
Northern England	934	26513 (81.3)	6094 (18.7)
Midlands	54	438 (26.7)	1200 (73.3)
Scotland	125	3200 (56.3)	2480 (43.7)
TOTAL, GB	1113	30151 (75.5)	9774 (24.5)

Source: E. Baines, *History of the Cotton Manufacture* (1835)

Table 5.8 Textile industries in the British Isles in 1838

Area	Employees in textile factories											
	Cotton		Wool		Worsted		Flax		Silk		All	
	Total	(%)	Total	(%)	Total	(%)	Total	(%)	Total	(%)	Total	(%)
Lancashire and Cheshire	188502	72.7	5121	9.3	925	2.9	2883	6.7	17423	57.1	214854	50.7
Yorkshire	12436	4.8	27548	50.0	26603	84.1	9654	22.2	1084	3.2	77325	18.2
Derby–Notts.–Staffs.	14000	5.4	196	0.4	550	1.7	96	0.2	5161	15.0	20003	4.7
Norfolk	130	0.1	121	0.2	385	1.2	–	–	2274	6.6	2910	0.7
Glouc.–Wilts.–Somerset	29	0.0	10866	19.7	–	–	566	1.3	2538	7.4	13999	3.8
Rest of England	3079	1.2	3413	6.2	3165	10.1	3374	7.8	5077	14.8	18108	4.3
Wales	1010	0.4	1507	2.8	–	–	–	–	–	–	2517	0.6
England and Wales	219186	84.5	48772	88.5	31628	100.0	16573	38.1	33557	97.8	349716	82.5
Clydeside	30464	11.7	1046	1.9	–	–	1132	2.6	725	2.1	33367	7.9
Eastern Scotland	3584	1.4	2421	4.4	–	–	16112	37.1	–	–	22117	5.2
Border Counties	–	–	1076	1.9	–	–	–	–	–	–	1076	0.2
Rest of Scotland	1528	0.6	533	1.0	–	–	653	1.5	38	0.1	2752	0.7
Scotland	35576	13.7	5076	9.2	–	–	17897	41.2	763	2.2	59312	14.0
Ireland	4622	1.8	1231	2.2	–	–	9017	20.7	–	–	14870	3.5
British Isles	259384	100	55079	100	31628	100	43487	100	34320	100	423898	100

Source: Returns of the Factory Inspectors, House of Commons Papers 1839, XLII
Clyde = Lanarkshire, Renfrewshire, Dumbartonshire and Ayrshire
Eastern Scotland = Aberdeen, Kincardine, Angus, Perth, Clackmannan and Fifeshire
Border Counties = Berwickshire, Roxburghshire and Selkirkshire
Total employees in each industry in the listed areas are also given as a percentage of the total in each industry in the British Isles.

the early 1800s, to steam, but it was the mid 1820s before William Horrocks of Stockport succeeded in developing a commercially competitive power loom. Its spread in the 1830s, together with the Roberts' self-acting mule patented in 1830, brought the cotton industry of the north west to its modern distribution. Large mills increasingly began to dominate both spinning and weaving. Big firms (such as Horrocks at Preston and Peels of Bury) had several mills with workforces of over 1,000 and were more highly capitalized than the smaller Arkwright roller watermills.

Technologically and commercially the cotton industry had substantial impacts on other textile manufacturers. Worsted spinning rapidly took up mule spinning and the Bradford area's mills rapidly absorbed the bulk of Britain's worsted industry in the early nineteenth century. Power looms were quickly modified for woollen and worsted cloth production. In Scotland the linen industry became largely focused on big mills in eastern Scotland, especially in the ports of Dundee, Arbroath and Kirkaldy.

The 1838 returns of the Factory Inspectorate (established in 1833) reveal a remarkable concentration of cotton (four-fifths in North West England) and worsted (nine-tenths in West Yorkshire) manufactures (Table 5.8). Other textiles, especially knitwear, were more widely dispersed and depended substantially on handicraft workers in rural locations (Figure 5.4). In the cotton industry, local specialisms within the main areas of production already gave towns a distinctive employment structure and business organization. The numerous warehouses of Manchester, the commercial capital, were the focus of marketing. It was also a major centre for finishing and was still involved in spinning and weaving in integrated mills. Bolton, and Oldham and Rochdale were predominantly spinners of fine and coarse yarns, respectively. Bury's combined mills, involved a good deal of weaving. Although the switch to power-loom weaving was incomplete, north Rossendale's industry was increasingly focused on new weaving sheds, often in combined mills, leading to a decline of handloom weaving and its virtual obliteration by the 1840s.

However, east Lancashire was never wholly dependent on cotton. Textiles had important links with other industries. First the wastefulness of coal consumption in early mill boilers drew mills to coalfield sites or, as in Preston, where canals gave access to cheap coal. Hence many textile firms had interests in collieries. Secondly, innovations in washing yarn and finishing cloth – bleaching, dyeing and printing – were important in textile manufacture, especially cotton, and helped to promote the modern chemicals industries (Chapter 10). Alkalis for soft soap were derived, in the eighteenth century, from imported potash or Scottish seaweed or Spanish barilla. The adoption of the Leblanc process, manufacturing soda by treating salt with sulphuric acid and roasting with coal and limestone, led to the chemical industries on Merseyside by the 1820s and to Charles Tennant's great works at St Rollex, Glasgow. Its by-product, hydrochloric acid, was equally important for bleach manufacture. Tennant's bleaching powder (1799) replaced the slow methods – using natural acids then weak sulphuric acid – of the eighteenth-century bleach fields and greatly assisted the finishing of cotton. Thirdly, machines and boilers, made by local millwrights and boilermakers, eventually promoted specialist textile machinery manu-

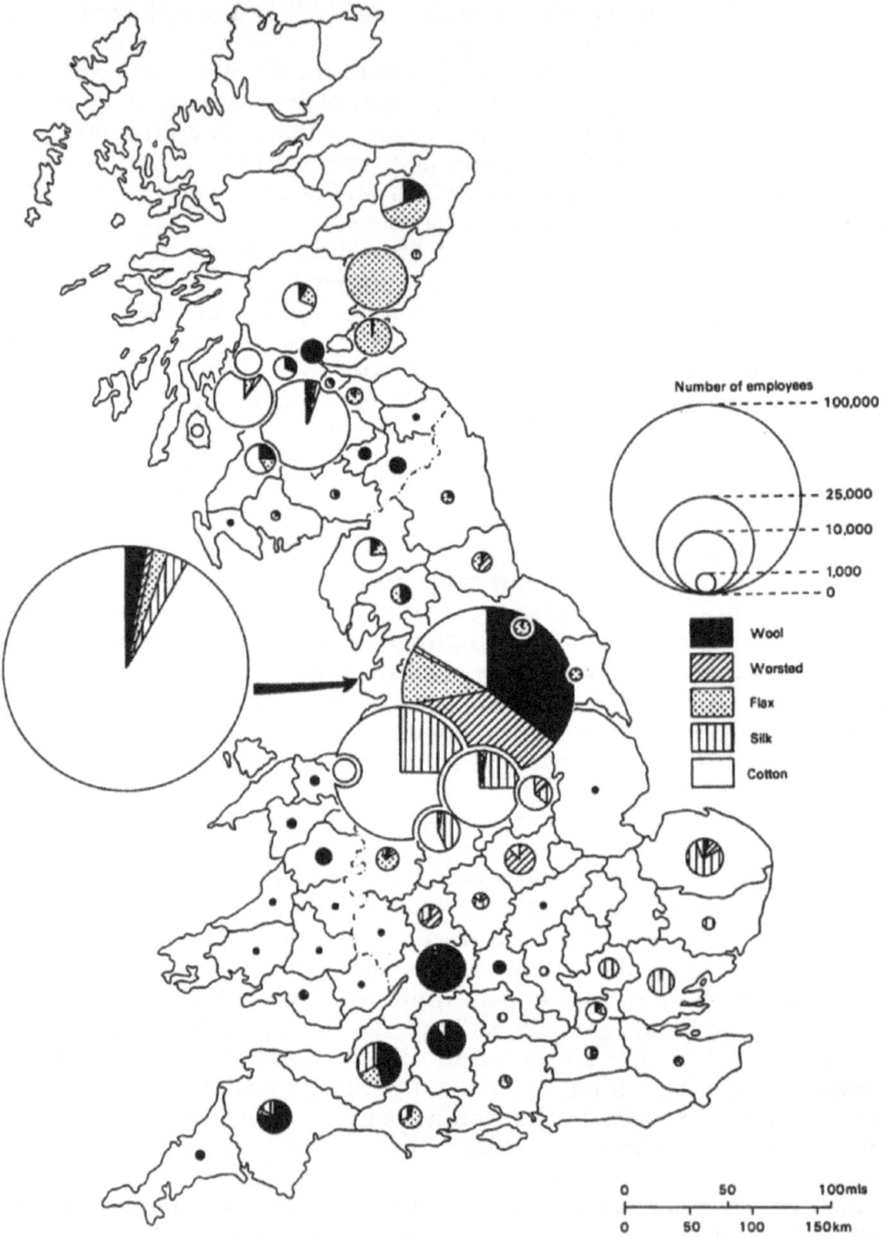

Figure 5.4 The distribution of textiles employment in Great Britain, 1838.
Source: Langton and Morris (1986) p. 110.

factures. In some mill towns of Lancashire and Yorkshire, and in Nottingham, Derby and Leicester general engineering developed from the stimulus given by textile machine manufacturers.

Thus, while the claim that textiles, particularly cotton, were the leading sector in the economic take-off in the eighteenth-century may be exaggerated, they certainly stimulated commercial activity, export income and technological and organizational innovation. They also brought substantial benefits to living standards through employment – in both country and town – and cheaper clothing. And their impacts on labour organization, especially in the factory, the provision of housing and infrastructure (for example, communications, especially canals, and public utilities) were significant in shaping new urban forms as well as life styles in the factory towns.

Conditions of work and standards of living

Unprecedented economic and technological change in the late eighteenth and early nineteenth centuries affected the working lives of ordinary men, women and children in different industries and regions in a variety of ways, the precise effects of which are difficult to measure. First, standards of living have to be related to both wage rates and family income and expenditure. Secondly, the calculation of family budgets tell us little about the changes in quality of life experienced by families moving from domestic to factory employment. Thirdly, national and regional statistics obscure the individual nature of varying work experience from factory to factory, for people employed in different jobs within one establishment, and over short periods of time.

The industrial revolution cannot be viewed as a simple transition from an agricultural and domestic economy to one dominated by factory régimes, but rather as a restructuring of economy and society equivalent to that of the 1920s and '30s, and which entailed the decline of old industries as much as the growth of new ones. For individual workers this meant the abandonment of old skills as well as the acquisition of new ones, while increasing regional specialization of industry created differing impacts from one locality to another.

Thus the impact of change in textile manufacture was felt not only in Lancashire and Yorkshire, but also in the old textile districts. The skills of textile workers in southern England were not immediately abandoned but diverted to more specialist and less competitive niches: in Essex the production of silk and bunting continued until at least the 1870s; and there were temporary increases in the manufacture of lace, gloves and straw plaiting in much of southern England as cloth manufacture declined.

Domestic industry had a number of characteristics which affected working conditions, including low wages and the potential for increasing the intensity of labour. Although family-based units of production conferred certain advantages over factory employment, they could only operate whilst low wages were tolerated and with the availability of low-cost and flexible child and female labour. Domestic spinning was mostly carried out by women who would also contribute to agricultural labour, take responsibility for child care and household chores, and help to train children for work in the family economy. But the clothiers, who controlled the supply of raw materials and who bought (and thus set the price for) finished cloth, and the majority of weavers were male.

Effectively, men controlled those tasks undertaken outside the home and those skills which were increasingly associated with more complex technology. Women were more likely to continue to weave on a simple hand-loom or to be hand- rather than framework-knitters: when the mule became available men took over the new spinning technology, leaving women to the traditional spinning wheel or jenny. Female skills were marginalized through a technology-related gender division, whilst women's work was also interlocked with responsibilities for childcare and domestic chores. As the domestic system gave way to factory production these characteristics became even more marked. Men dominated the most skilled jobs in mule spinning and, later, power weaving although by the 1830s there were substantial numbers of female mule spinners in factories despite attempts by male-dominated spinners' unions to exclude them. Most women remained domestic spinners, took up domestic weaving or filled lower-paid factory jobs such as reelers, warpers or tenters.

Despite exploitation, especially of women and children, the domestic system afforded all workers greater freedom than a factory régime. People could work to their natural rhythm, work all hours when necessary but take time off for fairs, festivals and family activities: 'Saint Monday' – traditionally a day for leisure and personal activities – was common in many handicraft trades. Such flexibility is illustrated in weavers' diaries from the 1780s. In October 1782 Cornelius Ashworth combined weaving on wet days with harvesting and threshing on fine. A full day's weaving might produce 8 or 9 yards of cloth but often he would make less, using the spare time for agricultural work, odd jobs in the village or to sell his cloth. Benjamin Shaw's father Joseph, who worked as a weaver in Garsdale (Yorks) in the 1770s, frequently become bored with weaving at home and took to dismantling and eventually teaching himself to repair clocks and spent an increasing amount of time travelling the locality mending clocks. As the demand for domestic weaving declined, he became an accomplished clock repairer and metal worker and eventually worked on the manufacture and repair of textile machinery. Such dual occupations – in agriculture and industry or in two different domestic trades – were typical of many rural craftsmen; their flexibility gave scope for new activities producing long-term benefits.

The scale and spatial organization of domestic manufacture varied considerably from area to area and between production processes. The eighteenth-century woollen districts of Yorkshire retained the traditional system of small independent clothiers. A typical domestic unit would entail the male head of the household buying wool from the local market to be carded and spun by his wife and children, with some wool put out for spinning by women in adjacent cottages. The clothier would dye and weave the wool himself, assisted by his sons and a small number of male apprentices or journeymen, and take the cloth to the nearest fulling mill and then to his stall in the local market town. He would produce only one or two pieces a week, but would probably also have a small acreage of farm land.

In contrast, in the worsted districts around Halifax manufacturing was organized on a large-scale capitalistic basis. Merchant-manufacturers would buy large quantities of wool to put out over a wide area for spinning and weaving by small domestic producers. Whereas the independent clothier/weaver operated over quite small distances, putting out to neighbours, travelling perhaps 5 miles to the nearest fulling mill and the large cloth hall in Leeds which had places for over 1000 small stall holders, the large worsted manufacturer had much wider horizons. He might distribute wool for manufacture over a distance of thirty miles and then market it at the cloth hall in Bradford where some 250 large manufacturers had stands (Figure 5.5).

In Lancashire small-scale capitalist middlemen developed by the middle of the eighteenth century. Fustian masters gave out raw cotton and linen threads for spinning and weaving on domestic premises over a wide

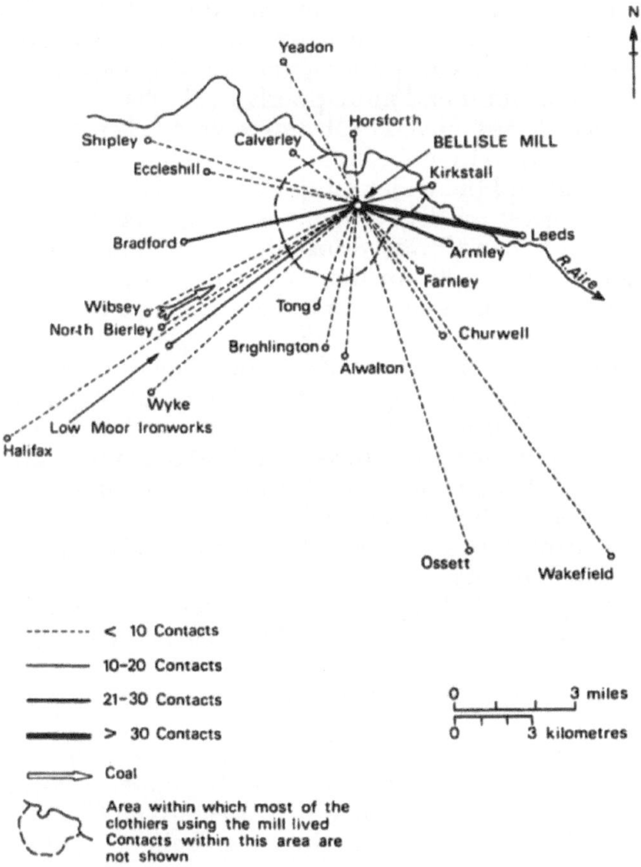

Figure 5.5 Area from which clothiers using the Bellisle Fulling Mill (Yorks) were drawn, *c.* 1811.
Source: Gregory (1982) p. 96.

geographical area and selling the cloth to separate merchants operating outside the production process. Some early cotton masters became very wealthy, especially following the spread of calico printing in Lancashire, and were responsible for establishing spinning factories in the late eighteenth century, although some spinning and a substantial amount of weaving remained in domestic premises and small workshops until the 1830s.

The development of large-scale factory production undoubtedly reduced the diversity of work experience found within domestic and proto-industrial processes. But by no means were all factories similar and there was a wide range of work experiences within any one factory unit. Many late eighteenth-century textile mills were rural and recruited labour from local domestic industry. Families often moved together to a new factory so that all members of the household could gain employment. A weaver used to working in a small weaving shed would be familiar with many aspects of the work environment – if not the scale – within a factory. Boys would probably be apprenticed to weaving, power-spinning or in the machine shop; girls might work in the carding room before moving on to other low-technology jobs within the mill. Generally, as new technology was adopted, men took control of the new processes in spinning and weaving whilst women were left with the older machines and more poorly paid jobs.

Increasingly, as factories moved to steam-powered sites, the labour force moved from rural mills to towns. While the new large urban mills offered greater opportunities and the wider range of employment in towns was an insurance against recession and unemployment, factory labour altered workers' lives in a number of ways. Most obvious was the loss of independence and freedom, especially for the man who had previously been his own master. Machine-makers who had travelled around the country building or mending machines at small premises throughout the textile districts found that in the large machine shops of factories they were tied to one place of work. Despite familiar technology, the factory weaver or spinner could no longer intersperse industrial work with agricultural labour or other activities. Many factory masters introduced rigid and draconian regulations to keep the workforce at their machines for long hours and to break their irregular work patterns. The very fact of being part of a continuous production process, where machines controlled the pace of work, was a fundamental change for many workers.

Children were a particularly attractive mill workforce. Their labour was cheap and they could be more easily controlled than independently-minded adult weavers. In rural areas of labour shortage, pauper children from urban poor houses were a common source of labour. In 1800 some 90 pauper children worked in the Greg's Quarry Bank mill, Styal (around 50 per cent of the workforce) and as late as 1816, when the system of parish apprentices was declining, 36 per cent of its labour force were pauper children from the Manchester area bound to seven-year apprenticeships and 70 per cent of the total workforce were children. In 1790 the Gregs built an apprentice house for the children who were paid a penny or two a week plus board and lodging for a variety of tasks around the mill. Older children learned to spin and card: some boys would become skilled mechanics; most would be

Table 5.9 Nominal annual earnings in selected occupations; England and Wales, 1710–1901 (Money wages in current pounds)

Year	Agricultural labourers	General labourers	Skilled engineers	Skilled building workers	Skilled textile workers	Skilled printing workers	Clerks (not government)	Teachers
1710	17.8	19.2	40.7	28.5	33.6	43.3	43.6	15.8
1737	17.2	20.2	41.6	29.1	34.3	44.2	68.3	15.0
1755	17.2	20.8	43.6	30.5	36.0	46.3	63.6	16.0
1781	21.1	23.1	50.8	35.6	41.9	54.0	101.6	16.5
1797	30.0	25.1	58.1	40.6	47.9	66.6	135.3	43.2
1815	40.0	43.9	94.9	66.4	67.6	79.2	200.8	51.1
1835	30.0	39.3	77.3	59.7	64.6	70.2	269.1	81.9
1861	36.0	44.2	88.8	72.9	63.3	74.7	248.5	93.8
1881	41.5	55.9	96.7	87.2	85.8	86.4	286.7	120.8
1901	46.1	68.9	116.2	103.2	101.4	92.7	286.9	147.5

Source: B. R. Mitchell, *British historical statistics* (Cambridge: Cambridge University Press, 1988) p.153

turned off when their term of apprenticeship was finished and new cheaper labour recruited. Non-pauper children at Styal, though paid a small weekly wage, were otherwise treated like the parish apprentices. Yet, despite long hours of work and uncomfortable living conditions, many children apprenticed at such a mill were better off than as paupers in the slums of Manchester.

There has been considerable debate about the extent to which industrialization improved or worsened the position of the working family. National wage series (Table 5.9) are imprecise and inconclusive, and debates about the standard of living can be tackled only at the local level (Chapter 1). For example, in Bath real wages declined from 1790 to 1812, rose again to 1832 but fell in the later 1830s so that the economic position of the labourer in 1830 was similar to that in 1790, though there might have been wider employment opportunities.

Quality of life (as opposed to standards of living) can only be measured for the individual. Then, as now, it was determined by variations in life cycle and family circumstances, by local housing and employment, by individual and family health, and by luck. Most individuals experienced sharp changes in quality of life over short periods, often for reasons quite outside their own control. During a twenty-year residence in early nineteenth-century Preston, Benjamin Shaw's weekly wages as a machine shop worker ranged from over £2 on piece rates in a time of full employment, through an average wage of around £1 to the poverty of unemployment. He was able to save enough to see his family through short periods of unemployment, but prolonged ill-health meant recourse to poor relief. Since his parish of Settlement was not Preston, he was removed by the Overseers of the Poor and his family temporarily split up.

Indeed health, affecting the ability to seek and maintain regular work, was a key factor affecting living standards. Although power-driven machinery in large factories increased the risk and severity of industrial accidents, working with poorly maintained and unguarded machinery in the home or small workshop was also dangerous. Overcrowding of both mills and domestic workshops increased the risk of accidents and encouraged the spread of infectious and industrial diseases, while the dust-laden atmosphere of many workshops and factories also induced respiratory disease. Children were particularly at risk in both workshops and mills, the lower level of control in the domestic system perhaps heightening the risk of accidents. Thus any analysis of the impact of industrial transformation on the lives of ordinary working people must take account of both continuity and change in working conditions which, with the diversity of regional and local economic circumstances, makes national generalizations particularly hazardous.

Bibliography and further reading for Chapter 5

For general trends in the British economy see the bibliographies for chapters 1 and 2.

Contemporary accounts include:

CAMPBELL, J. *A political survey of Great Britain* (London, 1774).

MACPHERSON, D. *Annals of commerce, manufactures, fisheries and navigation* (London, 1805).

PORTER, G. R. *The progress of the nation* (3 vols.) (London, 1836–38).

There are many studies of specific industries the most useful include:

Coal and power:

BUXTON, N. K. *The economic development of the British coal industry: from industrial revolution to the present day* (London: Batsford, 1980).

CROMAR, P. 'The coal industry on Tyneside 1771–1800: oligopoly and spatial change' *Economic Geography*, **53** (1977), pp. 80–93.

DICKINSON, H. W. *A short history of the steam engine* (Cambridge: Cambridge University Press, 2nd edition, 1963).

DUCKHAM, B. F. *A history of the Scottish coal industry* vol i, *1700–1815* (Newton Abbot: David and Charles, 1970).

FLINN, M. W. *The history of the British coal industry* vol 2: *1700–1830. The industrial revolution* (Oxford: Oxford University Press, 1984).

HILLS, R. L. *Power in the industrial revolution* (Manchester: Manchester University Press, 1970).

JOHN, A. H. *The industrial development of South Wales* (Cardiff: University of Wales Press, 1950).

LANGTON, J. *Geographical change and industrial revolution: coal mining in south-west Lancashire, 1590–1799* (Cambridge: Cambridge University Press, 1979).

LEWIS, B. *Coal mining in the eighteenth and nineteenth centuries* (London: Longman, 1971).

VON TUNZELMANN, G. N. *Steam power and British industrialization to 1860* (Oxford: Clarendon Press, 1978).

Iron and steel:

ASHTON, T. S. *Iron and steel in the industrial revolution* (Manchester: Manchester University Press, 3rd edition, 1963).

BIRCH, A. *The economic history of the British iron and steel industry, 1784–1879* (London: Cass, 1967).

DUTTON, H. I. and JONES, S. R. H. 'Invention and innovation in the British pin industry, 1790–1850' *Business History Review* **57** (1983) pp. 175–93.

GALE, W. K. V. *The British iron and steel industry. A technical history* (Newton Abbot: David and Charles, 1970).

HYDE, C. K. *Technological change and the British iron industry, 1700–1870* (Princeton: Princeton University Press, 1977).

ROEPKE, H. *Movements of the British iron and steel industry 1720–1951* (Urbana: University of Illinois Press, 1956).

The cotton industry:

BAINES, E. *History of the cotton manufacture of Great Britain* (1835, London: Cass, 2nd edition, 1966).

CHAPMAN, S. D. *The early factory masters* (Newton Abbot: David and Charles, 1967).

CHAPMAN, S. D. *The cotton industry in the industrial revolution* (London: Macmillan, 1972).

EDWARDS, M. M. *The growth of the British cotton trade 1780–1815* (Manchester: Manchester University Press, 1967).

FARNIE, D. A. *The English cotton industry and the world market 1815–1896* (Oxford: Clarendon Press, 1979).

FITTON, R. S. *The Arkwrights: spinners of fortune* (Manchester: Manchester University Press, 1989).

MARWICK, W. H. 'The cotton industry and the industrial revolution in Scotland' *Scottish Historical Review* **21** (1924).

RODGERS, H. B. 'The Lancashire cotton industry in 1840' *Transactions of the Institute of British Geographers* **28** (1960), pp. 135–53.
SMITH, D. M. 'The cotton industry in the East Midlands' *Geography* **47** (1962), pp. 256–69.
URE, A. *The cotton manufacture of Great Britain* (London, 1836).
WADSWORTH, A. P. and MANN, J. de L. *The cotton industry and industrial Lancashire, 1600–1780* (Manchester: Manchester University Press, 1931).

The woollen and worsted industries:
BISCHOFF, J. *A comprehensive history of worsted and woollen manufacture* (1857, reprinted London: Cass, 1968).
CLAPHAM, J. H. *The woollen and worsted industries* (London: Methuen, 1907).
CRUMP, W. B (ed.) *The Leeds woollen industry 1780–1820* (Leeds: Thoresby Society, 1931).
JAMES, J. *A history of the worsted manufacture in England* (London: Cass, 1968).
JENKINS, D. T. and PONTING, K. G. *The British wool textile industry 1770–1914* (London: Heinemann, 1982).
JENKINS, D. T. *The West Riding wool textile industry, 1770–1835: a study in fixed capital formation* (Edington: Pasold Research Fund, 1975).
JENKINS, J. G. *The Welsh woollen industry* (Cardiff: University of Wales Press, 1969).
PONTING, K. G (ed.) *Baines account of the woollen manufacture of England* (Newton Abbot: David and Charles, 1970).
PONTING, K. G. *The woollen industry of south west England* (Bath: Adams and Dart, 1971).

Other textile industries:
FELKIN, W. *History of the machine-wrought hosiery and lace manufactures* (1867, reprinted Newton Abbot: David and Charles, 1967).
HARTE, N. B. and PONTING, K (eds) *Textile history and economic history: essays in honour of Miss Julia de Lacy Mann* (Manchester: Manchester University Press, 1963).
SMITH, D. M. 'The silk industry in the East Midlands' *East Midland Geographer* **3** (1962), pp. 20–31.
SMITH, D. M. 'The British hosiery industry at the middle of the nineteenth century: an historical study in economic geography' *Transactions of the Institute of British Geographers* **32** (1963), pp. 125–42.
WILDE, P. D. 'Regional contrasts in the evolution of the Pennine silk industry before 1860' *East Midland Geographer* **17** (1976), pp. 217–29.

For the impact of industrialization on conditions of work, standards of living and regional landscapes see:
BERG, M. *The age of manufactures 1700–1820* (London: Fontana, 1985).
BERG, M., HUDSON, P. and SONENSCHER, M (eds) *Manufacture in town and country before the factory* (Cambridge: Cambridge University Press, 1983).
BYTHELL, D. *The handloom weavers* (Cambridge: Cambridge University Press, 1969).
COSSONS, N. *The B.P. book of industrial archaeology* (Newton Abbot, David and Charles, 1975).
FREEMAN, M. 'The industrial revolution and the regional geography of Britain: a comment' *Transactions of the Institute of British Geographers NS* **9** (1984) pp. 507–12.
GREGORY, D. *Regional transformation and industrial revolution* (London: Macmillan, 1982).
GREGORY, D. 'The production of regions in England's industrial revolution' *Journal of Historical Geography* **14** (1988), pp. 50–58.
HAMMOND, J. L. and B. *The skilled labourer 1760–1832* (London: Longman, 1919).

HUDSON, P. *The genesis of industrial capital: a study of the West Riding wool textile industry 1750–1850* (Cambridge: Cambridge University Press, 1986).
HUDSON, P (ed.) *Regions and industries: a perspective on the industrial revolution in Britain* (Cambridge: Cambridge University Press, 1989).
LANGTON, J. 'The industrial revolution and the regional geography of England' *Transactions of the Institute of British Geographers NS* **9** (1984), pp. 145–67.
ROSE, M. B. *The Gregs of Quarry Bank mill: the rise and decline of a family firm 1750–1914* (Cambridge: Cambridge University Press, 1986).
SMELSER, N. J. *Social change in the industrial revolution* (London: Routledge, 1959).
TAYLOR, A (ed.) *The standard of living in Britain in the industrial revolution* (London: Methuen, 1976).
THOMPSON, E. P. 'Time, work discipline and industrial capitalism' *Past and Present* **3** (1967), pp. 56–97.
TRINDER, B. *The industrial revolution in Shropshire* (Chichester: Phillimore, 1973).

6

Urbanization and urban life from the 1740s to the 1830s

The urban hierarchy

Urbanization, especially the growth of great cities, was one of the hallmarks of modernization and came to be one of the symbols of economic and social change during the industrial revolution. During the eighteenth century Britain's urban population growth overhauled that in Europe and by the later eighteenth century was twice the European rate. The aggregate number and proportion of urban-dwellers in England increased from some 0.85 million (17 per cent of the total population) in towns of over 5000 in 1700 to 1.2 million in 1750, 2.4m in 1801 and nearly 6 million in 1831 (21.0, 27.5 and over 40 per cent, respectively). In Scotland, rather later economic development produced substantial urban growth from c.100,000 (14.5 per cent) in towns in 1775, to 1.5 million (51.5 per cent) by the mid-nineteenth century (Table 6.1).

More significantly, towns moved from a relatively stable pre-modern system through a period of turmoil during which both the hierarchy of towns and their relative growth rate reflected fundamental changes in the economic forces shaping their development. Many old-established centres declined, new ones developed, and the rank order in a considerably expanded size-range changed fundamentally (Table 6.2). Of the 32 leading towns in England in the early part of the eighteenth century, only London had a population of over 30,000. Many towns had as few as 5,000 people: six cities – London, Norwich, York, Bristol, Newcastle and Exeter – had been among the leaders in 1600; by 1750 only four of these – London, Bristol, Norwich and Newcastle – had changed; by 1801 only London and Bristol remained amongst the leaders. Of the top 32 towns in 1701 only seven retained their place in the top half of the equivalent ranking list for 1801, six had moved up from the lower half and three were entirely new to that group of leading towns. Conversely, of the 1701 list ten were missing by 1801 and had been replaced by eight rapidly growing industrial centres (mostly in the midlands and northern England), one spa (Bath) and one garrison town (Colchester). By 1831 the list of the twenty largest towns was dominated by the major commercial cities, ports, industrial towns and resorts.

DOI: 10.4324/9781003174783-6

Table 6.1 Urbanization in Great Britain, 1801–1951

	Population (in millions)		Percentage of total population in towns						
	Total	Urban	All	<10,000	10–50,000	50–100,000	100–500,000	>500,000[1]	London
1801									
Eng. and Wales	8.9	3.0	33.8	9.9	9.5	2.2	–	–	12.2
Scotland	1.61	0.34	20.4	4.2	6.9	10.3	–	–	
1831									
Eng. and Wales	13.9	6.2	44.3	10.6	11.1	4.0	5.8	–	12.8
Scotland	2.37	0.75	31.7	2.4	10.6	2.4	16.2	–	
1851									
Eng. and Wales	17.9	9.7	54.0	9.9	13.4	5.8	11.0	–	13.9
Scotland	2.89	1.50	51.8	19.8	6.9	5.2	19.9	–	
1871									
Eng. and Wales	22.7	14.8	65.2	10.8	16.2	5.6	13.3	5.0	14.3
Scotland	3.36	1.95	57.4	18.7	7.1	4.3	10.6	16.7	
1891									
Eng. and Wales	29.0	21.6	74.5	10.2	16.2	8.6	20.0	4.8	14.7
Scotland	4.03	2.63	65.4	15.4	12.3	3.2	15.4	19.0	
1911									
Eng. and Wales	36.1	28.5	78.9	8.8	18.3	8.0	24.2	7.0	12.6
Scotland	4.76	3.33	69.9	11.2	16.4	4.9	15.7	21.7	
1951									
Eng. and Wales	43.8	35.6	81.2	3.9	14.5	10.1	13.1	20.9	18.7
Scotland	5.10	4.23	82.9	22.7	17.1	5.5	16.2	21.4	

[1]Includes (in date order) for England and Wales; Liverpool, Manchester and Salford (1871–); Birmingham (1911–); Sheffield (1931–); Leeds (1951); and, for Scotland, Glasgow only (1871–)

Sources: C. M. Law, (1967) for England and Wales, 1801–1911
T. C. Smout, (1969) for Scotland, 1801
A. F. Weber, (1969)
Censuses of Population: Great Britain, 1851; England and Wales and Scotland, 1871–1951

Table 6.2 The twenty largest towns in Britain, 1750–1951

Order	Towns by population (000s) in year: 1750	1801	1831	1891	1951
1	London (675)	London (959)	London (1 778)	London (4 266)	London (5 353)
2	Edinburgh (57)	Manchester (89)	Manchester (223)	Manchester (773)	Birmingham (1 113)
3	Bristol (50)	Liverpool (83)	Glasgow (202)	Glasgow (766)	Glasgow (1 090)
4	Norwich (36)	Edinburgh (83)	Liverpool (202)	Birmingham (634)	Manchester (881)
5	[2]Glasgow (32)	Glasgow (77)	Edinburgh (162)	Liverpool (632)	Liverpool (789)
6	Newcastle (29)	Birmingham (74)	Birmingham (144)	Leeds (368)	Sheffield (513)
7	Birmingham (24)	Bristol (61)	Leeds (123)	Edinburgh (342)	Leeds (505)
8	Liverpool (22)	Leeds (53)	Bristol (104)	Sheffield (324)	Edinburgh (467)
9	Manchester (18)	Sheffield (46)	Sheffield (92)	Bristol (289)	Bristol (443)
10	Leeds (16)	Plymouth (43)	Plymouth (66)	Bradford (266)	Nottingham (306)
11	Exeter (16)	Norwich (36)	Norwich (61)	Nottingham (214)	Hull (299)
12	[1]Aberdeen (16)	Newcastle (33)	Aberdeen (57)	Hull (200)	Newcastle (292)
13	Plymouth (15)	Bath (33)	Newcastle (54)	Stoke (193)	Bradford (292)
14	Chester (13)	Hull (30)	Hull (52)	Newcastle (186)	Leicester (285)
15	Coventry (13)	Nottingham (29)	Bath (51)	Leicester (175)	Stoke (275)
16	Nottingham (12)	Aberdeen (27)	Nottingham (50)	Portsmouth (159)	Coventry (258)
17	Sheffield (12)	Sunderland (26)	Dundee (45)	Bolton (156)	Cardiff (244)
18	[1]Dundee (12)	Dundee (26)	Bradford (44)	Dundee (154)	Portsmouth (234)
19	[1]York (11)	Stoke (23)	Bolton (42)	Plymouth (145)	Plymouth (208)
20	Great Yarmouth Portsmouth Sunderland Worcester (10)	Wolverhampton (21)	Brighton (41)	Sunderland (132)	Aberdeen (183)

Sources: London is the area equivalent to the County of London (1891–)
Figures for Manchester include Salford, Plymouth includes Devonport, and Edinburgh includes Leith. Stoke includes Burslem in 1801, Wolverhampton includes adjacent urbanized areas in 1801. For 1891, Glasgow includes the areas incorporated 1911–21, Birmingham those incorporated in 1881–1911; Liverpool those added in 1894 and Manchester/Salford those incorporated 1891–1911
1750 Wrigley, *People, Cities and Wealth*, Table 7.1; and [1]Smout, *A History of the Scottish People 1560–1830*, Table II (for 1755)
1801–1951 Censuses of Population

Scotland's more gradual urban revolution produced fewer such radical changes. In particular the rank order of the five leading towns – Edinburgh, Glasgow, Aberdeen, Dundee and Perth – remained until Glasgow forged ahead after 1821. But the emergence of ports in the eighteenth century and new industrial centres from the early nineteenth century brought about a rapid acceleration in urbanization and a profound change in Scotland's preindustrial burghal hierarchy.

In one further respect there was a substantial difference in the rank-size order between the two countries. There was no parallel for London's long-established primacy. The biggest city in Europe by 1700, it had been a driving force in the English economy from Elizabethan times. It provided a market for agricultural products from both its major hinterland and most parts of Britain; it dominated overseas trade, especially imports; and through migration it absorbed a disproportionately large part of the country's natural increase to offset its high mortality (Chapter 3). London was both the principal market and for many consumer and processing industries the largest manufacturing centre. By far the major focus of commerce and shipping it stood at the centre of a national communications network. In the pre-canal and pre-railway eras access to London via stage coach and carrier was faster than between most provincial cities even in adjacent regions and the extensive network of coastal and international sea routes was reflected in the primacy and rapid growth of the port and its shipbuilding and associated activities.

In contrast, although Edinburgh was the centre of Scottish government, commercial and cultural life it had only 2.5 per cent of the country's population in 1755, and the combined populations of the next few leading towns outnumbered it by 4:1. By 1831 Glasgow, with 202,000 people had surpassed Edinburgh and Leith's 162,000: the next three in the hierarchy (Aberdeen 57,000, Dundee 45,000 and Paisley 46,000) between them totalled almost as many as the capital.

Despite the continuing dominance of London, by the end of the eighteenth century the gap between it and the next group of large provincial cities was narrowing. In 1801 Greater London was over 10 times the size of Liverpool, Manchester, Edinburgh and Glasgow, as compared with a ratio of 20:1 over the next cities a century earlier. By 1831 that gap had narrowed to 8:1 (over Manchester/Salford). More importantly the rank order curve was straightening as the growth of smaller provincial cities and rapidly emerging industrial towns accelerated (Figure 6.1).

Britain's rapid late-Georgian urbanization was promoted by better transport and communications and combined forces of commerce and trade, industry, and residential and recreational development. From the 1770s to the 1820s the fastest-growing towns included many of the large provincial cities, the commercial and servicing centres of their regions whose manufacturing activity they shared in and complemented. In North West England, for example, Manchester was both the commercial centre for cotton and had important concentrations of spinning, finishing and other industries (including engineering) providing for the new factories. Liverpool was the major port for the cotton trade and, as its canal links with the textile regions of the north and the West Midlands metal-working regions expanded, from

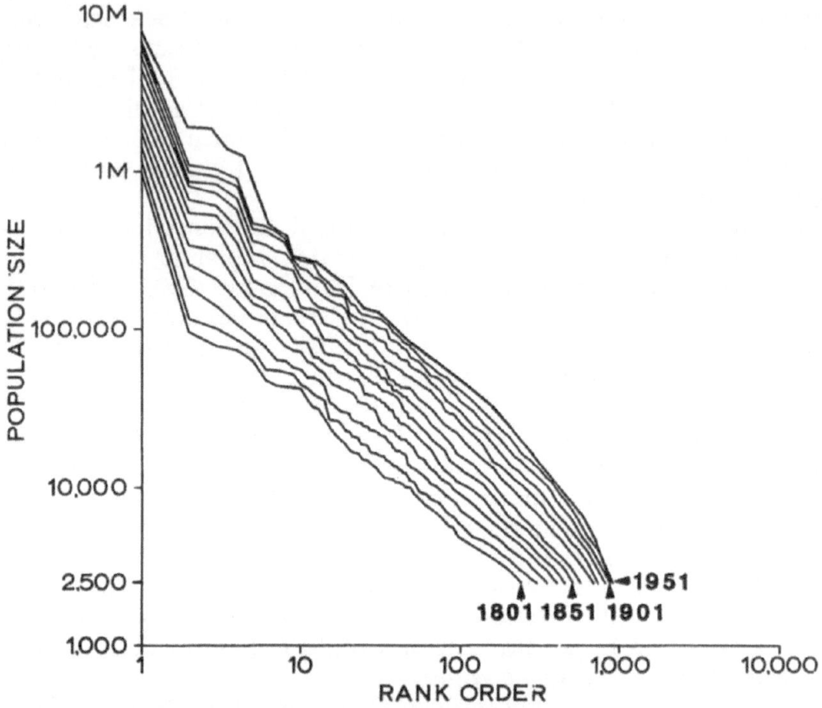

Figure 6.1 Urban rank-size relationships in England and Wales, 1801–1951. Based on B. Robson, *Urban Growth: An Approach*. (London: 1973) p. 30.

the 1760s, it became Britain's premier export port and through its varied import trade an important centre for the processing of food and other raw materials. Leeds, Birmingham, Nottingham and Leicester were, respectively, the centres for the woollen, metal, lace and knitwear industries of their hinterlands. Up to the 1830s many of those industries were still organized in small production units within a proto-industrial system (Chapter 5). Similarly Glasgow was not only a significant cotton manufacturing centre by the early nineteenth century but also the port and commercial centre for adjacent textile towns of Renfrewshire and north Ayrshire and the principal port for the cotton and other Atlantic trades. Tobacco and sugar had created much of its eighteenth-century prosperity and promoted development of 'outports' at Port Glasgow and Greenock to overcome the problem of navigation on the Clyde before its deepening in 1768–81.

Such juxtaposition of port and commercial capital in Britain's regions reflects the importance of expanding trade in urban growth. Many medium-sized towns were specialist ports; for example coal-exporting towns such as Sunderland, Maryport and Whitehaven. The importance of sea communications for internal and European trade was manifest in the dozens of busy coastal, estuarine and river ports which prospered during a phase of small sailing vessels and barges. Boston and King's Lynn handled much of their region's grain and produce exports and imports of coal; the port of Lancaster flourished with both coastal and trans-Atlantic trade; and small Welsh ports were the lifelines to Liverpool, Bristol and even London. With improvements in river navigation from the 1720s and the initiation and rapid development of a national canal network from the 1760s, inland ports developed at stategic nodes in the system (such as Stourport, Worcestershire), at termini (such as Port Dundas at the Glasgow end of the Forth and Clyde canal) and in company towns such as Goole on the Yorkshire Ouse, established in 1826.

Complementing Britain's growing commercial role and protecting the major Atlantic routeways against the French, naval dockyards and ports experienced very rapid growth and, in function and population structure, provided one of the most distinctive types of town. By 1801 Plymouth – also a regional capital – had advanced to eighth in the English urban hierarchy and Portsmouth – essentially a naval port – to eleventh. However, as strategic emphasis shifted, the Thames declined from its mid-eighteenth century importance when Chatham, like Portsmouth with some 10,000 population, was in the top twenty towns in the country.

The full flowering of the industrial town awaited the development of large-scale units of production and specialization characteristic of Victorian Britain (Chapters 10 and 11). But some of the most striking new urban

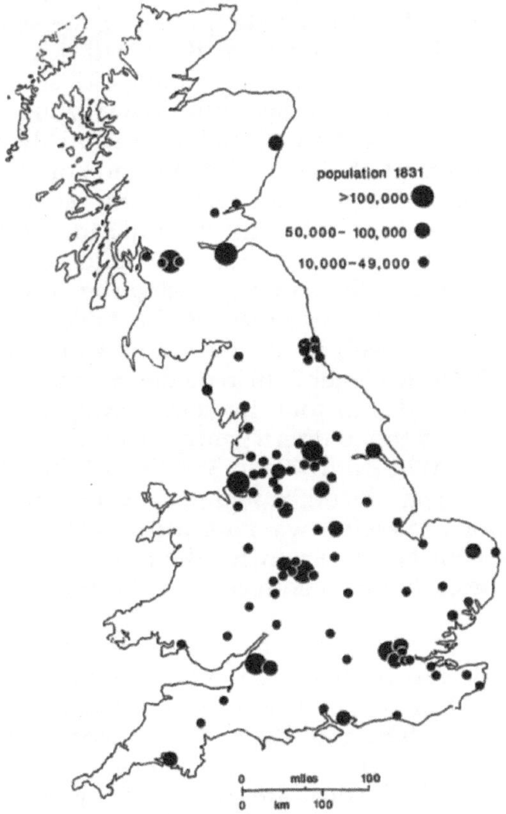

Figure 6.2 Major urban centres of Great Britain, 1831.
Source: Censuses of Great Britain, 1831 and 1851.

developments of the late eighteenth and early nineteenth centuries were in areas of early factory industry. Nowhere was this more evident than in the cotton- and worsted-manufacturing areas. From the 1770s to the early 1830s many places in spinning districts of south east Lancashire grew ten-fold to substantial towns of over 30,000, nowhere more so than Oldham, a mere village in the 1760s but a mill town of over 31,000 people in 1831 (Figure 6.2). Bradford grew even more spectacularly around its large worsted mills from negligible late eighteenth-century origins to 44,000 by 1831. Urbanization in the metal manufacturing areas is more difficult to estimate, since their industrialization embraced towns, mining villages and scattered handicraft communities. But by 1830 the Black Country – a region of mines, furnaces, forges and workshops with specialist manufacturing and market towns – had commercial centres which occupied a prominent place in both the regional and national hierarchy. Some of the fastest-growing industrial towns were in areas of least pre-industrial development. In South Wales the iron districts and ports alike sprang into prominence with the new coke-iron technology. The Dowlais works (1759) promoted the boom town of Merthyr Tydfil with a population of over 17,400 by 1831. Non-ferrous metal processing, along with port development, rapidly swelled Swansea's population to 20,000 by 1831. Similarly in Scotland the development of coke-iron smelting at Carron in 1760 promoted immediate growth of the small town of Stenhousemuir, although its full impact on the Coatbridge-Motherwell area awaited the new blast furnace technology of the 1830s.

One of the most distinctive and fastest-growing groups of late-Georgian towns, the spas and seaside resorts, had an essentially service function. Both reflected a new aspect of society, the provision of playgrounds for the rich and fashionable – more particularly the aristocracy and their entourages. 'The season' was a feature not only in London but of many county towns, as reflected in the considerable building of assembly rooms and theatres in the eighteenth century. 'Taking the cure' from spa waters (against promiscuity and the excesses of the table, in particular over-indulgence in port and claret) produced a new focus of activity for 'the fashion'. Bath, in 1700 little bigger than in medieval times, was designed and promoted by John Wood and took its social tone from Beau Brummel. Proximity to London and the southern gentry assured its success and it grew rapidly from some 1400 in the 1740s to 33,000 by 1801 and over 50,000 in 1831. The growth of spas in other regions – Cheltenham, Leamington, Harrogate and Tunbridge Wells – was later and slower. Indeed, only as they became residential centres (especially for the retired military and overseas merchants) from the early nineteenth century did they become substantial towns.

Similarly, despite eighteenth-century promotion of sea-bathing as health-giving, most seaside resorts began to appear significantly in the British urban hierarchy only from the early railway age (Chapter 11). Nevertheless Brighton, promoted by the Prince Regent's patronage had a population of 7,000 in 1801 and 41,000 by 1831. A string of similar resorts from Lyme Regis and Weymouth, along the Hampshire coast (Lymington and Christchurch) into Sussex and Kent were the foundation for the later distinctive south coast urbanization which owed much to its proximity to the capital.

Elsewhere, Scarborough, the 'Queen of Watering Places' was nurtured and later sustained by the railway to become one of the centres for that peculiarly British institution, the seaside holiday.

Although the most spectacular advances in urban development were in the large commercial and industrial cities, smaller English cities, including many country towns, also grew steadily. In Scotland though large centres of commerce and industry dominated, there was steady increase in a wide scatter of small burghs, occasionally densely clustered as along the Fifeshire coast. Some of these were old-established Royal Burghs; others, particularly from the seventeenth and eighteenth centuries, were 'trading' burghs; often set up by local landowners or promoted as industrial villages.

The population of smaller towns and of some older regional and sub-regional capitals such as Norwich, Exeter, Oxford, Chester or York grew at rates similar to those of their rural hinterlands, reflecting the prosperity of an expanding economy with increasing numbers of markets and a growing exchange of commodities. The processing of local products in mills, tanneries, sawpits, malthouses and breweries, and the provision of local services added to their importance. Many were also the focus for workshops and entrepreneurial organization of handicraft manufacturers. In a phase of high proto-industrial activity there were losers: for example the towns of the woollen districts of the West Country and the worsted and fancy-cloth areas of East Anglia. But there were also gainers: emerging centres in West Yorkshire, later to become factory towns such as Halifax, Huddersfield and Keighley; knitting and handloom weaving centres in the East Midlands, for example Hinckley, Shepshed or Ashby de la Zouche, and the Pennine and Lakeland Dales (Settle, Bentham or Kendal). The success of such commercial activity is clear from the way in which money raised by local bankers from landed and industrial investors was invested in regional and national enterprises and helped promote canals and turnpikes linking the countryside to regional centres, to their mutual benefit. They also created new and more widespread markets for their own specialities as well as access to a growing range of goods from British manufacturers and for such overseas goods as tea and sugar, and – for the gentry – wines and silks.

While the bigger county towns often lost ground at this time, many were laying foundations for industrial and commercial expansion or recovery later in the nineteenth century: Derby had its mixture of county functions and textiles; Norwich lost its textile industry but gained a growing range of agricultural processing and later engineering manufactures; while Lincoln and Shrewsbury had similar activities. For all such towns communications were vital. Not only did canals enhance the growth of existing towns but generated new ones. The canal network linked major regional centres and the chief manufacturing towns with their sources of raw materials and with internal and external markets that were crucial in promoting cheaper and better transport from the 1770s. The main foci for canals were the West Midlands and London, while major regional cities in both England and Scotland were linked to their own hinterland and, in combination with coastal trade, to each other (Figure 6.3). Transport was a key element in a new mobility of goods and people that permitted the specialization of production characteristic of the Industrial Revolution.

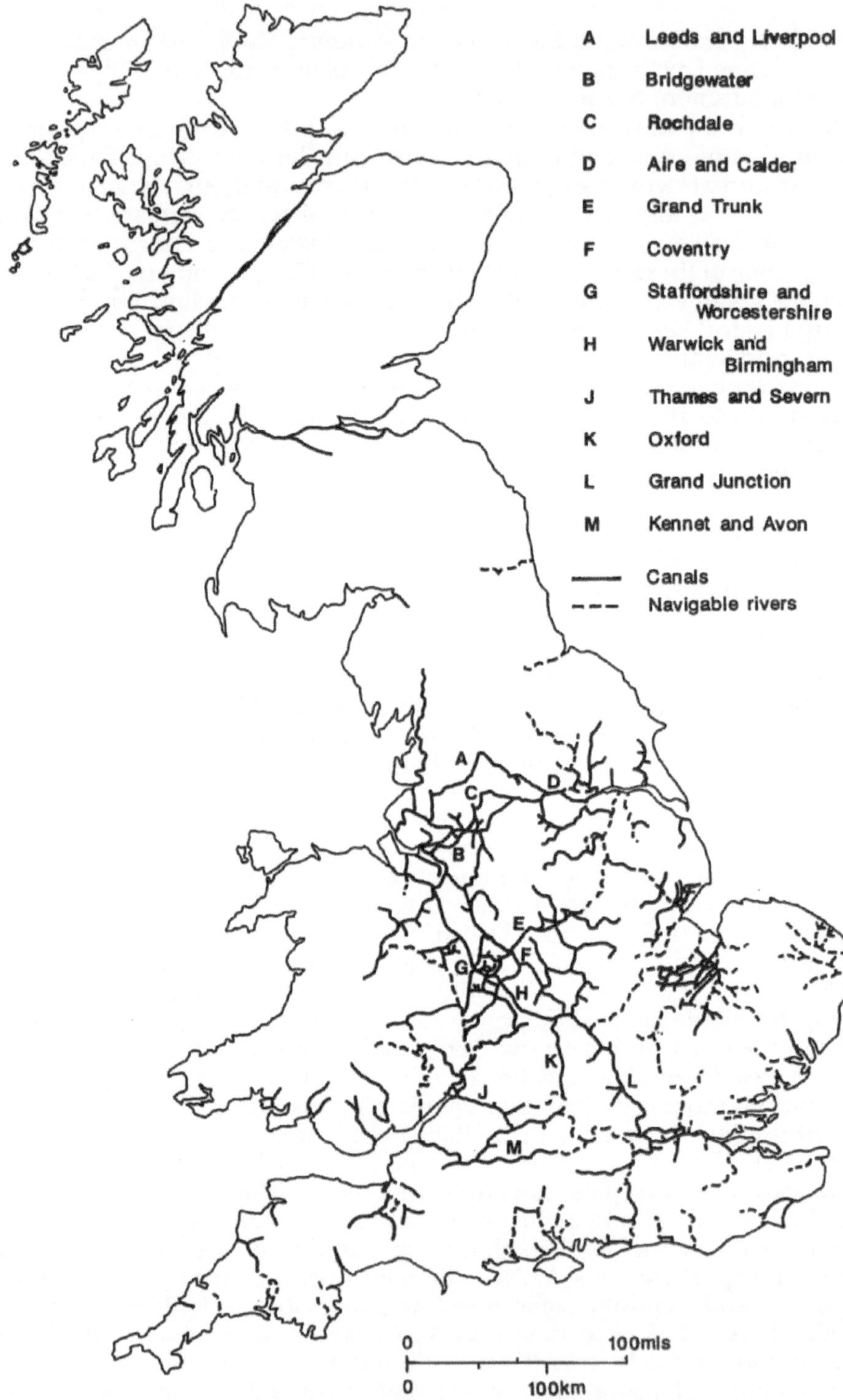

Figure 6.3 Inland waterways of Great Britain *c.* 1830. Based on Langton and Morris (1986) p. 85.

Population and housing

Migration played a key role in the growth of towns in the eighteenth century (Chapter 3) and influenced the nature and composition of urban populations. The majority of migrants travelled only short distances, especially the poor and young single migrants seeking work in the town. In the late eighteenth century 75 per cent of apprentices recruited by a Sheffield cutlers' company came from within Sheffield or from settlements within five miles of the city, and a survey of adult beggars in London in 1790 showed that only 17 per cent had travelled from English parishes more than ten miles from London, although 34 per cent had originated in Ireland and 3 per cent in Scotland. Long-distance movement from Ireland and Scotland was increasing in the eighteenth century: most Scottish migrants were skilled and professional men (often moving with their families); Irish migration was disproportionately of poor Catholics. Although some were seasonal migrants, many settled in towns like London, Glasgow, Liverpool and Bristol, a foretaste of the vast mid nineteenth-century population inflow from Ireland.

With such exceptions, most rural to urban migrants were moving within a familiar regional context. In theory the Laws of Settlement, could return anyone likely to become a burden on the poor rates to their parish of original settlement and could have acted as a brake on migration: but the volume of movement was so great, and identification of parishes of origin so difficult, that they were operated only sporadically. The weakest members of the community were most frequently subjected to the laws: single, especially pregnant, women and children were most likely to be harshly treated by the Overseers of the Poor, though this did not prevent them migrating to towns: indeed most eighteenth-century towns had an excess of females.

The inflow of population to eighteenth-century towns was accommodated both by massive new building and through subdivision and change of use in existing buildings. New building was provided in a variety of different ways by many separate builders and developers. There was usually no shortage of building land in and around towns, although certain locations may have been restricted, as in the case of Nottingham where late enclosure of the common fields prevented expansion to the north-west and south until after 1830 and greatly raised land prices.

The landowner was not generally the builder, although small infill developments in gardens and yards could be carried out by the original owner. Usually land was sold to a middleman or developer who would finance and organize the building process. Landowners included municipal corporations (for example in Bath, Leicester, Liverpool and Newcastle); charities, schools and churches; large private landowners; and professional and businessmen with small parcels of land. Developers could include local merchants, tradesmen, professional men (especially lawyers) and builders who would raise capital locally to finance house construction. Land was conveyed freehold in perhaps 50 per cent of transactions, or through a building lease. In either case the landowner might try to influence the nature of development through specific clauses or covenants. Thus Liverpool Corporation took a close interest in the development of land leased for building from the 1730s: by the 1790s it closely controlled the layout and amenities of its leasehold estate through restrictive convenants. Developers, who raised capital usually from local tradesmen or landowners, expected a return on investment of 5 to 10 per cent. From the late eighteenth century small groups of builder-owners were beginning to form joint terminating building societies for the erection of dwellings though most workers rented their housing in the eighteenth century (Chapter 11).

Urban housebuilding in the eighteenth century took many forms. Most cities had some grand housing for the rich and leisured classes, perhaps best typified by the sweeping terraces built by John Wood in Bath and in the squares, terraces and crescents of London's West End. In Bath such houses typically had a street frontage of 20 to 25 feet with two substantial rooms on each of 4 floors, together with a basement kitchen: even in the 1770s they sold for £1100 (Figure 6.4). Similarly the grand housing constructed in Edinburgh's New Town from the 1750s was part of a planned redevelopment of the city by James Craig, whose plans resulted from the vision and energy of Provost George Drummond who aimed to create a city which could compete with London. With the council keeping tight control over development of the New Town through building regulations, and the appointment of Robert Adam to complete Craig's work, by the end of the eighteenth century Edinburgh had become a model of Georgian town planning. Glasgow too added the substantial Georgian town houses around George Square for its merchant élite between 1750 and 1775, and Aberdeen developed plans for a New Town from the 1790s.

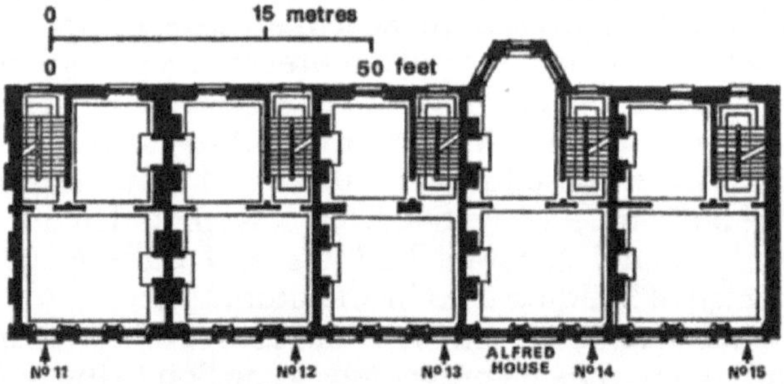

Figure 6.4 Plans of Georgian town houses in Bath *c.* 1772.
Source: Chalklin (1974) p. 220.

Artisans' housing ranged from substantial terraced houses to small courts. In Birmingham a relatively affluent skilled craftsman might live in a two-or three-storey house with two rooms on each floor and an associated yard and workshop. Such a property could cost up to £200 to buy in the mid-eighteenth century and would be rented for at least £8 a year. Since only a minority of workers could afford such rent many houses were multi-occupied by the 1780s. More commonly, workers' housing in Birmingham consisted of two or three rooms with associated workshops, often built in the yards of bigger houses, while from the 1770s rows of back-to-back houses were being constructed. Typically these had one room on each floor, cost less than £60 and were rented for less than £5 per year. However, as

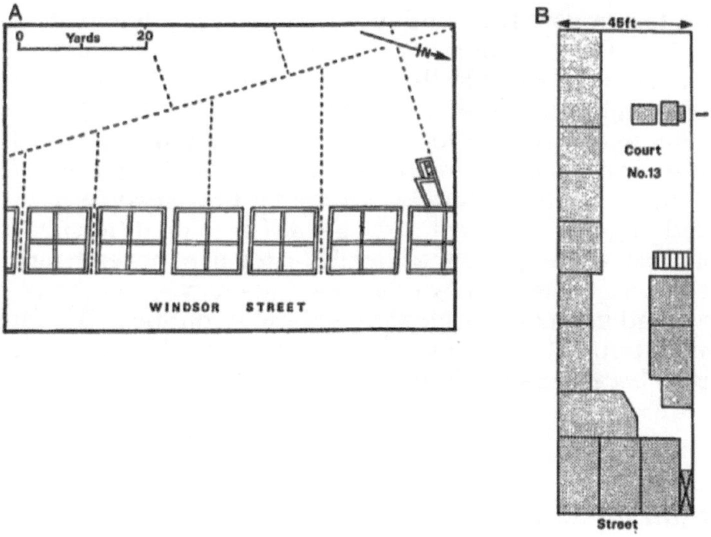

Figure 6.5 Plans of artisans' housing in Birmingham (A) *c.* 1790; (B) *c.* 1820.
Source: Chalklin (1974) pp. 200–1.

pressure on space increased many of these were also multi-occupied (Figure 6.5).

In Liverpool too, workers increasingly lived in back-to-back houses crammed into courts built from the 1780s. Simmons's enumeration of 1790 lists 1608 'back houses' in the city containing some 7,955 people, 14.8 per cent of the city's population. Approached through a narrow tunnel entrance, such houses typically had three storeys with one room measuring perhaps 12 feet square on each floor. They rapidly became the slums of the Victorian city (Figure 11.4). A large proportion of Liverpool's houses had cellars and, as pressure of population on living space grew in the 1780s, these were separately inhabited. While similar conditions were also found in Manchester and London, Liverpool's cellar dwellings were notorious. By 1790, 1,728 inhabited cellars enumerated in the city provided accommodation for 6,780 people, 12.6 per cent of the city's population, at an average density of 3.9 persons per cellar. Although many cellars, often housing single women, were no more overcrowded than other dwellings, in Liverpool their special sanitary and health problems singled them out for attention from late eighteenth- and nineteenth-century reformers.

However, some of the worst housing conditions in eighteenth- and early nineteenth-century Britain were found in London where population pressures and constraints on space were far more acute than in provincial towns. The old courts and tenements which had grown up around the City of London in the seventeenth century and earlier were by the eighteenth century the poorest districts. Typically, working-class families lived in a single room or cellar, without any proper sanitation or water supply, and paying 2–3 shillings per week rent. Lodging houses were also common in London, and in the poorest districts as many as 15 people would sleep in one room, each paying 1 or 2 pence for a night's shelter.

It is hard to compare rural and urban housing conditions in eighteenth-century Britain. Contemporary descriptions tend to accentuate the horrors of urban living experienced by the very poor who also lived in very cramped and uncomfortable conditions in rural areas (Chapter 4). Many of the migrants renting a couple of rooms in a newly-built house probably had as much living space as they would have enjoyed in the countryside. There were certainly more opportunities for young townspeople to leave home and find lodgings than in most rural areas. The main difference was in the density of urban housing. Living literally on top of or beneath neighbours in a multi-occupied tenement was a new experience for many requiring considerable adjustments in lifestyle and daily routines. The intense concentration of population spread disease more quickly, so sanitation and water supply became increasingly important issues from the late eighteenth century.

The spatial structure of towns

In comparison with Victorian cities we know relatively little about the internal structure of Georgian towns. This is particularly unfortunate since, in the eighteenth and early nineteenth centuries, cities were experiencing rapid growth, around an industrial economy, whilst being transformed from their 'pre industrial' state to a 'modern' urban form.

The extent to which Georgian towns developed a clearly defined internal structure is partly related to size. Eighteenth-century London had a well-defined internal geography: a series of self-contained communities separated by class and occupational differences, by the difficulties of moving over large distances within the Metropolis, and by the perception of some districts as being dangerous. By 1798 the middle class was concentrated especially in central, west and north-west London, with a considerable degree of social segregation. However, segregation was not total. The wide spread of manufacturing meant that some wealthy households were found in most districts of London: but they did not form more than half of the resident population in any one district. As in all pre-industrial towns there was still a mix of poverty and wealth in most areas: the rich generated their own hinterlands for service and retail workers many of whom lived in slums adjacent to the rich families they served, thus restricting the extent to which large-scale class segregation could occur.

Although much smaller than London, Liverpool was one of the largest provincial towns with a population of 83,000 by 1801 and a well-developed commercial and industrial structure. These factors were reflected in its social geography by strongly differentiated housing, occupations and social class. As in London there were no large homogenous districts, but considerable social segregation between adjacent streets and between front houses and courts. Similarly in late eighteenth-century Bristol both rich and poor were found in all parishes, but the wealthy were concentrated in six central parishes and, increasingly, in high-status suburban areas (Figure 6.6). At the other end of the urban hierarchy, Chorley (Lancashire) a small industrial

A

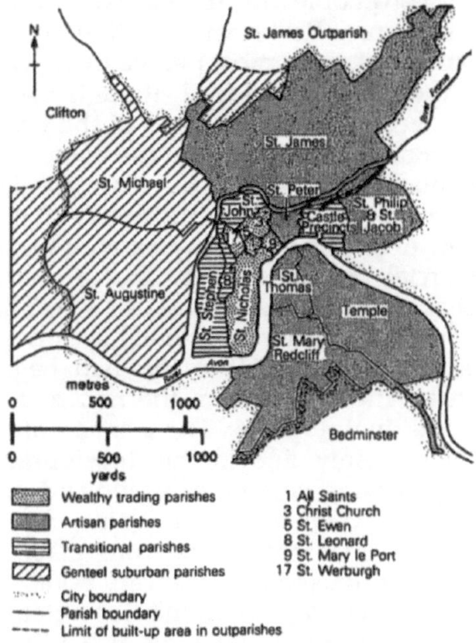

B

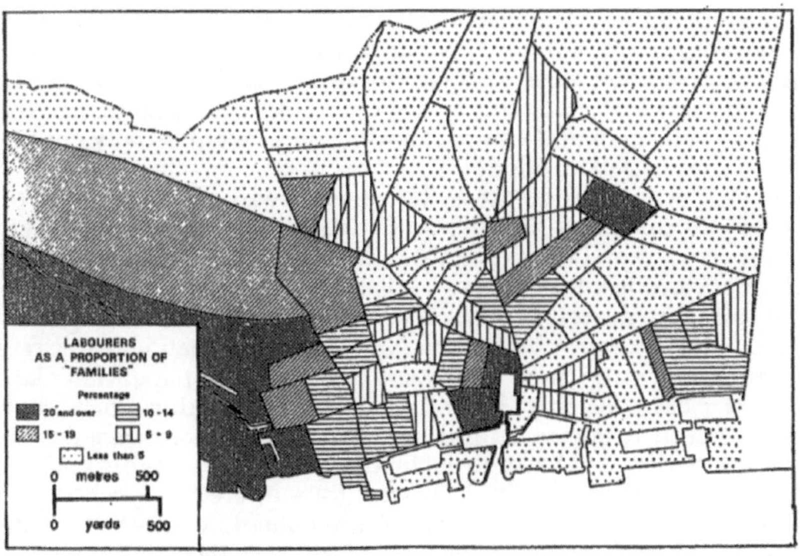

Figure 6.6 Tne spatial structure of late eighteenth-century Bristol (A) and Liverpool (B).
Sources: (A) Baigent (1988) p. 119 and (B) Laxton (1981) p. 91.

town of no more than 2,400 people in 1816, had a discontinuous and largely linear morphological structure. Residential location was largely controlled by the nature of employment, with most people living close to the mills or workshops in which they were employed, though there were also clear differences in the prosperity of different parts of the town as rich families were beginning to separate themselves from the poor.

Most social geographies have been based on a single source of information, and thus fail to capture the dynamic nature of eighteenth-century urban life. Rapid inflow of population, coupled with massive expansion of industrial and commercial premises meant that a district could quickly acquire a new social character. The Avon Street district of Bath, developed in the early 1730s as a medium-quality residential area designed to provide fashionable lodgings for visitors to the city, had begun to acquire an unsavoury reputation by the 1760s when the rich had mostly moved to newer houses elsewhere in the town. The demise of Avon Street as a fashionable district was mainly due to the development of small-scale industry nearby, the increasing volume of traffic and people that flowed through the area, and the frequency of flooding. By the 1820s it was considered the worst slum in Bath: a dangerous place which harboured villains, spread disease and housed only the very poor.

The dynamic nature of everyday urban life, the extent of daily mobility and the role of neighbourhood communities are also difficult to capture. The daily 'action space' of most people seems to have been relatively circumscribed. The rich had spare time and money, but their perception of undesirable areas determined their routes through the city and the movement of women, in particular, was constrained by social norms. Most working-class activities were constrained by long working hours and low incomes: people lived close to their workplace, and food and other items were bought in nearby markets and shops. Thus most eighteenth-century towns were still small and compact and contacts were usually confined to the immediate neighbourhood.

In comparison with rural areas, however, towns offered greater opportunities to join clubs and societies and engage in organized and often self-improving recreation. Better urban standards of living facilitated the uptake of such opportunities. For instance, aristocratic clubs were available for the relatively affluent, as well as political societies and organized religion which could tap a much larger population. News and information also circulated more quickly in towns (there were some 130 local newspapers circulating in 55 different urban centres by 1760). Opportunities for education were better in urban than in rural areas, and largely-urban friendly societies, which provided mutuality and basic social insurance for the working classes, had over 600,000 members by 1800. Towns thus offered a greater range of possibilities of material and moral self-improvement, for interaction, discussion and broadening of horizons. The extent to which ordinary working people actually used these facilities regularly is poorly understood. Many would have insufficient time or money to do more than buy an occasional newspaper or attend a political gathering. Moreover, participation in such activities, and the associated mobility around town, was generally a male activity. Many women, restricted by the double responsibility of work and child care, found most of their everyday contacts and relaxation in and around the workplace or home.

Public health and municipal reform

The unhealthy nature of overcrowded eighteenth-century towns attracted increasing contemporary attention. Such diseases as smallpox, typhus and influenza were both endemic and epidemic: they killed large numbers of both rural- and urban-dwellers, but disproportionately affected the young and malnourished of the urban slums (Chapter 3). Largely urban outbreaks of smallpox occurred in 1740–42, 1747, 1781 and 1796: in Edinburgh (total population 40,000) there were 2,700 smallpox deaths in 1740–42 alone, and in Glasgow 19 per cent of all deaths in the last quarter of the eighteenth century were due to smallpox. Smallpox, a highly infectious disease spread by direct contact with a carrier, quickly took hold in overcrowded urban homes and workplaces.

Typhus fever was endemic in London, and epidemics occurred in other towns in 1741–2, 1772–3, 1799–1800, 1817–19, 1826–27 and 1831–2. In the 1741 epidemic 25 per cent of all deaths in London were attributed to typhus and 20 per cent in Edinburgh. Typhus is transmitted from person to person by the body louse and thrives in overcrowded urban conditions. In 1773 the death rate from typhus in Manchester was estimated as twice that in the surrounding countryside. Influenza epidemics occurred particularly in 1743, 1755, 1781–82, 1803 and 1831. The 1781 epidemic spread from Newcastle upon Tyne to the rest of Britain during the following year eventually affecting up to 80 per cent of the adult population. Although most people recovered, it caused great distress and considerable loss of income.

Children were particularly vulnerable to most infectious diseases, but infants suffered particularly from the effects of diarrhoea, scarlet fever and measles. In 1773 50 per cent of Manchester's children died before the age of five and it has been suggested that between 1730 and 1749 75 per cent of London children died before their fifth birthday. The main nineteenth-century killer of adults, tuberculosis, was also on the increase in the second half of the eighteenth century. For both adults and children the combination of poverty, malnutrition and overcrowded living and working conditions were the main factors encouraging the spread of disease. As towns grew, polluted water became an increasingly pressing problem and was the cause of many diseases such as infantile diarrhoea and typhoid fever, which increased throughout the eighteenth century, and finally cholera which was introduced into Britain in 1831 from its Asian breeding grounds. Reaching Sunderland through carriers coming into the port, cholera quickly spread to both towns and villages, particularly to areas where water was heavily contaminated by sewage, and the disease was a significant spur to public health legislation after 1830 (Chapter 11).

Awareness of the spread of epidemic disease in eighteenth-century towns created pressures for improvement of the environment. Local inhabitants, especially doctors, petitioned the local corporation to provide such facilities as fresh water, street cleansing and lighting, refuse and sewage disposal, and to establish a watch or police force. The usual response of English town councils was the establishment of Improvement Commissions under a local Act of Parliament, with powers to levy local rates to carry out improvements. The earliest Commissions were mostly concerned with street cleansing and lighting, but by 1800 over 100 different Commissions had been established: by 1830 only four municipal boroughs with over 11,000 people did not have an Improvement Commission.

The effectiveness of most Improvement Commissions (and in Scotland of the Dean of Guild Courts and intervention by individual borough councils) was limited, and calls for more direct intervention in the urban environment often met with considerable opposition. For instance in late eighteenth-century Liverpool a number of medical men, especially Dr Currie of the Liverpool Dispensary, argued the case for sanitary reform and in 1801, following an epidemic, the Council was persuaded to consult medical opinion about the scope of a proposed Improvement Act. Currie's report argued for building controls, proper sanitation and water supply to all courts, and the preparation of a town plan to regulate building. None of these suggestions was included in the 1802 bill which was in any case withdrawn due to local opposition. In the following year an Improvement Act was again proposed, encompassing some of Currie's proposals, but in the face of even stiffer opposition it again failed. Such was the experience of most towns. Even partially effective intervention in the urban environment was delayed until the mid-nineteenth century, by which time the problems had multiplied considerably (Chapter 11).

Preventative measures were limited in the eighteenth century but, as medical practice became better organized and more professional, many towns established hospitals and dispensaries, usually as charitable institutions. By 1800 London had at least 17 medical dispensaries. There were some 31 institutions in the provinces: the Liverpool Dispensary was founded in 1778; the Manchester Dispensary was built onto the existing infirmary service (founded in 1752) in 1781 and Manchester was the first town to establish a Local Board of Health, founded by a group of volunteers in 1794–6. However, the dispensaries failed to reduce mortality because of both limitations in medical knowledge and growing environmental hazards. The only really significant medical breakthrough was through late eighteenth-century developments in vaccination against smallpox.

Apart from limited knowledge about the causes of disease and the reluctance of property owners to spend money on improving housing for the poor, many attempts at environmental betterment were ineffective because of poor local government organization. Most local councils were fiercely independent, and within Britain a variety of different local government structures made coordination between national, county and local administrative levels difficult. Many substantial industrial towns (for example Manchester and Birmingham) were unincorporated in the eighteenth century and, in theory, controlled by the county. Although they gained control over their own affairs through local Improvement Acts the system did not lend itself to effective local government. Unincorporated industrial towns had no direct representation in Parliament and found it difficult to petition for change. Incorporated towns varied greatly in the way in which local government was organized. 'Closed' corporations, such as Bath, Leeds, Liverpool, Coventry and Leicester, were often run by a small oligarchy appointed for life. Under these circumstances it is not surprising that local government was slow to respond to the increasingly serious problems of urban life until after the Reform Act of 1835 (Chapter 11).

Bibliography and further reading for Chapter 6

For an overview of urbanization and urban life:

ABRAMS, P. and WRIGLEY, E. A. (eds) *Towns in societies* (Cambridge: Cambridge University Press, 1978).

ADAMS, I. *The making of urban Scotland* (London: Croom Helm, 1978).

CHALKLIN, C. *The provincial towns of Georgian England: a study of the building process 1740–1820* (London: Arnold, 1974).

CLARK, P. (ed.) *The transformation of English provincial towns 1600–1800* (London: Hutchinson, 1984).

CORFIELD, P. *The impact of English towns 1700–1800* (Oxford: Oxford University Press, 1982).

HAMMOND, J. L. and B. *The town labourer, 1760–1832* (London: Longman, 1917).

WILLIAMSON, J. G. *Coping with city growth during the British industrial revolution* (Cambridge: Cambridge University Press, 1989).

General studies of specific towns include:

ARMSTRONG, A. *Stability and change in an English county town: York 1800–1850* (Cambridge: Cambridge University Press, 1974).

BARKER, T. C. and HARRIS, J. R. *A Merseyside town in the industrial revolution: St Helens 1750–1900* (Liverpool: Liverpool University Press, 1954).

GEORGE, M. D. *London life in the eighteenth century* (London: Kegan Paul, 2nd edition, 1930).

MARSHALL, J. D. 'The rise and transformation of the Cumbrian market towns 1660–1900' *Northern History*. **19** (1983), pp. 128–209.

NEALE, R. S. *Bath: a social history 1680–1850* (London: Routledge, 1981).

YOUNGSON, A. J. *The making of classical Edinburgh 1750–1840* (Edinburgh: Edinburgh University Press, 1966).

On health, housing and the internal structure of towns see:

BAIGENT, E. 'Economy and society in eighteenth-century English towns: Bristol in the 1770s' in Denecke, D. and Shaw, G, (eds) *Urban Historical Geography. Recent progress in Britain and Germany* (Cambridge: Cambridge University Press, 1988) pp. 109–124.

DAVIES, G. 'Beyond the Georgian facade: the Avon street district of Bath' in Gaskell, S. M. (ed.) *Slums* (Leicester: Leicester University Press, 1990).

HOWE, G. M. *Man, environment and disease in Britain* (Harmondsworth: Penguin, 1976).

LANGTON, J. and LAXTON, P. 'Parish registers and urban structure: the example of late eighteenth-century Liverpool' *Urban History Yearbook* (Leicester: Leicester University Press, 1978), pp. 74–84.

LAXTON, P. 'Liverpool in 1801' *Transactions of the Historic Society of Lancashire and Cheshire* **130** (1981), pp. 73–113.

MORRIS, R. J. *Cholera, 1832* (London: Croom Helm, 1976).

SCHWARTZ, L. D. 'Social class and social geography: the middle classes in London at the end of the eighteenth century' *Social History* **7** (1982), pp. 167–85.

TAYLOR, I. C. 'The court and cellar dwellings: The eighteenth-century origins of the Liverpool slums' *Transactions of the Historic Society of Lancashire and Cheshire*. **122** (1970), pp. 67–90.

WARNES, A. 'Residential patterns in an emerging industrial town' in Clarke, B. D. and Gleave, M. B. (eds) *Social patterns in cities* (London: Institute of British Geographers, 1972), pp. 169–89.

SECTION II:

Britain from the 1830s to the 1890s

7

The political, economic and social context, 1830–1890

The concept of Victorian Britain conjures up a whole series of popular images. Margaret Thatcher's government of the 1980s spoke repeatedly of a return to 'Victorian Values': self help; a *laissez-faire* economy; the role of the nuclear family in social control; and national self-confidence and pride bred by imperialism. However, Victorian Britain was more complex than this. Such values were not necessarily dominant, and certainly did not equally affect all regions and individuals. This chapter outlines some of the main political, technological, economic and cultural changes affecting Britain from the 1830s to the 1890s and how these may have affected particular regions and social groups. Succeeding chapters examine specific themes nationally and regionally, showing how they were affected by such changes.

The period 1830 to 1890 was one of both great national stability and change. Queen Victoria's long reign (1837–1901) provided underlying continuity to profound economic and social change. The country had largely recovered from the effects of the Napoleonic Wars (1793–1815) and, with the exception of the Crimean War (1854–56) which claimed 25,000 British lives and cost the British Government approximately £70,000,000, Britain remained free from conflicts in Europe until the First World War (1914–18). Elsewhere, with the exception of some 'colonial' wars and the Boer War (1899–1902) Britain remained at peace. This almost unparalleled period of peace and Britain's dominance as a world power created stability which assisted economic growth and social change.

This sixty-year period was one of almost continuous national economic growth. The 'main motif' of the years 1815 to 1914 was, according to Thomson (1950), the 'remarkable accumulation of material wealth and power which the English people achieved during the century'. The wealth however was unevenly distributed between regions and social groups, and

DOI: 10.4324/9781003174783-7

by the last quarter of the nineteenth century Britain's economic and imperial power began to be called into question, though the widely held belief in economic growth and international power persuaded many in the better-off section of society that it could cope with the problems of urbanization, industrialization and economic growth. Carefully prepared political propaganda and improving standards of living for most people persuaded most social classes in Britain to share these beliefs and values. Belief in material progress and what it could achieve was epitomized by the Great Exhibition at Crystal Palace in 1851: '. . . a sight the like of which has never happened before . . .' as *The Times* wrote. Great self-belief created the substantial internal stability of Victorian times.

The materialism which underpinned mid-Victorian prosperity was tempered by a set of religious and moral values which both legitimated the accumulation of wealth and, in some at least, generated a moral consciousness which contributed towards nineteenth-century social reform. The extent to which religion was important for all social classes, and the degree to which moral consciousness rather than self-interest was a spur to social reform are open to debate. But shared values between the church, the state and the business world must have increased the self-confidence and stability of the leaders of society.

Yet this deceptively reassuring framework was shot through with changes, challenges and new ideas which substantially altered the social and economic geography of Britain. Politically there were twenty changes of Prime Minister, involving ten individuals between 1830 and 1890; Liberal-Whig administrations held power for almost twice as long as Tory-Conservative governments. Potentially the most important political change, which eventually had repercussions everywhere, was the initiation of Parliamentary Reform and the progressive extension of the franchise. The 1832 Reform Act, a small step along the road to universal suffrage which retained the principle that property was the main qualification for the vote, increased the total electorate by only some 217,000 males (less than a 50 per cent increase). There was, however, a significant redistribution of Parliamentary seats. Some 86 boroughs either lost or had their representation reduced, whilst 43 new boroughs – mostly northern industrial and commercial towns – gained representation in Parliament. The dominance of the rural and landed interest was weakened, and nationally England experienced a slight reduction in seats as Wales and, especially, Scotland gained.

The Second Reform Acts further extended male franchise in England in 1867 and 1868 in Scotland. In the counties, occupiers of houses rated at £12 or more and leaseholders with property of at least £4 in annual value were given the vote; in boroughs householders who had been in residence for at least one year and lodgers paying £10 or more per annum were enfranchised. Although the total electorate – still entirely male – was under 10 per cent of the population, these changes, together with a further redistribution of seats away from small towns and towards the counties and large urban centres, strengthened the political voice of the middle class and some of the skilled working class. The secret ballot introduced for Parliamentary elections in 1872 reduced opportunities for intimidation, and further Reform

Bills of 1884 and 1885 extended the vote and redistributed more seats to industrial towns.

Thus by the 1890s a significant proportion of ordinary men had the opportunity to express their views through the ballot box; the changing geography of Britain had been reflected in a redistribution of Parliamentary seats; and national politicians were beginning to realize that the views of an increasingly working-class electorate should be taken seriously. Such changes as the re-organization of the Poor Law (1834), the Municipal Reform Act (1835) and the 1867 Reform Bill, which also applied to municipal elections, meant that similar pressures were experienced at a local level. Throughout, local government was increasingly subject to national trends and legislation, and local politicians were having to respond to both local and national opinion to protect their political interests. One effect was that the political and legislative responses to change increasingly became similar from one part of the country to another.

A second set of changes and challenges was stimulated by urbanization and economic expansion. Urban growth and industrialization had a major impact on all regions and all strata of society. Whilst all women and many ordinary working men were excluded from the effects of Parliamentary Reform, all were to some extent touched by the massive economic and social changes. These may be viewed on two levels: on the one hand economic growth offered new opportunities and opened new horizons; on the other hand it provided new constraints and condemned many to penury and hardship in rapidly growing industrial towns. The extent to which urban growth and industrial change was beneficial or detrimental varied from region to region and between social groups, and varied over time depending on national and local economic circumstances and their effect on the life-cycle of families and individuals. For example the textile industry, pre-eminent in the British economy throughout this period (Chapter 10), provided employment for over half a million workers, but was increasingly concentrated in Lancashire and West Yorkshire. For the mill owner economic expansion offered the opportunity to make a fortune and many factory workers were well placed to benefit from the industry's growth, despite hard conditions of work and periodic depression. Those in less-skilled or technologically dying trades, such as hand-spinning or weaving, experienced poorer working conditions and low and irregular wages, especially for women and children. For such people the benefits of industrialization were distinctly limited. For all, the dangers of ill-health in poor urban environments or industrial accidents in an era of poor safety precautions could have catastrophic implications for a family. The social and geographical effects of the Victorian economic 'miracle' were complex, unstable and uneven.

Victorian economic development is epitomized by the growth of the railways (Figures 7.1 and 7.2). By mid-century a basic network had emerged; by the 1860s most significant towns were connected to the railway system; and by 1914 Britain had a network of over 20,000 miles of railway. During the same period travel time shrank rapidly: in the pre-railway era six hours travel from London would not take one beyond South-East England; by 1845 Manchester could be reached in six hours; whilst by 1910 six hours would

Figure 7.1 The railway system in Great Britain, 1852. Based on Langton and Morris (1986) p. 89 and Pope (1989) p. 106.

take the traveller from London to Scotland. The railways offered those who could afford it new and growing opportunities to move between regions, to travel long distances for business or pleasure and to commute longer distances to work. They also facilitated and cheapened the movement of news and goods over long distances, helping to reduce differences between places. However, not everyone benefited from the growth of the railways. Railway construction extracted a substantial toll of misery and death; the growth of new routes quickened rural out-migration and assisted the long-term decline of many communities; in towns it led to extensive demolition of

Figure 7.2 The railway system in Great Britain, 1900. Based on Langton and Morris (1986) p. 89 and Pope (1989) p. 107.

houses, increased overcrowding and contributed to an increasingly noisy and polluted environment. Many people were simply too poor to enjoy the benefits which the railways brought. Their impact depended very much on who you were, where and when you lived, and how much your earned.

A third set of changes, in attitudes and values, created greater national uniformity in some respects while perpetuating or creating regional or local differences. Minority groups and minority beliefs were especially important in the changing social geography of Victorian Britain. Although attitudes towards Protestant dissenters and Roman Catholics had gradually become more tolerant during the late eighteenth century, they remained barred from public office and Roman Catholics could still experience persecution for practising their religion. Anglicanism, the official state religion, dominated England and Wales despite substantial Nonconformist and Catholic minorities. In 1828 the Test and Corporation Acts which discriminated against Protestant dissenters were repealed and in 1829 the Roman Catholic Relief Act achieved Catholic emancipation. In theory Catholics, dissenters and Anglicans should have had equal opportunities from the 1830s onwards.

Yet the effects of these acts varied considerably since the balance of religious adherence varied from region to region. The 1851 religious census shows that Wesleyan Methodists were numerous in much of industrial northern England, the South West and parts of the South East; Calvinistic Methodism was concentrated in north and west Wales; while the Free Church and United Presbyterians dominated the Scottish Highlands. Roman Catholics were particularly important in Lancashire, a feature accentuated by Irish migration especially after the Irish famine. Whilst Protestant dissenters were readily accepted into society, Roman Catholics were often discriminated against, especially where the Catholic threat was perceived to be significant, or where Catholicism was equated with other negative attributes, such as among southern Irish communities. Thus, sectarian discrimination and violence was common in Glasgow, Liverpool and other industrial towns in the nineteenth century and played an important part in shaping their social geography.

Attitudes to other minority groups were also taking shape during the Victorian period. British imperialism, empire migration and British involvement in the slave trade (abolished in the British Empire in 1807) and in slavery itself (abolished in theory in the colonies in 1833, but lingering until 1840) promoted stereotypes of immigrants. The small population of Blacks in Victorian Britain (concentrated in London and port cities like Bristol and Liverpool) were discriminated against far more severely than even the Irish. Jews, migrating in substantial numbers from Eastern Europe from the 1880s, encountered similar segregation and racism that continued to the present day. Victorian Britain was also a male-dominated society and sexism was commonplace in all aspects of economy and society. Not only were women excluded from most positions of political and economic power, but attitudes severely restricted their mobility and day to day activities.

In many other areas of national life there appears, on the one hand, to have been a convergence of values but, on the other hand, there were considerable local and regional variations in the ways in which those values were expressed. Although a Smilesian vision of self-help is said to characterize Victorian attitudes to work, thrift and community life, a vast range of popular institutions such as co-operatives, working-men's clubs and friendly societies implemented these values in different ways in different localities. Despite a gradual movement towards a national education system, especially following the Education Act of 1870, there were great variations in quality of schooling and in the level of school attendance in different localities. Working-class areas were destined generally to have the poorest schools, the greatest levels of truancy and the least success in breaking through the bonds of class and privilege and the barriers between skilled and unskilled work. At the other end of the educational spectrum scientific and technological developments in the Victorian period rested on a progressively wider scientific knowledge which was of considerable significance in many industries of the period (Chapter 10). Their particular impact, through the development of Universities and Technical Colleges, had very different effects in particular regions and localities.

While between the 1830s and the 1890s significant changes were projected onto a backcloth of apparent national stability and security, there was not one Victorian Britain but many. Trends towards greater uniformity at the national level could be cross-cut by increasing diversity in the ways in which these trends worked through in particular localities as the following sections show.

Bibliography and further reading for Chapter 7

ASHWORTH, W. *A short history of the international economy since 1850* (London: Longman, 3rd edition, 1975).

BEST, G. *Mid-Victorian Britain 1851–75* (London, Weidenfeld, 1971).

BRIGGS, A. *Victorian people* (London, Odhams Press, 1954. Penguin edition, 1965).

CHECKLAND, S. G. *The rise of industrial society in England 1815–85* (London: Longman, 1964).

CROUZET, F (Trans. A. S. Forster) *The Victorian economy* (London: Methuen, 1982).

DYOS, H. J. and WOLFF, M (eds) *The Victorian City: Images and realities* (London: Routledge, 1973).

DYOS, H. J. and ALDCROFT, D. H. *British Transport: an economic survey from the seventeenth century to the twentieth* (Harmondsworth: Penguin, 1974).

EVANS, E. J. *The forging of the modern state: early industrial Britain 1783–1870* (London: Longman, 1983).

FLOUD, R. and McCLOSKEY, D (eds) *The economic history of Britain since 1700:* vol 2, *1860 to the 1970s* (Cambridge: Cambridge University Press, 1981).

GILBERT, A. D. *Religion and society in industrial England: church, chapel and social change 1740–1914* (London: Longman, 1976).

HARRISON, J. *The early Victorians 1832–1851* (London: Weidenfeld and Nicolson, 1971).

HOBSBAWM, E. J. *The age of Capital, 1848–75* (London, Weidenfeld and Nicolson, 1975).

HOBSBAWM, E. J. *The age of Empire, 1875–1914* (London: Weidenfeld and Nicolson, 1987).

HUNT, E. H. *British labour history, 1815–1914* (London: Weidenfeld and Nicolson, 1981).

McCLOSKEY, D. *Enterprise and trade in Victorian Britain: essays in historical economics* (London: Allen and Unwin, 1981).

MARSHALL, M. *Long waves of regional development* (London: Macmillan, 1987).

MINGAY, G. E. *The transformation of Britain, 1830–1939* (London: Routledge, 1986).

PERKIN, H. *The origins of modern English society 1780–1880* (London: Routledge, 1969).

PERKIN, H. *The age of the railway* (London: Panther, 1970).

PERRY, P. J. *A geography of nineteenth-century Britain* (London: Batsford, 1975).

ROBBINS, K. *The eclipse of a great power: modern Britain 1870–1975* (London: Longman, 1983).

ROEBUCK, J. *The making of modern English society from 1850* (London: Routledge, 1973).
ROUTH, G. *Occupations of the people of Britain, 1801–1981* (London: Macmillan, 1987).
THOMSON, D. *England in the nineteenth century 1815–1914* (Harmondsworth: Penguin, 1950).
WALTON, J. *The English seaside resort: a social history 1750–1914* (Leicester: Leicester University Press, 1983).
WALTON, J. *Lancashire: a social history 1558–1939* (Manchester: Manchester University Press, 1987).
WALVIN, J. *Victorian values* (London: Deutsch, 1987).
WHITTINGTON, G. and WHITE, I. *An historical geography of Scotland* (London: Academic Press, 1983).

8

Demographic change from the 1830s to the 1890s

Introduction

Between 1831 and 1900 Britain's population growth continued steadily. Although rates of increase slackened from mid-Victorian times, especially in Scotland, annual increments rose from around 250,000 in the 1830s to nearly 400,000 in the 1890s (Figure 8.2), mostly in towns and industrial areas. From the 1830s rural population growth slackened and progressive migrational losses led to continuous decline of population in virtually all agricultural townships between mid-century and the 1900s. All regions were affected, although the greatest relative impact was on the marginal areas of upland Britain and on largely farming economies, both lowland and upland. Despite substantial overseas emigration most of the rural surplus was absorbed in urban labour markets.

The broad balance of population distribution shifted, between the mid-nineteenth and the early twentieth century, not only from rural to urban/industrial Britain but between and within major industrial regions. Many of the pioneering areas of the early industrial revolution grew more slowly from mid-Victorian times, for example the West Midlands (Table 3.3 and Figure 8.1). Yorkshire, north-western and northern England, South Wales and central Scotland attracted a continuing increase in their share of the national population. But the major regional increase was in Greater London and the South East reflecting their role in promoting the 'new' industries of the late nineteenth century and growing dominance of the tertiary sector, hinting at the core-periphery redistribution which was to become such a feature of twentieth-century Britain.

By the end of the nineteenth century the economic influences on population growth had changed. The dominant influence of food prices, previously the major element in real wages, on marriage rates and levels of fertility was no longer the key determinant of population growth. Earnings were driven mainly by the secondary and tertiary sectors which employed most people and shaped labour demand and population mobility (Figure 3.1). But the mortality decline, which had contributed much to accelerating population growth from the 1740s, was halted between the 1820s and early

 DOI: 10.4324/9781003174783-8

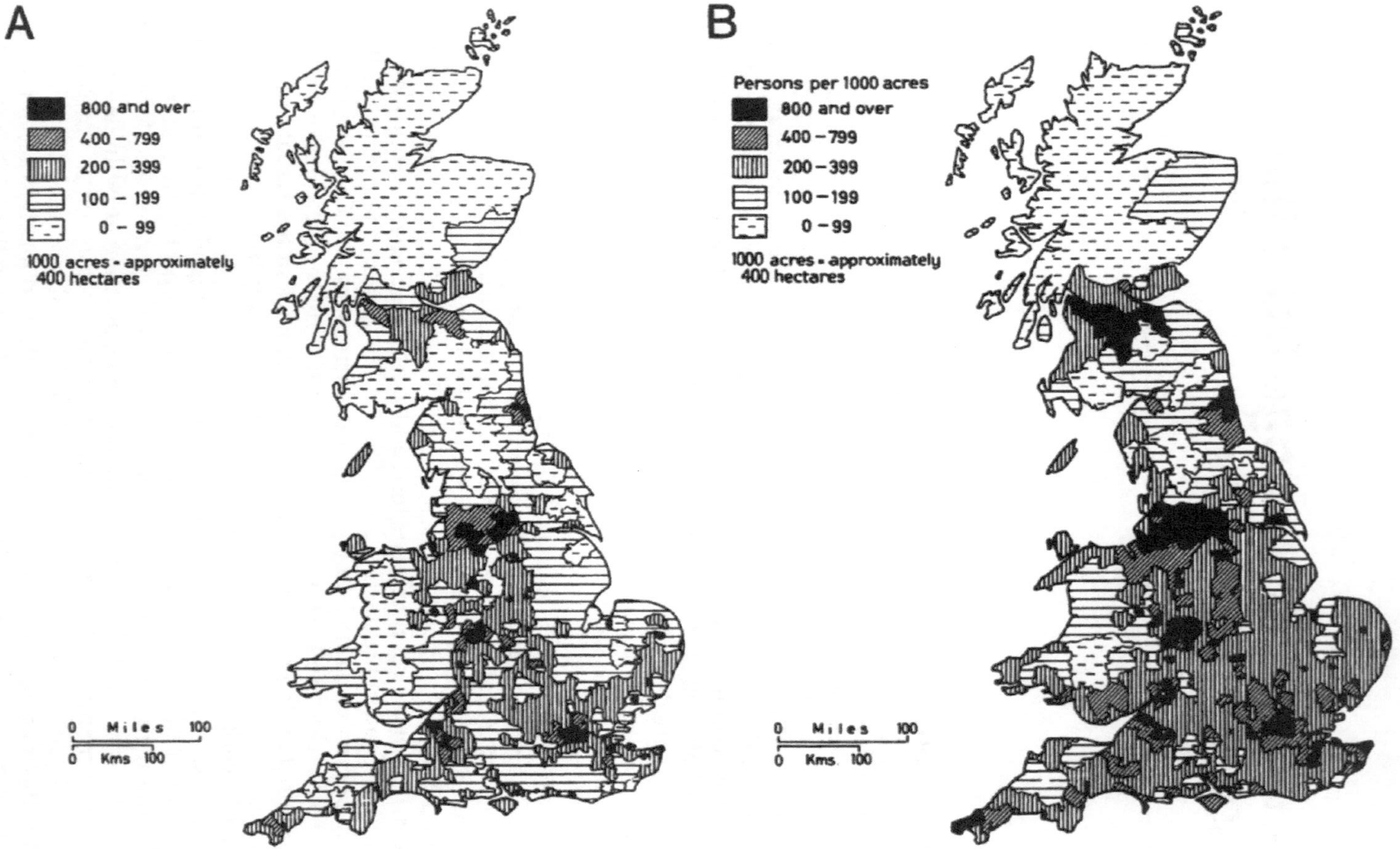

Figure 8.1 The distribution of population in Great Britain, 1801 (A) and 1851 (B). Densities are given for registration districts in England and Wales and counties in Scotland.
Source: Census of Great Britain, 1851.

1870s by the toll exacted by high urban mortality. Not until improvements in urban health, in child and then – from the end of the century – infant mortality did mortality resume its downward trend (Figure 3.2).

Overseas migration played an important part in reduced rates of population increase throughout the Victorian era (Figure 8.2). The declining share of Scotland in the national population was largely due to greater emigration and net losses of 0.3 to 0.6 per cent per annum from the 1840s to 1914. In England and Wales net losses were substantially lower, 0.04 to 0.2 per cent

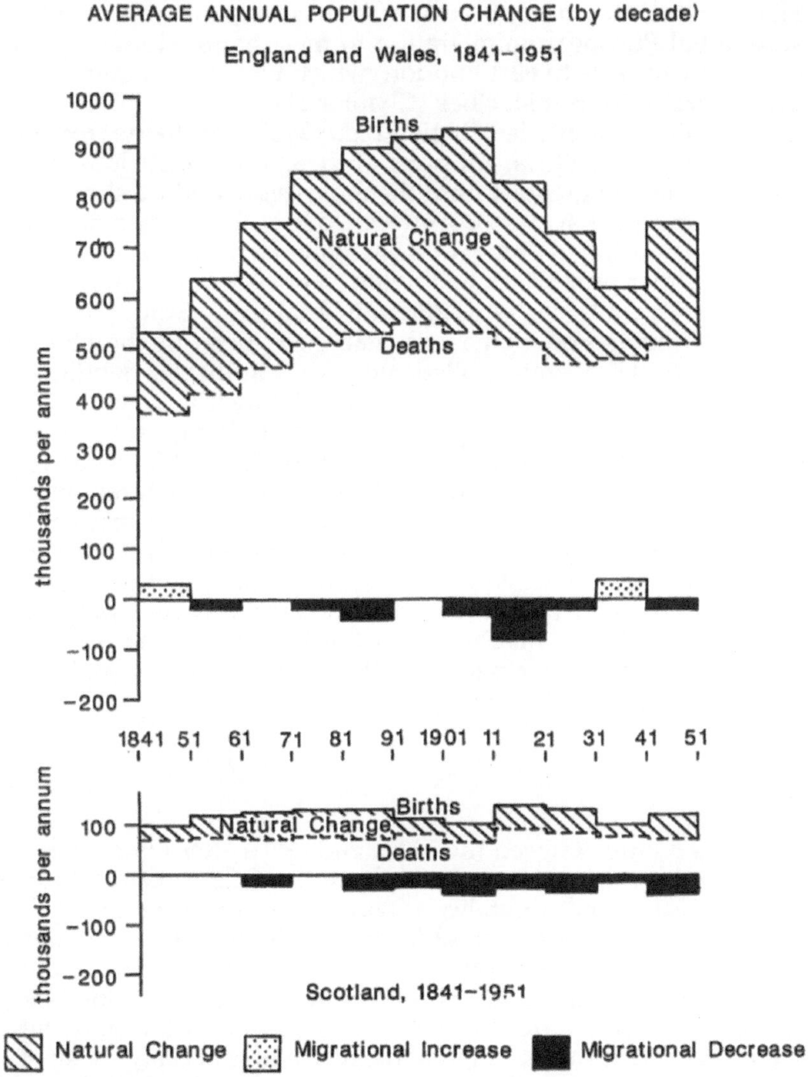

Figure 8.2 Average annual population change (by decade) in England and Wales, 1841–1951 and Scotland, 1861–1951.
Source: Censuses of Great Britain, 1851, and of England and Wales and Scotland, 1861–91; annual reports of the Registrars General.

per annum, though with gains during the Irish influx of the 1840s. In all some 10 million emigrants left Britain between 1815 and 1914 as compared with a total population increase of 29 million and an estimated natural increase of 30 million. That the balance was not more adverse is due to substantial immigration, especially from Ireland. Catastrophic emigration of some 2.4 million Irish after the famines of the late 1840s led to sharp increases in the numbers of Irish-born in Britain. By 1861 602,000 Irish-born were enumerated in England and Wales as against 291,000 in 1841; in Scotland their numbers increased from 126,000 to a peak of 219,000 in 1881. In the late nineteenth century the tide of Irish immigrants subsided but there was substantial European immigration to most major cities, especially of eastern European Jews to east London, which had a substantial impact on the social character of British cities (Chapter 11).

Over 1.8 million people left England and Wales in the depressed 1880s and 1.9 million in the 1900s. All parts of the country contributed, though losses were relatively greatest from the most depressed rural counties and declining mining areas such as Cornwall. Up to mid-Victorian times half the emigrants were unskilled (many of them agricultural labourers): by the end of the century that proportion had fallen to one-third and four out of five emigrants were from large towns and industrial areas. Studies by Charlotte Erickson have suggested that more than half those in the ship lists of emigrants to the USA were skilled men from industry, handicrafts, the services and professions, often accompanied by their families.

The structure of Britain's population did not change substantially between the early and the late nineteenth century (Table 3.2). Relatively high fertility, limited falls in mortality and moderate net migrational losses, mainly of younger people, retained a youthful population structure. Despite a high dependency rate (the ratio of young and elderly to the working population) an expanding workforce absorbed the growth. As birth rates and family size began to fall from the 1880s the proportion of persons under five years began to fall. Similarly, assisted by improving life expectancy, the proportions of older middle-age (45–64) and then elderly people began to increase.

Mortality

Levels changed little between the 1820s and 1870s after which they moved hesitantly downwards to the turn of the century. Of the major sets of factors influencing health and mortality (Figure 8.3), socio-economic forces – particularly rising real wages and improving living standards and diet – offered some improvement, though not for the urban poor. Bio-medical factors offered few major break-throughs in curative medicine before the late nineteenth century despite better hospital provision and improved treatment and containment of epidemic diseases especially those of childhood such as scarlet fever, diphtheria and measles. Great, at times almost intolerable, pressure from adverse environmental conditions in the large towns in which an increasing proportion of the population lived, restricted improvement. Only with effective legislation to improve sanitation, water

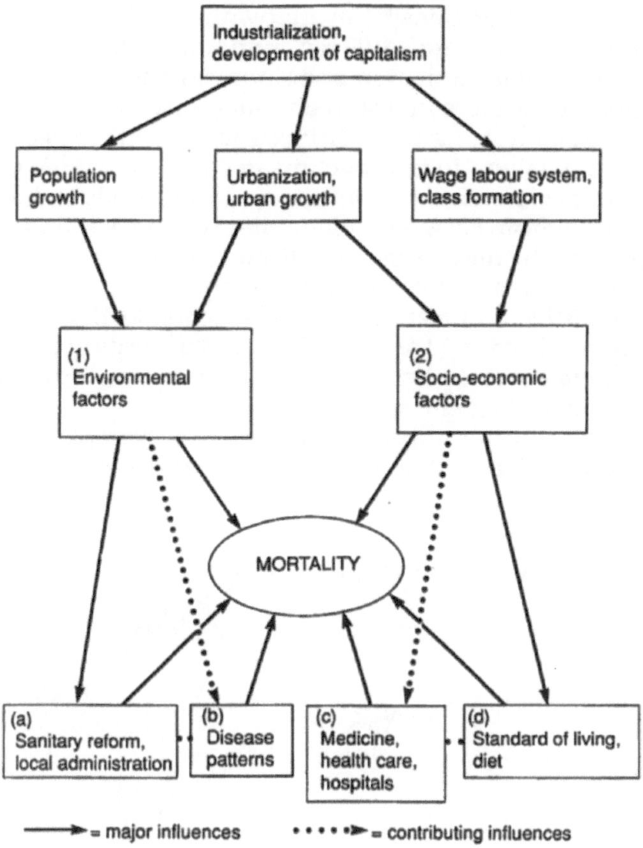

Figure 8.3 Variables influencing the level of mortality in the nineteenth century. Based on R. Woods and J. Woodward, *Urban disease and mortality in nineteenth-century England*. (London: 1984) p. 21.

supply and housing, and to apply effective measures of preventive medicine – especially in the control of epidemic diseases – were these gradually abated (Chapter 11). While medical science may have changed slowly, improving public and private medicine and, from mid-Victorian times, more and better-run hospitals improved health and life expectancy, especially among the middle classes.

While most epidemic diseases resisted cure, prevention and treatment could ameliorate their impact. Such epidemic years as 1831, 1847–49, and the mid-1860s saw the average mortality of around 22 per thousand increase to 24–25 per thousand (Figure 3.2). Excess mortality in the large cities and industrial areas adversely affected natural increase as reflected in the contrast, identified by William Farr, between the 'Healthy Districts' (the rural and suburban areas) that had an average life expectancy at birth (e_o) of 51.5 years in the late 1830s, and the 'Poor Districts' (unhealthy inner cities and many industrial areas) where it was less than 29. This gap narrowed from the 1880s when it began a slow fall to figures for County Boroughs and Rural Districts of 47.5 and 66.3 years respectively, in 1911. The direct link between high population density, overcrowding and death rates at all ages, especially among infants and children, underlines the continuing importance of environmental (sanitation and housing) and socio-economic (poverty and ignorance) factors in health and mortality.

The major reason for the wide discrepancies in life expectancy and the principal cause of failure to improve this in early and mid-Victorian times, was the failure to conquer infant mortality. Whereas child deaths began to decline erratically from early Victorian times (Figure 8.4) then more swiftly and steadily from the mid-1860s, only from 1900 was there a parallel fall in infant mortality (Chapter 13). Infant mortality remained at 140 to 150 per thousand live births in England and Wales (Scotland 120–130), accounting for 20 per cent of all deaths. Trends and spatial contrasts in the levels of child and infant mortality underline the significant contrasts between densely populated urban areas and rural districts.

Infant mortality in the unhealthiest cities (for example, Liverpool) was more than double that in healthy rural areas and twice that of suburban

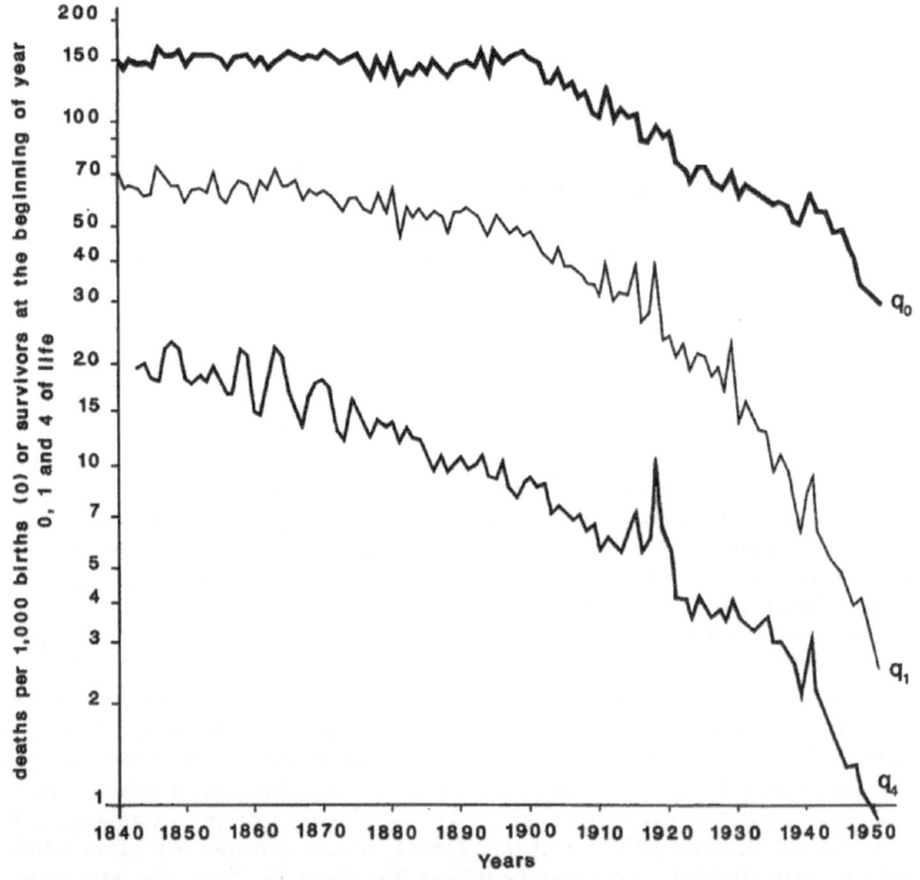

Figure 8.4 Infant mortality (q_0)and selected child mortality (q_1 and q_4) in England and Wales, 1840–1950. Based on R. Woods, P. Watterson and J. Woodward 'The causes of rapid infant mortality decline in England and Wales, 1861–1921' *Population Studies* 42 (1988) p. 351.

areas. During the crisis years of the late 1840s, with famine Irish migrants flooding in, fever rampant and a cholera visitation in 1849, the general crude death rate in the Borough of Liverpool was 34.2 per thousand as against 26.3 per thousand in the healthier suburbs. Within the Borough there were wide contrasts between the crowded working class North End, with its predominantly Irish-born population, and middle-class and suburban areas. In 1847 deaths from 'fever', including typhoid and typhus (the 'Irish disease'), and diarrhoea and dysentery ranged from under 6 per thousand in high-class residential areas of the Borough to over 51 in dockside Vauxhall. In August 1849, at the height of the cholera epidemic, the annual equivalent death rate was over 140 per thousand in central and north dockside areas but only one-fifth of that (28.7) in good residential areas. Such wide discrepancies persisted into the early 1880s when the death rates in the healthier wards in good years and the worst wards in bad years ranged from 18 to 50 per thousand (Figure 11.7). Glasgow's intra-urban mortality experience in the 1870s was not dissimilar – ranging from 21 to 46 per thousand in 1871 and from 16 to 38 per thousand in 1881 – with even wider discrepancies in child mortality between Wards of 69 to 166 per thousand in 1871 and 53 to 139 per thousand in 1881.

Steep gradients in mortality underline the importance of environmental factors in the protracted mortality decline from the mid-eighteenth century in Britain. Despite a general awareness of the links between high mortality from infectious and bronchial diseases (especially pulmonary tuberculosis), the precise causes of their development and transmission were not understood until they were revealed by the advances in bacteriology in the late nineteenth century, although advances in urban public health, water supplies and sanitation and, more belatedly, housing began to effect some improvements (Chapter 11).

As McKeown (1976) has shown, over three-quarters of the fall in mortality between 1848 and 1901 was brought about by a decline in such airborne diseases as scarlet fever, diphtheria and measles and those caused by infected water and food such as typhoid, cholera and – most significantly – dysentery and diarrhoea. These two groups accounted for 44 and 33 per cent respectively of the decline in mortality over these years (other micro-organisms a further 15 per cent). Considerable improvement was also effected in prevention of respiratory tuberculosis, that other great scourge of city-dwellers, deaths from which fell by 56.3 per cent between mid- and late-century thanks to better housing, nutrition and nursing. There was no amelioration of other bronchial deaths (including pneumonia and influenza) to which growing air pollution undoubtedly contributed, though their fatality was beginning to be reduced by the turn of the century, long before the miracle drugs of the 1940s and 1950s (Chapter 13).

Even within the countryside substantial differences in mortality, especially child mortality, may reflect environmental and nutritional contrasts. In some districts, such as the Fenlands, damp and humid summer heat tainted food and increased mortality in areas where babies were weaned early. Where children were breast fed and/or there was access to abundant fresh milk (as in many parts of upland Britain) infant mortality was often below average. However, the precise roles played by better living conditions and improved water supply and sanitation as against improvements in the quality of diet and better food hygiene are difficult to establish. Too much has probably been made of purer milk supplies from the late nineteenth century in improved infant mortality: but alongside a growing public health movement and awareness of the importance of cleanliness in the home (helped by the popular press) it contributed to a final phase of rapid decline in mortality from the 1890s.

Fertility

The fertility transition from late-Victorian times (Chapter 13) was more rapid than that in mortality. Fertility levels had already stabilized by the 1830s. The lower marriage age which had contributed to the increased natural growth of the early industrial revolution (Chapter 3) gave way after the depressed 1820s and 1830s to later marriage, a slight increase in the proportion of women who never married and lower birth rates of 35–36 per thousand women in the 1840s as compared with over 40 per thousand in the 1800s.

While marriage practices are not a direct guide to either general or, more especially, regional trends in fertility, social factors such as the availability of marriage partners in areas of high emigration, as in the western Highlands and Islands, or persistent out-migration (throughout rural Britain) limited both marriage levels and affected births. Limitations on marriage in certain occupational groups, for example living-in domestic servants and farm labourers, also affected local fertility patterns. Low marriage rates typical of rural Scotland up to the end of the nineteenth century reflected, in part, 'living-in' of farm labourers (especially in north eastern Scotland) and was related to high levels of bastardy. The general increase in the mean age of marriage to around 25.8 years for women and some two years higher for men by mid-century, with a further increase from the 1870s to the First World War, also reflects changing economic circumstances and the desire for more spending power and independence that characterized late-Victorian middle- and, to some extent, working-class society.

There were considerable differences between the industrial areas (where there were more and earlier marriages) the rural areas (where marriage tended to be later), and between the different social classes: urban labourers and coal miners married young; prudent white-collar workers, shopkeepers and the middle class postponed marriage until they felt able to afford it. Similarly single children leaving home for the city – whether as a domestic servant or an industrial of office worker – often lived for a time in lodgings before taking on family responsibilities: hence the high proportion of households with lodgers reflected in both census enumerators' books and the pages of Dickens's novels.

While the first signs of a downward secular trend in fertility came with the adoption of mechanical birth control within marriage among some middle-class groups, the rapid spread of family limitation to all social classes owed more to traditional methods. Despite religious and cultural beliefs which delayed its adoption in some sectors of society, increasing secularization of society caused barriers to family limitation to vanish quickly in the twentieth century. The argument that family limitation represents the diffusion of birth control from the professional and upper middle classes – the maid learning from the mistress, as it were – to the lower classes (the last to adopt being unskilled urban and farm labourers) does not stand up to close examination. Among the first to limit family size were 'skilled' non-manual and commercial workers (shopkeepers, clerks and the like) who were also prominent amongst cautious late-marriers.

Economic incentives to limiting the number and spacing of births were strong where women were prominent in the workforce. In the mills of Lancashire and West Yorkshire or in the Potteries women might delay having children, or have a smaller family and return to work as soon as possible. Increasing numbers of women involved in shop and, from the 1890s, office work might also have deferred marriage and limited their families. Amongst the middle classes, the increasing expense of raising children with rising costs of domestic servants and school fees, as well as a growing desire for greater freedom and more money to spend on luxuries and entertainment, were obvious incentives to have fewer children. Among working-class parents the loss of children's income as the use of child labour was restricted was also a disincentive to large families, especially after the Attendance Officers of the new School Boards set up under the 1870 Education Act saw to it that more and more children did attend school. The growing independence of women was also important in the onset of the fertility transition, especially among middle-class wives. As child mortality declined, more survived to adult life so that there was less need for large families and more incentive to put more space between births so as to avoid excessive pressures on mothers and households.

Thus a series of complex factors underlie the beginnings of the fertility decline from the late 1870s and its spread over the next fifty years. While the rapid fall to the one- and two-child family 'norm' was mainly achieved in the early twentieth century (Chapter 13), the fall in average family size from 6.2, for women between the ages of 20 and 24 married in the 1860s, to 4.1 for those marrying in the 1890s indicates the substantial impact of family limitation by the late nineteenth century. Moreover the rapid decline in the average age at which the mother's last child was born (from the age of 41 to 34 over that period) is a clear reflection of deliberate spacing and limitation of births within marriage.

Nevertheless the speed of adoption of family limitation varied considerably between different occupational groups and reflects considerable differences in marital fertility between different types of area in 1891. Relatively low birth rates in textile districts and residential towns, with large numbers of single women in domestic service and middle-class households, contrast with earlier and more universal marriages with larger families among iron- and steel-making and coalmining communities where abundant demand for boys and young men in the mines and early peaking of earnings reduced incentives to limit families, while fewer opportunities for female employment and the stereotyping of women – to look after men on shift work – meant that girls tended to marry young. Rural fertility also remained high

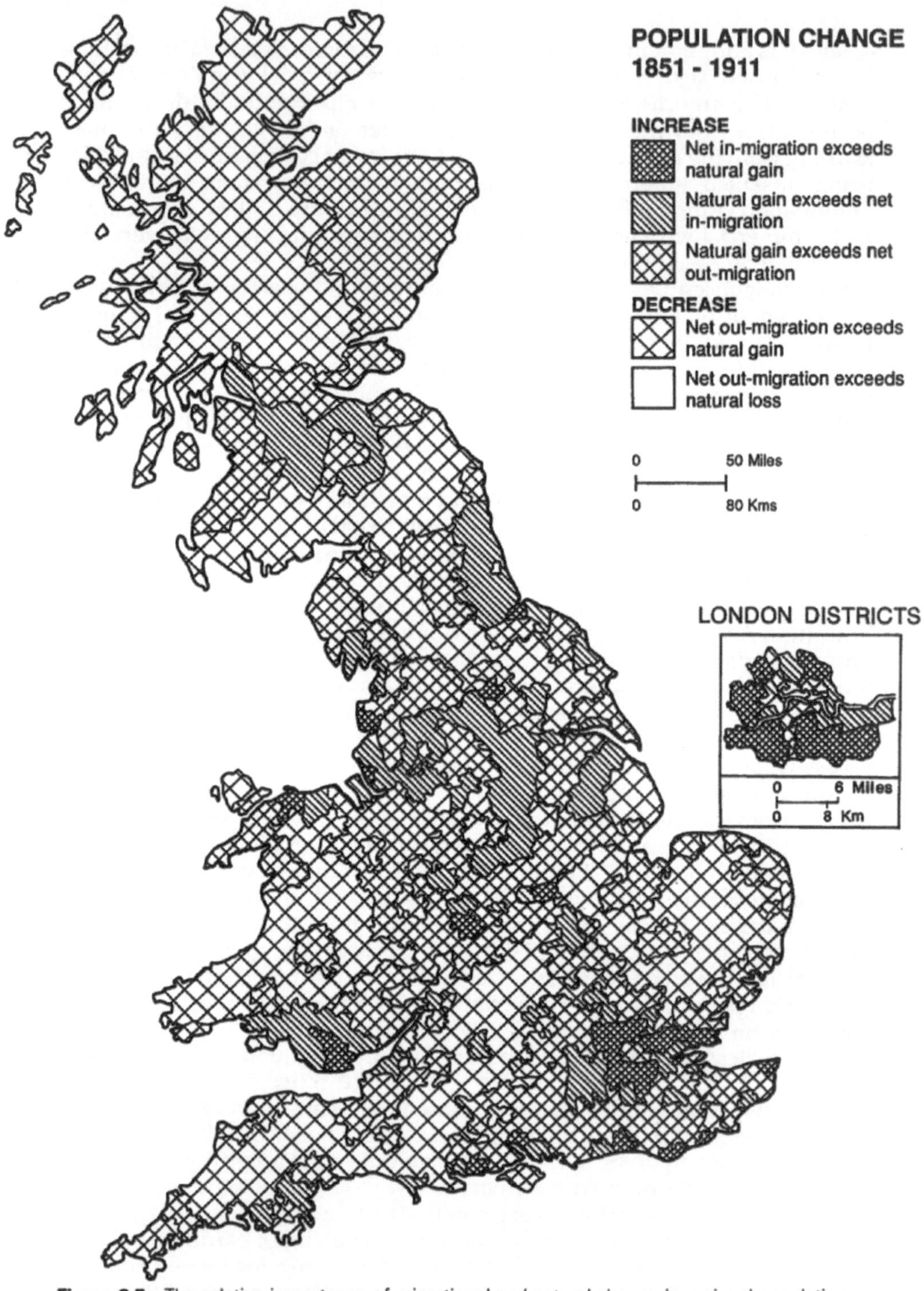

Figure 8.5 The relative importance of migrational and natural change in regional population trends in Britain, 1851–1911.
Source: Censuses of Great Britain, 1851, and of England and Wales and Scotland, 1861–1911; annual Reports of the Registrars General.

although birth rates declined as out-migration created an ageing population, especially in some areas of highland Britain.

Migration

Population migration, one of the most significant of demographic phenomena, is among the most difficult to measure. Births and deaths are finite events which, from 1837 in England and Wales and 1855 in Scotland, were recorded with increasing accuracy. They are also immediate in their impact on family, friends and neighbours. In contrast migration, which has never been recorded systematically in Britain, ranges from short-term transitory mobility, the motives for and effects of which are frequently unclear, to permanent moves.

Many people see rural to urban movement as the dominant feature of migration in the period 1830 to 1890. Redford (1926) highlighted the importance of rural to urban migration and emphasized the demographic importance of migration for the early nineteenth-century growth of towns. This view is misleading since at the aggregate level and in most places, even by the 1840s, net in-migration was less important than natural increase in urban growth. Only newer settlements – resort towns, residential suburbs (especially around London), and newly established industrial towns – depended mainly on migration for growth. While many left the countryside for towns, inducing nation-wide rural depopulation, and there was substantial and changing movement between towns, natural increase was of growing importance in urban population development (Figure 8.5).

The motives for and effects of migration were very varied and increased in complexity from the 1820s and '30s as the geography of Britain itself became more complex and highly differentiated and as people had the information and resources to move freely around the country. Figure 8.6 suggests a classification of migratory moves in Britain in this period emphasizing the complexity of migratory experiences and the way in which inter-urban and urban-suburban movement became increasingly important as the century progressed. Figure 8.7 suggests that many of these features may have been linked together with a rural to urban move not a discrete event unconnected to other migratory experience but part of an overall life history of individual migration from the countryside, to an adjacent village and then to a local town. Thereafter the individual might move several times up the urban hierarchy perhaps reaching a large city in the 1850s and subsequently moving between and within cities to end in an outer suburb in the 1890s. Such a migration history may be closer to reality than the simple stereotype of rural-urban migration.

Before the fuller birthplace data of the 1851 census direct information on regional patterns of migration is limited. Even then it is impossible to produce a comprehensive picture of regional migration trends which in general show a high volume of migration in all periods and a relatively constant pattern of internal migration since the late eighteenth century, but with a gradual broadening of the range and quantity of migration. Three principal points on early nineteenth-century migration can be highlighted.

Type of Move	Long Distance	Short Distance
Rural-Rural	Temporary harvest migration.	Local inter-village movement including marriage migration.
Rural-Urban	Usually a series of 'stepwise' moves up the urban hierarchy.	Most common during early phase of urban growth - short distance movement to nearest town.
Urban-Urban	Increasingly common between large cities after about 1860.	Movement up urban hierarchy from small town to nearby city.
Urban-Rural	-	Movement of high-status households to rural suburbs.
Intra-Urban	-	Frequent short distance moves mostly in same area of city.

Figure 8.6 Classification of major categories of migration found in nineteenth-century Britain.

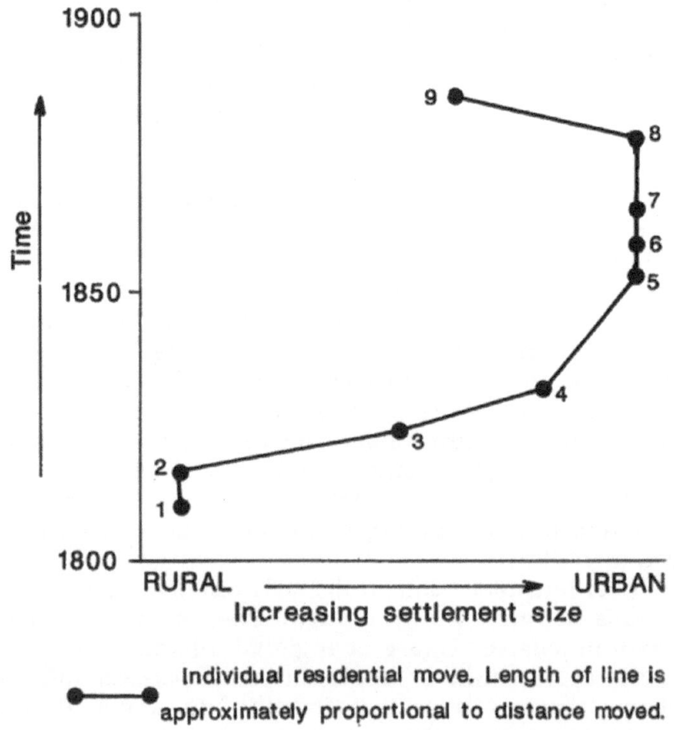

Figure 8.7 Model of hypothetical individual life-time migration in nineteenth-century Britain.

First, most migration was directed at those counties and regions experiencing rapid industrialization and urbanization: thus Nicholas and Shergold (1987) emphasize the pull of London and the high level of movement between industrial counties (not only rural to urban movement) between 1818 and 1839; while Withers (1985) demonstrates the attractiveness of the expanding Dundee textile industry for migrants from the eastern Highlands (especially Perthshire) between 1821 and 1854. Secondly, the majority of individual moves were relatively short-distance through a series of stepwise movements. Thirdly, longer-distance moves tended to be selective by occupation and age: such migrants were often significantly more skilled and literate than non-movers.

Some migrants moved frequently and over long distances for specific occupational opportunities. Southall's (1986) study of tramping artisans in

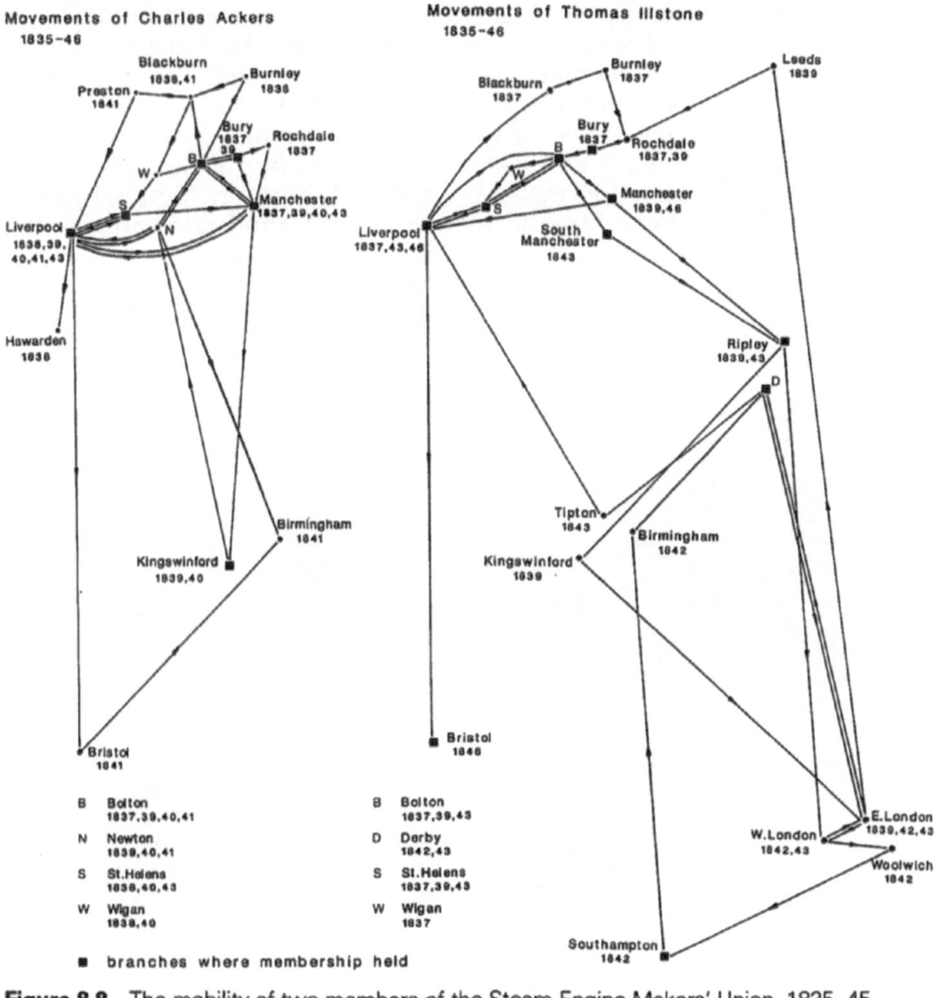

Figure 8.8 The mobility of two members of the Steam Engine Makers' Union, 1835–45. Based on Southall (1986) p. 33.

the period 1835–45 shows that engineers moved frequently and that long-distance moves were commonplace: such artisans averaged 2.5 moves over six years and many travelled several hundred miles in search of work, especially during the depression years of 1841–2 (Figure 8.8). This group is atypical: élite craftsmen, members of a well-organized union which provided information about job opportunities and arranged accommodation, they mostly moved alone, leaving family behind and sending home remittances when money was available. But their example demonstrates that long-distance mobility – mostly on foot – was not unusual among skilled artisans.

Although census birthplace data are imperfect they offer a series of reasonably comprehensive pictures of life-time inter-county population movement after 1851. Most studies emphasize the dominance of London as the target of long-distance migration. Industrial areas outside London initially tended to rely heavily on regional networks of migration, though exchanges of migrants between industrial regions increased as the century progressed (Figure 8.9). Between 1851 and 1891 the limited occupational opportunities of a small town like Lancaster attracted migrants over only a short distance and it relied mainly on its rural hinterland. Larger towns of North West England, such as Preston and Bolton, had wider spheres of attraction which increased in the later nineteenth century. Although competing with Liverpool and Manchester they offered significant and specific employment opportunities and exchanged migrants with the larger towns. In contrast, Liverpool and Manchester attracted migrants from all parts of Britain, Liverpool having marginally the larger migration field (Figure 8.10).

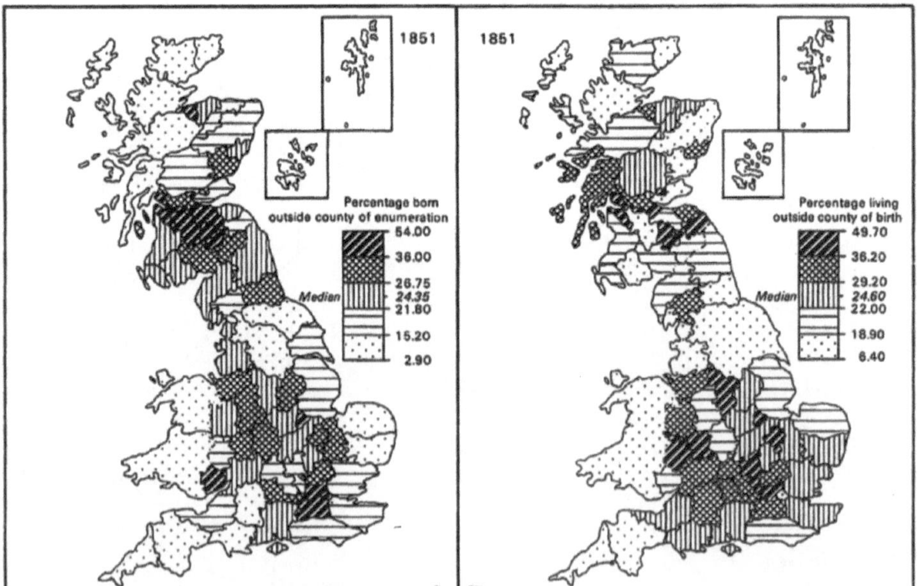

Figure 8.9 In-migration (A) and out-migration (B) by county in Great Britain, 1851.
Source: Census of Great Britain, 1851.

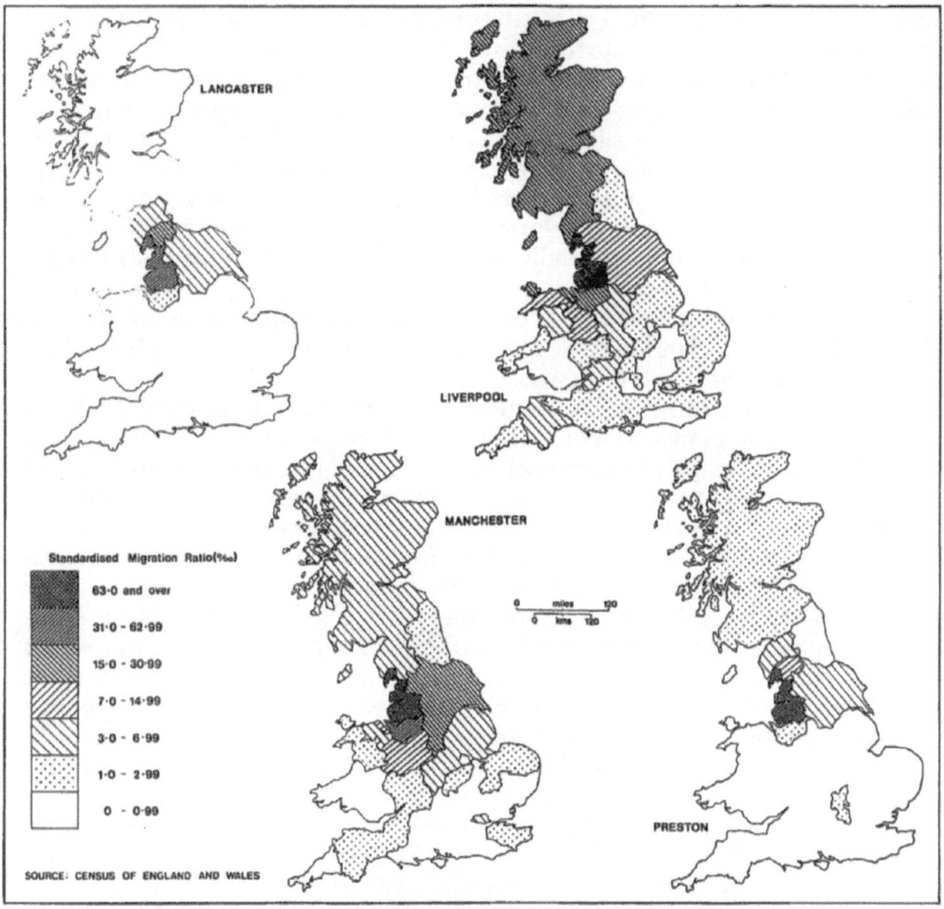

Figure 8.10 Migration to selected towns in north-west England, 1871.
Source: Census of England and Wales, 1871.

However, even in these cities most migrants had been born close by, though increasingly they came from nearby towns rather than the countryside. Liverpool's and Glasgow's large migrant populations, encouraged by easy access by sea, were typical of many nineteenth-century port towns.

Specialist employment opportunities attracted particular migrant streams over long distances. The substantial Welsh-born population of Middlesbrough in the 1870s was drawn almost exclusively from South Wales and reflected a well-established migration stream between areas with a similar industrial structure. Several Welsh iron masters moved to Middlesbrough to exploit expanding economic opportunities on Teeside and continued to recruit Welsh labour over a considerable period of time. Similarly, St Helens' expanding glass industry recruited many skilled workers from other glass-making areas, supporting the view that longer-distance migrants were often more skilled, better-educated and of higher social status.

Most rural areas lost population during the later nineteenth century, but in those districts closest to expanding towns population rapidly stabilized and, by the end of the nineteenth century, in some cases increased as urban growth spilled over the surrounding countryside. Around London improvements in communications and pressure on space within the metropolis led to early and rapid suburbanization: around Liverpool, villages on Deeside were attracting residential migrants by the 1890s after decades of population loss. In remoter rural areas population losses were more general and continuous and were exacerbated in some cases by the coming of the railway. Initially providing employment, railways subsequently offered an easy outlet for migrants, especially with the demise of rural industry in the face of competition from urban-based factories (Chapter 10). Reorganization by large landowners also led to displacement of population as in the Scottish Highlands where clearances led to mass migration to Scottish cities – especially Glasgow – and overseas.

What impacts did such movement have on the communities left by migrants and the places to which they moved: how did it affect the lives of individuals and families? In areas of marked imbalance between inflow and outflow – in numbers, demographic or social characteristics – the effects on communities were significant: the impact of population clearances on the Scottish Highlands was devastating; the effect of Irish famine migration into Glasgow or Liverpool was considerable, fundamentally changing their social geography. Elsewhere results were more subtle. Most large cities cumulatively lost and gained vast numbers through migration, but social and economic characteristics were not radically changed as leavers were mainly replaced by people of similar socio-economic characteristics and culturally distinctive migrant groups were proportionately small except in particular localities (Chapter 11). In smaller settlements atypical migratory change is most striking: the short-term impact of navvy gangs involved in railway construction on villages and small market towns; the recruits to defence establishments in such towns as Portsmouth, Plymouth and Aldershot; the short-term impact of rapid industrial growth in new towns like Middlesbrough and Crewe or in expanding mining areas. But in most places the long-term impact of migration was gradual and more easily assimilated.

For the individual, mobility was not necessarily traumatic. A short-distance move to a local town through well-known territory, utilizing well-established friendship or kinship networks to find work and accommodation, eased the leaving of home for many young migrants. Longer-distance moves could be smoothed by contacts providing help and access to a larger labour market. Welsh migrants from Merthyr Tydfil to Middlesbrough knew what sort of place they were going to from previous letters and contacts, and iron workers could be confident of employment in an industry which they understood and where they would be assimilated into a substantial Welsh community with a minimum of alienation and disruption. Family migration was also more important than previously thought. Rather than individuals arriving lost and friendless in a strange city, in Victorian Britain many families moved along well-established paths into known territory. Even where the move was into unfamiliar territory, mutual support within a kin or cultural group could lessen difficulties.

In Victorian Britain population migration affected many people at some time in their lives and was taken for granted as a part of lifetime experience, without severe long-term effects on either individual or community. The search for work was the dominant motive, especially in longer-distance movement targeted on specific labour markets. Shorter-distance moves which remained dominant throughout the nineteenth century involved a host of individual reasons, leading migrants towards familiar territory and leaving open the possibility of return. Indeed, although evidence from detailed longitudinal studies is limited, return migration or repeated movement between specific locations was probably more common than once thought.

Regional population structures

Varying rates of fertility, mortality and migration produced substantial contrasts in the structure of population which, in turn, gave rise to considerable differences in the potential for regional population growth. Britain's mid-nineteenth century population was youthful (Table 3.2). Rural districts were the least youthful, especially where – as in the Scottish Highlands, upland Wales, northern England and the West Country – strong out-movement was combined with falling birth rates (Figure 8.11). Where fertility was high – in the East Midlands, South East England, west Wales, and in the mining counties – there were above-average proportions of children under 15. High proportions of younger working-age population (aged 15–44) characterized industrial areas and large cities, including London. Ageing populations, with above average proportions over 65, in general complemented areas with low proportions of under-15s: for the most part they were predominantly areas of rural out-migration.

Sex differentials in migration are reflected in the substantial deficit of men in rural Scotland, especially the Highlands, and in many parts of rural England, for example East Anglia and the South West (Figure 8.12). But shortage of work for girls was even more marked than for men in most rural areas, as reflected in relatively high male:female ratios in much of south eastern England from which many women went to work in London and along the Welsh border whence there was substantial migration to domestic service in Liverpool, Manchester and the West Midlands. In contrast the above-average maleness of Clydeside, Staffordshire and the South Wales and North East coalfields clearly reflects their male-dominant employment. By the end of the century that position had changed substantially. The phase of male-dominant Highland migration had passed and the lowest male:female ratios were in north eastern Scotland and the Borders. High female employment in mill towns and the cities of North West England and the continuing demand for domestic servants and female service workers in London were reflected in their female-dominant populations. The lowest ratios were by then in coastal counties, especially at resorts and in Cornwall where there had been heavy and persistent mining emigration. Male dominance continued in heavy industrial areas and some of the arable counties (for example Lincolnshire) and was most marked in the coalfields and areas with substantial defence establishments.

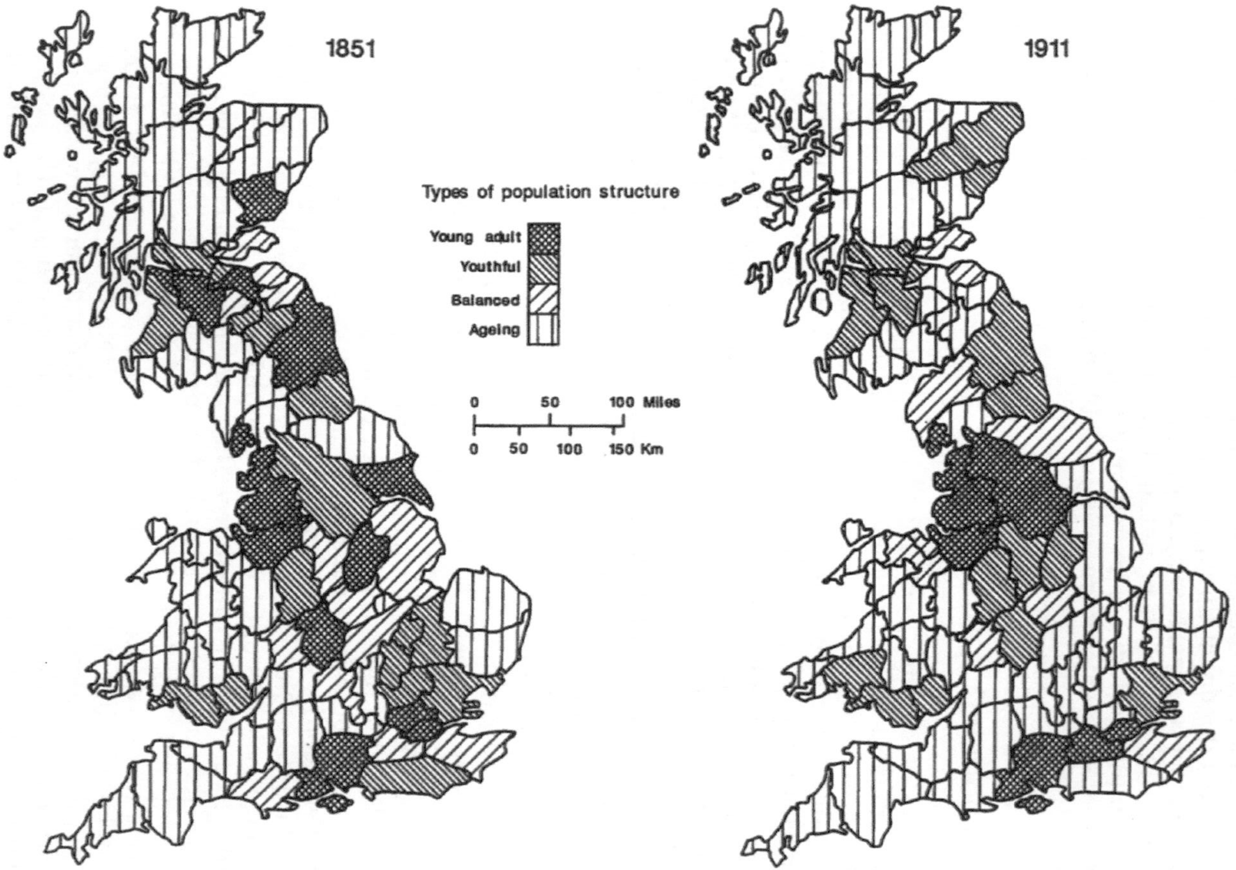

Figure 8.11 The age structure in Great Britain in 1851 and 1911. 'Young adult' counties are those with a substantial excess of people in age-groups 15–44; 'youthful' populations have a surplus of ages 0–14 with near- or above-average 15–44; 'balanced' populations have three groups close to the national average; 'ageing' counties have an excess of population over 65 years and a deficiency of working-age population, especially the 15–44 age-group.

Sources: Census of Great Britain, 1851; Census of England and Wales and of Scotland, 1911.

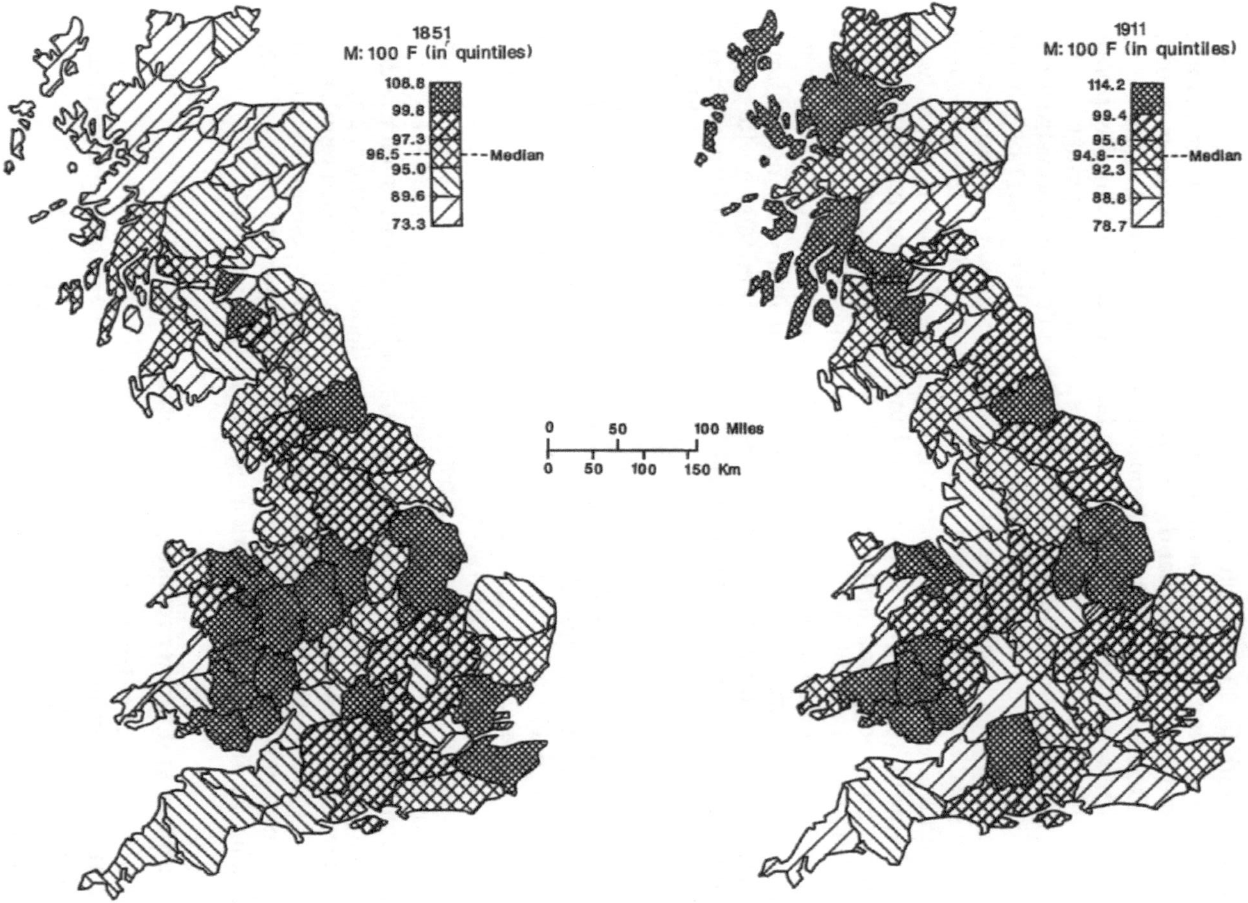

Figure 8.12 The sex structure of the population of Great Britain in 1851 and 1911.
Sources: Census of Great Britain, 1851; Census of England and Wales and of Scotland, 1911.

In Britain as a whole, rising life expectancy and the beginnings of decline in the birth rate were reflected in smaller proportions of young people by the turn of the century (Figure 8.11). Movement of families from the land and from inner city centres and suburbs are apparent by 1900 in the fall in the proportion of under 15s in such areas. In contrast active migration of labour to the towns and industrial areas of central Scotland, North East and North West England, the West Midlands, South Wales and, to a lesser extent, Greater London led to a substantial concentration of younger adults (aged 15–44). By then, however, an increasingly familiar pattern of ageing populations in the countryside, in coastal resorts and some residential towns, including the spas, was becoming a feature of Britain's population structure.

The effects of demographic, social and economic change on households were more complex. Most areas of heavy and prolonged out-movement, especially rural areas, experienced gradual reductions in the size of household. Within cities there were substantial differences between one- or two-person households living in often cramped and multiply-occupied buildings in central 'reception areas' and larger family-based households, often with many lodgers, in established working-class communities. Occupational structure was reflected in the retention of both sons and, to a lesser extent, daughters in the large households of mining and iron-working communities. In contrast middle-class households often declined in size with fewer children and fewer servants between the 1850s and the turn of the century.

The proportion of households headed by widow(er)s reflected higher mortality in Victorian England than in the twentieth century and together with the large numbers of people living alone or left alone (whether by children moving from home or being orphaned) produced greater numbers of single-person households than in the mid-twentieth century. Nevertheless the two-generation, nuclear family household – not the extended household – remained the norm in both rural and urban areas of Victorian Britain as it had always been.

Changes in the character of late nineteenth-century British society were reflected in its changing demographic character and population structure. Environmental conditions, especially in the polluted and overcrowded towns and industrial areas, and economic forces (labour migration, improving real wages and better food and housing) were still important regulators. But social forces – for example marital behaviour, the role and status of women, universal education – were increasingly important in population behaviour. In these the 1890s mark a significant turning point.

Bibliography and further reading for Chapter 8

Standard works on population change are:

FLINN, M. W. et al. *Scottish population history from the seventeenth century to the 1930s* (Cambridge: Cambridge University Press, 1977).

McKEOWN, T. R. *The modern rise of population* (London: Arnold, 1976).

TRANTER, N. *Population and society, 1750–1940* (London: Longman, 1985).

WRIGLEY, E. A. and SCHOFIELD, R. *The population history of England 1541–1871: a reconstruction* (London: Arnold, 1981).

Other studies on changes in fertility and mortality include:

MITCHINSON, R. *British population change since 1860* (London: Macmillan, 1977).

SZRETER, S. 'The importance of social intervention in Britain's mortality decline 1850–1914: a reinterpretation of the role of public health' *Social History of Medicine* **1** (1988), pp. 1–38.

WOODS, R. and SMITH, C. 'The decline of marital fertility in the late-nineteenth century' *Population Studies* **37** (1983), pp. 207–25.

WOODS, R. 'Approaches to the fertility transition in Victorian England' *Population Studies* **41** (1987), pp. 283–311.

WOODS, R. and HINDE, R. 'Mortality in Victorian England: models and patterns' *Journal of Interdisciplinary History* **18** (1987), pp. 27–54.

On spatial trends see:

JONES, H. 'Population from circa 1600', Chapter 5 of Whittington, G. and Whyte, I. (eds) *An historical geography of Scotland* (London: Academic Press, 1983).

LAWTON, R. 'Population changes in England and Wales in the later nineteenth century' *Transactions of the Institute of British Geographers* **44** (1968), pp. 55–74.

LAWTON, R. 'Population trends in Lancashire and Cheshire from 1801' *Transactions of the Historic Society of Lancashire and Cheshire* **114** (1962), pp. 189–214.

SAVILLE, J. *Rural depopulation in England and Wales, 1851–1951* (London: Routledge, 1957).

WOODS, R. 'The structure of mortality in mid-nineteenth century England and Wales' *Journal of Historical Geography* **8** (1982), pp. 373–94.

On migration see:

BAINES, D. *Migration in a mature economy: emigration and internal migration in England and Wales 1861–1900* (Cambridge: Cambridge University Press, 1985).

ERICKSON, C. J. 'Who were the English and Scottish emigrants to the United States in the late nineteenth century?' in Glass, D. and Revelle, R. (eds) *Population and social change* (London: Arnold, 1972), pp. 347–81.

ERICKSON, C. J. 'Emigration from the British Isles to the USA in 1841. Part 1: Emigration from the British Isles.' *Population Studies* **43** (1989), pp. 347–67, and 'Part 2: Who were the English emigrants?' *Population Studies* **44** (1990), pp. 21–40.

FRIEDLANDER, D. and ROSHIER, J. 'A study of internal migration in England and Wales' *Population Studies* **19** (1966), pp. 239–79.

HANDLEY, J. E. *The Irish in modern Scotland* (Cork: Cork University Press, 1947).

HOLMES, C. *Immigration and British Society*, 1871–1971 (London: Macmillan, 1987).

JACKSON, J. A. *The Irish in Britain* (London: Routledge, 1963).

JACKSON, J. T. 'Long distance migrant workers in nineteenth century Britain: a case study of the St. Helens glassmakers' *Transactions of the Historic Society of Lancashire and Cheshire* **131** (1982), pp. 113–37.

NICHOLAS, S. and SHERGOLD, P. 'Internal migration in England, 1818–1839' *Journal of Historical Geography* **13** (1987), pp. 155–68.

OSBORNE, R. H. 'The movement of people in Scotland 1851–1951.' *Scottish Studies* **2** (1958), pp. 1–46.

POOLEY, C. G. 'Welsh migration to England in the nineteenth century' *Journal of Historical Geography* **9** (1983), pp. 287–305.

POOLEY, C. G. and WHYTE, I. D (eds) *Migrants, emigrants and immigrants: a social history of migration* (London: Routledge, 1991).

REDFORD, R. *Labour migration in England 1800–1850* (Manchester: Manchester University Press, 1926).

SOUTHALL, H. 'The tramping artisan revisits: the spatial structure of early trade union organization' in Withers, C. (ed.) *Geography of population and mobility in nineteenth century Britain* (Cheltenham: HGRG, 1986).

SWIFT, R. and GILLEY, S. *The Irish in Britain, 1815–1939* (London: Pinter Press, 1989).

WITHERS, C. 'Highland migration to Dundee, Perth and Stirling, 1753–1891' *Journal of Historical Geography* **11** (1985), pp. 395–418.

WITHERS, C. and WESTERN, A. J. 'Stepwise migration and Highland migration to Glasgow, 1852–1898' *Journal of Historical Geography* **17** (1991), pp. 35–55.

9

The countryside from the 1830s to the 1890s

Introduction

The Victorian countryside was increasingly influenced by external as much as internal forces. Recovery from the poor harvests and deep depression of the 1820s and '30s together with rising home markets took agriculture into a Golden Age of British farming from the late 1840s to the early 1870s. But the eventual consequences of free trade ended the dominance of wheat growing as grain prices collapsed under a flood of cheap imports from the New World (Table 9.1). Although markets for stock and dairy products and perishable cash and fruit crops benefited from rising real wages and increasing demand, they too experienced external competition with the development of refrigeration and canning in late-Victorian times. The agricultural depression of the late 1880s and '90s was widespread and crippling as reflected in the decline of agriculture's share of the national income from one-fifth in mid-century to one-twelfth by the 1890s.

By then many agricultural labourers and craftsmen had left the countryside (Chapter 8). The push of low wages, increasing mechanization of both agriculture and industry, the pull of job opportunities and a livelier urban life style drew the young and ambitious. Better and cheaper transport progressively drew local trade towards larger country towns and regional centres and with it many professional and service activities. Seen against a background of rising demand, a relative then absolute fall in labour (Table 9.2) and growing exposure to foreign competition the achievements of British farming from the 1830s to the mid 1870s were impressive. More widespread adoption of existing agrarian improvements and new techniques of dealing with difficult soils on claylands, poor light land and the more intransigent marshlands saw improvement spread to all types of farming. More intensive farming, using chemical as well as natural fertilizers, produced substantially greater yields. The processes were assisted by more scientific farming which reached many farmers through farming clubs, agricultural shows, newspapers and periodicals, and by increased energy inputs from machinery, animal and steam power.

 DOI: 10.4324/9781003174783-9

Table 9.1 Value of British food, drink and selected raw materials imports, 1792–1955

	Average annual value (£ million)								
	Corn	Meat and animals	Butter and margarine	Hides and skins	Tea, coffee, sugar	Wines and tobacco	Oil seeds and fats	Wool	Timber
1792–96	1.03	–	–	–	6.12	0.92	–	0.23	0.51
1811–15[1]	0.61	–	–	0.54	12.21	0.95	1.14	0.52	0.58
1841–45	3.85	–	–	2.25	14.65	1.26	3.44	1.57	1.34
1871–75	51.96[2]	12.94	7.50	7.12	39.08	11.54	17.86	20.64	16.74
1901–05	64.10	49.40	23.22	7.62	29.06	9.06	19.00	22.18	24.76
1931–35	44.28	82.30	39.32	13.50	44.62	18.30	19.58	35.22	31.92
1951–55	231.86	275.78	96.24	49.72	213.50	86.28	146.90	202.54	182.75

[1]Data for 1813 missing
[2]Including flour
Figures from 1871–75 are for UK
In addition fruit and vegetable imports, some £50 m.p.a. in the late 1930s averaged £182.48 m. in 1951–55
Source: Mitchell and Deane (1962), pp. 290–5 and 298–301; Mitchell and Jones (1971), pp. 131–32

Table 9.2 Agricultural labour in England and Wales, 1851–1931

	Males aged over 20 (000s)				Males aged over 12 (000s)
	1851	1871	1891	1911	1931
Farmers (and family)	299.3	272.4	243.9	271.4	248.2
Bailiffs and foremen	10.5	16.3	18.0	22.0	16.7
Shepherds and labourers	735.1	657.8	546.6	498.2	494.8
Gardeners,* gamekeepers and others	82.4	129.4	175.0	264.2	274.6
Total	1317.3	1075.8	983.4	1055.8	1148.3

*includes domestic gardeners
Source: Censuses of England and Wales

Land use and farming systems

As James Caird argued in 1850–51, in an increasingly market economy farming could only succeed as a business, maximising profits by increased yields from high inputs and/or reducing the real cost of working the land. British agriculture, faced by increasing foreign competition, needed flexibility in cropping and land use and convertibility in farming enterprise so that it could respond to changing economic and agrarian conditions. The key to success was to adapt varying soil and climatic conditions to profitable products. The crucial elements which determined different cropping and farming systems and the margin of profitability were, first, the costs of working the land – rent, inputs of capital, labour, fertilizer, seed, stock, etc. – and of getting the product to market (transport and marketing costs); secondly, income which reflected the quality of land and cultivation, the nature and location of demand and access to the market.

Despite increasing adaptability of farming, weather and soil conditions were a powerful influence on Britain's three major land-use and farming systems. First, upland farmers raised sheep and store cattle (with some dairying near to the main industrial markets) in an overwhelmingly pastoral landscape and on small family farms with access to extensive commons, though (after mid-century enclosures) these competed with big hill flocks on large estates, especially in the Scottish Highlands. Secondly, the drier lowland areas of eastern Britain were dominated by arable farming though its precise end-products, not least the balance between crop and stock, varied with changing market conditions, local soils and climate, and farmers' aptitudes. On the wetter lowlands and heavy, water-retentive land a third type of farming increasingly focused on grassland for fattening and dairying as the price of grain dropped in the 1870s (Table 9.3). In later Victorian times a fourth system of intensive cash-crop arable farming characterized high-quality, high-yielding soils close to major urban markets: the Fenland for London; the mosslands for south Lancashire.

From the 1840s the recovery from agricultural depression, considerable urban population growth, improved railway access to markets and continu-

Table 9.3 Food price movements in Great Britain, 1851–1880

	1851–55	1856–60	1861–65	1866–70	1871–75	1876–80	1870–80 % / 1851–55
Wheat	103	98	87	100	100	87	−7
Barley	82	98	86	101	103	95	+10
Oats	90	87	87	101	104	96	+7
Beef	77	85	87	94	110	103	+31
Mutton	80	88	93	93	108	105	+27
Cheese	75	86	84	102	97	85	+13
Milk	65	84	82	89	91	111	+36

1865–74 = 100
Source: E. L. Jones (1968)

ing agricultural innovation were reflected in the mosaic of regional patterns of farming depicted in the Prize Essays on the agriculture of counties published in the *Journal of the Royal Agricultural Society* between 1844 and 1874. The files and maps of the Tithe Commutation Commissioners of the 1840s and the parish statistics of crops and stock gathered in the Annual Agricultural Returns from 1866 provide a good picture of changing land use. Up to the early 1870s wheat dominated all types of soil in arable lowland Britain, occupying one-quarter to one-half of tillage and combined, especially on lighter land, with barley (Figure 9.1). In wetter, cooler western and upland areas oats with barley were usual. In both systems roots (turnips and swedes on lighter land and mangolds and brassicas, especially kale, on heavier), rotation grasses and clovers were the standard fodder crops,

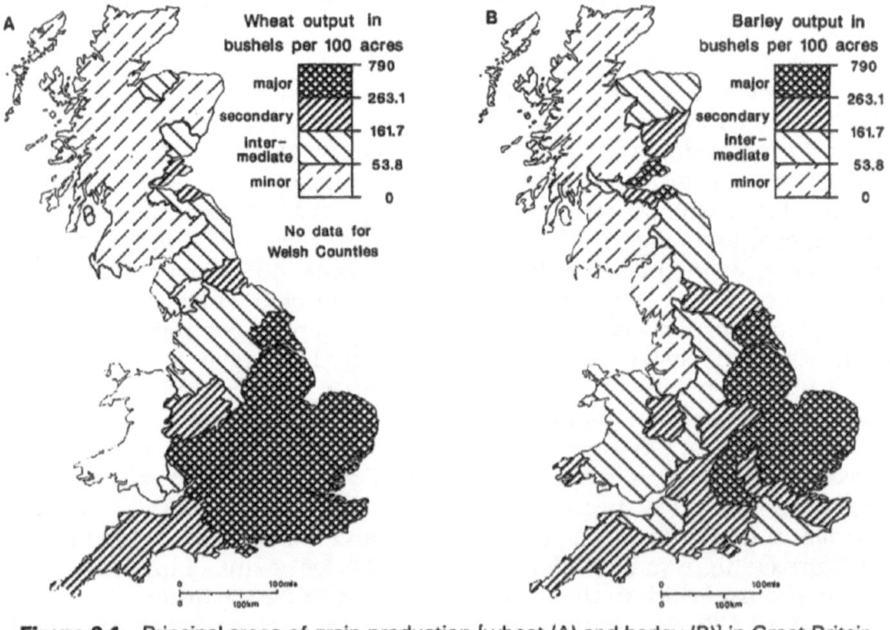

Figure 9.1 Principal areas of grain production [wheat (A) and barley (B)] in Great Britain, 1871. Based on Langton and Morris (1986) pp. 37–9.

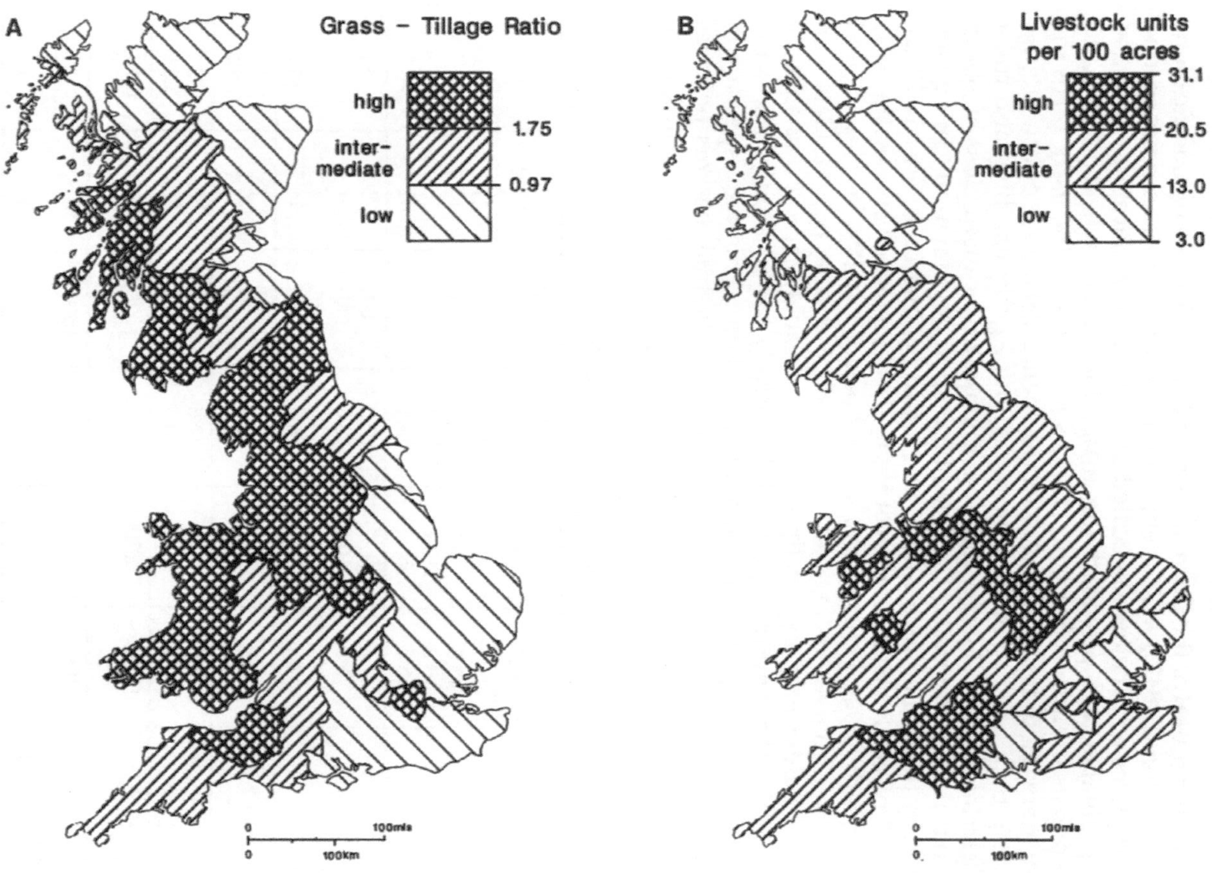

Figure 9.2 Principal areas of grassland (A) and livestock (B) in Great Britain, 1871. Based on Langton and Morris (1986) p. 47.

though in some intensive farming systems these were supplemented by oil seeds, valuable for both cattle cake and lamp oil, and other green crops. Though stock were important throughout an essentially mixed farming system, they dominated in western, upland and some midland areas (Figure 9.2).

Despite the repeal of the Corn Laws in 1846 (after a series of modifications between 1791 and 1842), which removed the restrictions on imports so long as wheat prices were below a specified upper limit, grain farming continued to be profitable and provided the major food crop: wheat in lowland England; oats in upland Scotland (Table 9.4). But from the early 1870s the price of grain fell dramatically as railways opened up the American prairies and cheap bulk ocean transport promoted huge grain handling and roller milling facilities in most large ports – in themselves a cause of loss of jobs in numerous rural water mills. The cost of transport of grain from Chicago fell to 3s 11d by the end of the century from 15s 1d per quarter in the late 1860s. Wheat prices, fell from an average of 55s to 28s per quarter between the early 1870s and early 1890s (Figure 9.3). Barley, much of it malted for brewing and distilling, fell from an average 40s to 24s over the same period. Oats – an important fodder crop especially for an increasingly large population of horses for draught and transport – did relatively better; though prices fell from 26s to 18s per quarter its share of cultivation increased, especially in midland and western areas. In northern and industrial markets potatoes were increasingly an alternative to bread in the diet and became a substantial crop in both general rotations and cash crop farming.

Table 9.4 Crops as a percentage of arable in England and Wales, 1801–1945

Crops	Percentage of total arable 1801	1836	1871	1911	1938	1945
Wheat	22.7	26.8	23.0	16.3	21.2	15.0
Barley	13.6	15.7	14.3	12.6	10.3	13.7
Oats	18.8	12.6	11.4	18.1	15.1	15.9
All cereals	55.5	55.1	49.1	47.4	47.9	48.0
Potatoes	1.6	N.K.	3.0	3.8	5.5	6.8
Fodder roots	5.9	10.2	13.5	11.3	7.4	5.3
Sugar beet	–	–	–	–	3.8	2.8
All roots	7.5	N.K.	16.7	16.0	16.7	14.9
Peas and beans	6.9	4.7	6.0	4.1	2.0	1.7
Fallow	20.3	9.4	3.5	2.9	4.1	2.3
All tillage	89.8	79.5	79.5	76.9	76.4	76.1
Temporary grass	10.2(e)	20.5	20.5	23.1	23.6	23.9
TOTAL ARABLE (000 acres)	11350	12700	14946	11299	8935	12916
TOTAL PERMANENT GRASS (000 acres)	N.K.	N.K.	11376	15950	15604	11804

N.K. = not known (e) = estimated

Data for 1801 are estimated from the 1801 Crop Returns and for 1836 from the Tithe Commutation Survey; for 1871–1945 from the Annual Agricultural Returns

Sources: Grigg, (1989), Tables 4.4 and 5.3; Lord Ernle (1961 Edition), Appendix VIII, Table 1

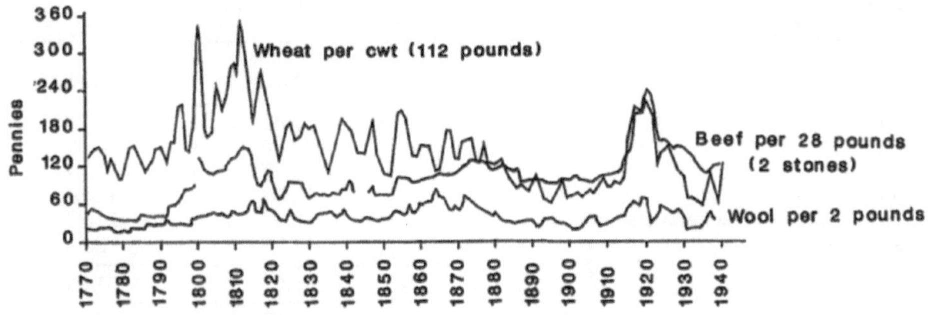

Figure 9.3 Wheat, beef and wool prices in England, 1770–1940. Based on Grigg (1989) p. 19.

Although some animal products, especially wool and hides, were increasingly imported, it was only with the development of canning from the late 1860s and compressed air refrigeration from 1879 that imports of frozen beef and, especially, mutton from the Americas and Australasia began to threaten the livelihood of stock farmers, though poor quality and loss of flavour in refrigeration and storage restricted competition to the poorer end of the market. Therefore, although prices moved against wool and mutton producers, livestock farming in general survived the post-1870 depression better than arable.

Changing relative values had considerable impact on cropping (Tables 9.5 and 9.6). Within arable rotations every opportunity was taken to increase stock food, though the parallel fall in the cost of imported fodder grains and oil cake focused this chiefly on rotation roots and green crops. Where land could be converted to grass, as on many heavy soils, permanent grassland and extended grass leys became the basis of livestock farming, effectively shifting the economic margin of arable cultivation eastwards and onto lower land. The number and density of stock increased substantially in most areas

Table 9.5 British agricultural output, 1831–1951

	Percentage of total agricultural output[1]					
Type of product	1831	1856	1867–73	1908	1925	1951
All crops	54.5	48.7	41	32.4	31.4	30.0
Field crops	–	–	36	29.5	20.5	20.4
Horticulture	–	–	5	2.9	10.9	9.6
All livestock	45.5	51.3	55	66.0	67.3	68.1
Milk and dairy	–	–	15	19.1	25.6	31.5
Fatstock	–	–	37	43.7	35.0	22.7
Poultry and eggs	–	–	3	3.2	6.7	13.9
Other (fruit etc.)	–	–	4	1.6	1.3	1.9

[1]For GB (1831–1873), England and Wales (1908 and 1925) and UK (1951)
Source: Grigg (1989), Table 2.7

Table 9.6 Gross output of agriculture in the United Kingdom, 1867–1903

Product	Value of output (£ million) 1867–69	1870–76	1894–1903
Wheat	35.38	27.56	7.72
Barley	16.78	17.56	9.43
Oats	10.54	9.07	8.07
Potatoes	14.02	13.82	11.34
Hay and straw, fruit, vegetables	20.11	19.40	21.75
Other crops	7.34	7.58	3.64
Arable (subtotal)	104.17	94.99	61.95
Beef	34.90	45.67	42.05
Mutton	25.92	30.51	25.20
Pig meat	18.60	22.95	19.13
Horses	1.50	2.00	3.00
Milk	33.78	38.51	43.56
Wool	7.49	8.27	3.24
Poultry and eggs	4.57	6.96	10.00
Livestock (subtotal)	126.76	154.87	146.18
Total	230.93	249.86	208.13
Index (1867–69 = 100)	100	108	90

Source: T. W. Fletcher (1961) reprinted in P. J. Perry (ed) (1973) p.54

except those least suited to grazing such as the light lands of East Anglia or the heavy clays of Essex. Improvements in drainage and grassland management from mid-century added to the shift to stock, a process assisted by increasing demand for stock products in a growing urban industrial market with rising real wages and improving diets. In particular, consumption of milk and dairy products increased substantially. New methods of factory cheese-making and, in the 1880s, of processing butter favoured low-cost producers, but these were challenged by imports from Denmark and Australasia by the late nineteenth-century. The growing market in liquid milk – marked by a *per capita* increase in daily consumption from one-quarter to one-third of a pint between 1878 and 1904 – was entirely supplied by home producers. The best quality milk came from town dairies: Liverpool's 'cow-keepers' were often dalesmen from the northern Pennines; the Welsh were prominent in the London milk trade. Dairying often dominated pastures around cities and, as the barrier of distance in transport of this perishable commodity was overcome by rail transport, the 'milk-shed' of large markets was substantially extended. By the later nineteenth century London's milk supply was drawn not only from farms in the Home Counties and the Weald but also from East Anglia and the south Midlands, north Wiltshire and, later, South West England, especially in the lower-yield winter months. The traditional cheese-makers of Cheshire's permanent grass pastures began to supply 'train-milk' to Liverpool and Manchester, and stockmen in northern dales shifted from a primary concern with fattening to milk supply for east Lancashire and west Yorkshire where many small farms also had their own milk rounds.

Changing patterns of diet, improved transport and technical improvements in food-processing provided an increasing market for vegetables and fruit from high-yielding soils within reach of the big markets. By the 1890s canning, jam-making and preserving were using both home and imported products for processing at large ports – Hartleys in Liverpool, Keillers at Dundee: but many manufacturers were in specialist fruit or vegetable districts – e.g. Wilkins at Tiptree, Essex, and Chivers at Histon, Cambridgeshire. In many established arable districts – such as the Fenland, south west Lancashire, the Lothians and coastal Ayrshire – cash crops of potatoes, green vegetables and legumes were added to previously grain-dominated rotations and some areas – including such specialist fruit-growing districts as the Vale of Evesham, the Wisbech district of the Fens and the Carse of Gowrie – incorporated market gardening, with extensive areas of glass-house cultivation in such areas as the Hull valley, or production of highly specialized crops such as early rhubarb (near Leeds) or celery (on mosslands near Manchester).

Improvements and innovation

Specialization reflected the farmer's ability to adapt to the market by investment in improved techniques and equipment. From the 1840s the flow of capital into farming was helped by the spread of poor relief over Poor Law Unions under the 1834 Act, the Commutation of Tithes in 1836 and recovery of prices. More stable conditions saw grain prices rise by 0.2 per cent p.a. 1846–70 and those of stock products by 0.7 per cent p.a. James Caird, the advocate of 'High Farming – the Best Substitute for Protection', as he entitled his 1848 pamphlet – argued for maximisation of profit through high investment in land management, buildings and machinery. These brought marked increases in yields of crops and stock between the 1830s and early 1870s that were little improved upon over the next half century.

One major improvement came about through more effective drainage. On heavy claylands the high water table inhibited cultivation by restricting the working season and hindering the intake of nutrients by plants due to waterlogging. Although the precise benefits are difficult to quantify, 15–16 million acres – around half the cultivated land in England – was estimated to be in need of drainage in the mid-nineteenth century. Earlier attempts using tile or hollow drains had limited success (Chapter 4): traditional methods of ridging heavy lands by ploughing were often more effective. It was not until the perfecting of cheap, machine-made pipes (e.g. the Scragg's pipe) in the 1840s and the availability of £2m in loans under the Public Draining Acts of 1846 and 1850, and provision for borrowing under the Private Money Draining Act of 1849 (incorporated into the 1864 Improvement of Land Act) that underdraining of heavy soils took off. Some 4–5 million acres were tackled between 1846–76 affecting one-third to over one-half the arable in such areas as North East England, the Welsh border and parts of the east and south Midlands. While investment and take-up of underdrainage was earliest and greatest on large estates, many tenant farmers invested in or paid loan-interest on higher rents to finance it. In all, around £9m was loaned for drainage between 1846 and 1899: the estimated total cost of £27.5m was not far short of the £29m expenditure on all Parliamentary enclosure in England and Wales.

Although underdraining slackened with the decline in farm prices from the 1870s, and there was limited investment in it on grassland areas, contemporaries believed that it benefited yields of all crops on strong clays and loams. Hitherto these were less affected by new rotations and methods of cultivation than light land. Many retained old methods of cultivation. Fallows persisted on very heavy land such as Holderness and south Essex into the early twentieth century. Grain, with beans as a nitrogenous crop, remained the focus of much clayland arable though wider adoption of rest crops, including mangolds and brassicas such as kale, cabbage and kohl rabi, gave greater flexibility to clayland cultivation and increased stocking densities from mid-century.

Many large-scale 'improvers' were also 'fanciers' (whether of bloodstock, dogs, cattle or sheep) who improved stock through selective breeding and maintenance of stock books, but ordinary farmers took little care in breeding or purchase of stock even by the end of the century. However, improved fodder supply both from the farm and from cheap imported feedstuffs enhanced meat and milk yields and stocking densities from mid-Victorian times.

It is difficult to assess the relative contribution of underdraining, improved cropping and stock, and other changes in farming methods to increased yields and productivity. Fertilizers had an important contribution to make. The continuing folding of sheep on light soils, bigger flocks on upland and lowland grazing's, more muck from bullock yards and dairies and access to town waste – for example night soil from Manchester carried by barge in the reclamation of Chat Moss in the 1830s – increased the supply of natural manures. Animal by-products such as bones and bone-meal, hair, hoof and horn, and fish meal provided slow-release nitrogen essential for grain cultivation. Greater use was made of industrial wastes such as rags and shoddy and, from the 1880s, basic (phosphoric) slag from blast furnaces was heavily used on 'sour' land. Heavy liming was helped by cheaper transport and larger-scale working. Some of the biggest fertilizer inputs were imported natural potash, which from 1851 largely replaced kelp from the Highlands, and nitrates from home-produced coprolite and Peruvian guano, first shipped to Liverpool in 1835 and, with Chilean nitrate, a big import from the late 1840s. From negligible levels in 1840, consumption of nitrate increased to 34,000 tons by 1874 and general use of chemical fertilizers increased substantially by the end of the century.

Although arguably slow to take root in standard practice, there was a significant trend towards more scientific farming in which the role of soil chemistry was crucial. Despite pioneering efforts by Sir Humphrey Davy (e.g. in his Board of Agriculture lectures, 1803–13) it was the theoretical work of Justus von Liebig (in particular his 1840 book on *Chemistry in its Application to Agriculture and Physiology*) that laid the foundations of experimental work begun by Sir John Lawes at the Rothampstead Agricultural Research Station from 1843.

The means by and extent to which scientific farming was promulgated and adopted is uncertain. The spate of agricultural writing of the late eighteenth and early nineteenth century slackened in the 1820s and 30s but by the 1840s it was more scientific and more widely disseminated;

for example in the Journal of the Royal Agricultural Society (1839–) and via agricultural shows and exhibitions and farmers' clubs. To what extent a scientific, 'improving' mentality spread from advanced to ordinary farmers is debatable. The membership of societies widened, though it is doubtful whether many ordinary farmers regularly read farming books and periodicals. They were more likely to be influenced by requirements of leases and contact with other farmers at shows, clubs or on market days.

Although the use of machinery was advocated by agricultural improvers, effective mechanization was slow to develop and spread. There were substantial improvements in the quality and design of implements to suit particular soil conditions from the late eighteenth century: replacement of wooden by iron ploughs, harrows and rollers; perfecting of drills and horse hoes and their adoption in grain as well as turnip husbandry. All improved cultivation, while better carts and wagons reduced production and transport costs. Improved draught horses and their harnessing to horse gins for threshing and other purposes in the wheel-houses that became a feature of farms in arable districts increased energy efficiency. So did such humble innovations as the replacement of the sickle by the scythe for hay cutting and harvesting, or the use of turnip-slicers. As long as implement manufacture was a small-scale business of local smithies and foundries, the pace of change was slow. Cheap labour, not least in harvest gangs of women, children and itinerant Irish, gave little incentive to invest in machinery. Moreover, the machine-breaking of the 'Captain Swing' riots of 1830 reflected widespread opposition by labour to their use. As agriculture recovered that opposition faded. The Great Exhibition of 1851 was a splendid showplace for agricultural machinery, but general adoption as judged from catalogues, farm sales and inventories was limited before the High Farming of the 1850s and '60s.

Early attempts to mechanize harvesting date from the Scottish Bell reaper of the late eighteenth century, but modern reapers came after 1848 with large-scale production of the American McCormick reaper (patented in 1834). Hand haymaking continued until the adoption of horse-rakes and American-inspired cutters (which repaid their cost in a couple of seasons) from the late 1850s and hay-making machines were commonplace by 1880. Many developments awaited improvements in motive power. Plough horses had ousted oxen from all but the heaviest land but steam did not achieve general success in harvesting or ploughing, though the spectacular grubbing-up and ploughing by stationary steam engines of the remnants of Hainult Forest, Essex in 1851 showed its potential. Threshing – a slow, costly operation initially partly mechanized through de-husking on wheelhouse drums and hand-operated winnowing – saw steam power applied by Ransome's of Ipswich in 1842 after some thirty years of experimentation. Many of the large makers of agricultural machinery and steam engines were in the arable areas of eastern England (Chapter 10). Richard Hornsby, a Grantham blacksmith, started on threshing and winnowing machinery in 1815 then built seed drills and, from 1849, steam engines. The Lincoln firm of Rushton, Procter and Co. (amalgamated with Hornsby's in 1918), who were big manufacturers of steam engines from 1857, turned to oil and diesel engines in the late nineteenth century and, like other Lincolnshire and East Anglian firms, developed substantial international markets.

However, many farmers (especially on small family farms) could not afford mechanization. Harvesting was not generally mechanized until movement from the land and restriction of child labour (Chapter 8) together with rising wages made hand methods increasingly uneconomic. Although the 1880s depression saw fewer innovations, there was more widespread use of machinery: in 1871 only one-quarter of British grain was harvested by machine, by 1900 four-fifths was, and while it took 10 or 11 worker-days to produce one acre of corn in the 1840s by 1900 – using reaper, binder and steam thresher – it took only 7.5.

There was large-scale investment in farm buildings in Victorian times, particularly on large estates. Parliamentary enclosure had created many new farms (Chapter 4), though rebuilding on the new holdings was often spread over decades. By mid-Victorian times big farms sometimes had a boilerhouse and engine to drive equipment, process stock food or run a sawmill. Stock farms increasingly provided winter shelter for both milking and fatstock, as well as piggeries, hen-houses and the like. Big dairies were often very elaborate, with patterned tiles, special insulation and hygienic ventilation, such as that on the royal estate at Windsor built by Prince Albert in 1858. Though exceptional, such farms were part of a considerable rebuilding in the 1850s and '60s that, functionally and architecturally, used ideas of designers such as J. C. Loudon (1783–1818) and these provided both the farmer's home and his production unit – the rural equivalent of the factory as James Caird put it. While most still used local materials, cheap rail transport introduced slate, machine-made brick and cast-iron into the buildings and their equipping, especially on large estates. But most poorer farmers could neither afford new buildings nor improvements to existing ones: thus only with enforcement of the Milk and Dairies Order of 1926 was new stock accommodation provided on many small dairy farms. The decline in farming from the 1870s saw much neglect of both farm buildings and of labourers' cottages, which were already a major problem in the countryside.

Enclosure, reclamation and tenure

Most of Britain's open field arable and much of its common and waste land had been enclosed by 1830 (Chapter 4). A further 0.6 million acres of waste were enclosed under General Acts of 1836 and 1845 between 1845 and 1870 and transformed alongside earlier reclamations. On Exmoor, for example, it was only possible to reclaim peat moor and support Cheviot and Scottish Blackface flocks as well as Scotch, North Devon and Shorthorn cattle after deep ploughing in the 1840s, to break the iron pan, and eventually by draining and fertilizing. Long-lease tenant farmers helped complete the transformation in the 1850s and '60s, though the final cost of £6,000 to establish a 300-acre farm (and rents of 15s per acre) was expensive and in the depression of the 1880s and '90s many leases fell vacant. But the reclamation transformed a wilderness, planted a new village and parish (1856) of some

400 people, clothed the lower slopes with beech hedge and bank, and provided substantial shelter belts.

Exmoor epitomizes the reclamation of upland wastes. On many upland commons of Wales, Cumbria and the northern Pennines, family sheep farms (often rented from large estates) were extended. In Scotland, the spread of sheep raising and, later, deer forest on very large, absentee landlord estates was a major social and economic problem. Yet with the collapse of wool and, later, lamb prices, hill farming became a very marginal enterprise, profitable only during wartime and kept going through family labour and ancillary activities in quarrying, residual craft industries such as Welsh rural woollen mills and, later, forestry (Chapter 14). It was a way of life progressively eroded by depopulation in the later nineteenth century and, in many areas, virtually abandoned in the twentieth.

Other difficult environments which previously resisted complete reclamation included substantial lowland heaths on very light, highly leached and water-deficient sands and coarse gravels. In the Breckland of south-western Norfolk (Arthur Young's 'Poor Sands' region), traditionally sheep walk and rabbit warren, over 100,000 acres were enclosed between 1800–17. But transformation to mixed farming on large farms of big estates was completed only in the 1850s and '60s under extended Norfolk rotations with barley, rye or oats rather than wheat and up to five out of ten years in leys. Despite massive investment on Lord Iveagh's Elvedon Estate near Thetford half the cultivated area reverted to waste after 1870 or was afforested, mainly with exotic conifers, a foretaste of later Forestry Commission activities (Chapter 14).

On soils of intrinsically high natural fertility, such as lowland fen and marsh, it was a very different story. Despite draining from the mid seventeenth century (Chapter 4), extensive areas of the Fenland were waste, especially in south Lincolnshire. On drained peatland, as the surface was lowered through shrinkage and chemical wastage of organic matter under cultivation, banks and outfalls became increasingly vulnerable and there was widespread flooding. Only the gradual replacement of windmills by steam engines from the 1820s solved the problem, though 30 per cent of the Fenland's windmills were still in use in 1852 when centrifugal pumps were used to drain Whittlesey Mere (one of the deeper meres) and rapidly replaced the old 'lifting' engines and scoop wheels. In the 1840s the peat fens were a grain-growing area. The underlying clay brought up into the peaty top soil by ploughing and better drainage suited grain, and autumn-sown wheat often replaced spring oats and cole-seed. Grazing – a traditional Fenland pursuit – remained in the silt belt around the Wash but that changed with rail links and the extension from the 1850s of cash-crop farming to include oilseeds, mustard, woad (for dyes), pulses and potatoes, as well as wheat. Through crops competitive in the home market Fenland farmers weathered the post-1870s depression and developed intensive fruit and vegetable cash-cropping, market gardening, and, later, bulb cultivation in the Boston-Spalding area.

Near to the northern cities, the mosslands of south Lancashire offer a similar story of slow initial reclamation of their acid peat soils and rapid advances in farming from the 1830s to 1880s. The draining of Martin Mere was completed with the aid of steam engines in 1850 and, along with Chat and other smaller mosslands, became major suppliers of brassicas, specialist salad vegetables, potatoes and grain crops to nearby industrial areas.

The progressive switch to a market-orientated farming economy in an era of free trade was reflected in changes in tenure and size of farms and estates. Eighteenth-century enclosures had increased the numbers of small holders and had produced a three-fold structure of landlord, tenant farmer and landless labourers – the proportions of which varied substantially from one region to another – and between different systems of farming: England and, especially, Scotland were dominated by large estates but the balance between tenant and family farms was more equal in Wales. Extensive estates in grain-growing areas of eastern England and the Lothians had big tenant farms with many labourers; fatstock farms of the English Midlands were bigger than those of dairying areas; Welsh stock farmers had small areas of improved land, with access to commons, and depended on family labour in a low-income economy. There were great contrasts in size between tenancies on the vast sheep estates of the Scottish Highlands and the tiny crofts of the western coasts and islands.

Whereas changes in farming methods contributed to substantial falls in agricultural labour from mid-century (Table 9.2) the number of holdings increased, though many farmers went out of business in the post-1870s' depression. Although mixed, stock and cash-crop farmers adapted well, bankruptcies of grain producers in the late 1870s and early '80s and of small stock farmers in the early 1890s reflected poor harvests, increasing grain imports and the fall in wheat, wool, then meat prices. Great estates reached their peak of dominance in the 1870s when in England and Wales some 1700 of 270,000 landowners held over 43 per cent of the land and in Scotland the 25 largest owners held one-third of the land. At mid-century small farms held only one-fifth of the cultivated area on over three-fifths of the holdings. One-third of the land of England and Wales was farmed on holdings over 300 acres (8 per cent) reflecting the dominance of arable farming in south eastern Britain where capital inputs increasingly demanded substantial acreages for maximum viability. In contrast Victorian dairy herds and farms were often small, one factor in the greater incidence of farms of under 100 acres in parts of western Britain.

While very small units were largely removed from many areas by enclosure and in later nineteenth-century depopulation, small acreages typified intensive cash-crop production in the Fens and south-west Lancashire for example, and in early vegetable, fruit and market gardening production (e.g. in Kent, the Vale of Evesham and parts of South West England). From the depressed 1880s, there was a drive to increase small holdings: Acts to extend allotments (in 1882) and – under the pressure of Joseph Arch's 1885 compaign slogan 'three acres and a cow' – in 1887 empowered local councils to compulsorily purchase land for that purpose. But, while allotments were a valued feature of two-thirds of English villages, small holdings (of 1–5 acres) were limited.

In the Scottish Highlands the long conflict between lord and tenant going back to the break-up of the clan system after the '45 rebellion and the shift to a rental (or feu) system delegated to large tenants (often initially kinsmen of the laird) continued. With the decline of fishing and kelping the fragmented holdings of many coastal and island communities were no longer viable. As Victorian landowners (many absentees) sought to maximize profits on their estates through extensive sheep farming the termination of small tenures effectively completed the clearance of the glens. Many emigrated (Chapter 8) but unrest in the early 1880s produced a full investigation, of considerable cultural as well as economic importance, leading to the Crofters' Act of 1886. This empowered purchase of land for resettlement of beleagured Gaelic communities in the western Highlands and Islands. But limited tillage for oats and potatoes, high costs of transport in marketing stock products and of importing necessities and a decline in traditional supplements to income from fishing and textiles have seen continuing depopulation (Chapter 14).

Responses to agricultural change

The landscapes and economy of rural Britain were substantially reshaped during Victorian times. Despite substantial advances in 'new farming', many parts of Britain were still agriculturally backward in the 1830s and '40s. The stimulus of High Farming transformed grain growing and enhanced commercial meat then milk production, and was reflected in investment in soil improvement, stock, machinery and farm buildings. These had greatest impact on the bigger estates and large tenant farms of lowland arable and grazing districts and on milk and vegetable producers near the urban markets. Remote, small, under-capitalized farms were still decades behind the leaders.

By 1874 the High Farming boom was over. The depression to the 1890s was initially largely a grain crisis that most affected high-cost clayland cultivators (Figure 9.4). Large-scale cereal growers on lighter soils could adjust by growing more fodder, keeping more stock or producing alternative cash crops (Table 9.6). Cheap imported maize and cattle cake may have damaged the arable side of mixed farming but were advantageous to dairying (Figure 9.5) and fatstock production. Farmers growing perishable commodities coped well: those with neither opportunity nor ability to adapt went under. For example, conservative corn farmers on heavy land in south Essex were badly hit, notwithstanding an apparently large market for stock products in nearby London; Scottish farmers coming onto the same land managed to make a living. Small farms in both arable and hill districts were worst affected, many having neither the resources nor skills to adjust.

Land use changed substantially from the late 1870s. Marginal land went out of cultivation on a massive scale: some went over to forestry (e.g. in Breckland); some to sporting estates (whether on northern grouse moors or Scottish deer forests); some tumbled down to weedy pasture and scrub. There was a good deal of conversion of arable – especially cereal – to grassland with permanent pasture on heavier soils and longer temporary grass leys within arable rotations (Figure 9.6). The grassland/arable boundary was displaced eastward, as reflected in the changed pattern of land use and stock by the early twentieth century (Chapter 14).

Measures of agricultural prosperity are relative but changes in tax assess-

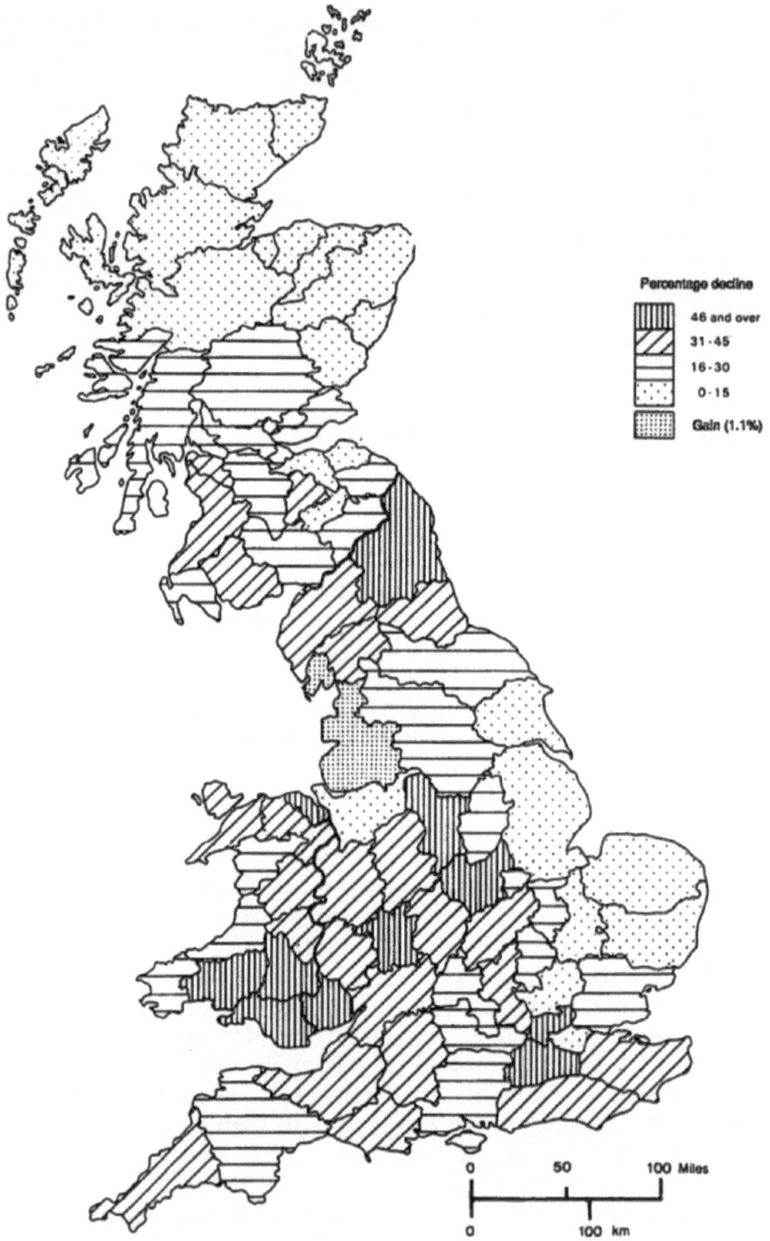

Figure 9.4 Decline in the corn acreage (wheat, barley, oats, rye) of Great Britain, 1870–1914.
Source: Pope (1989) p. 9.

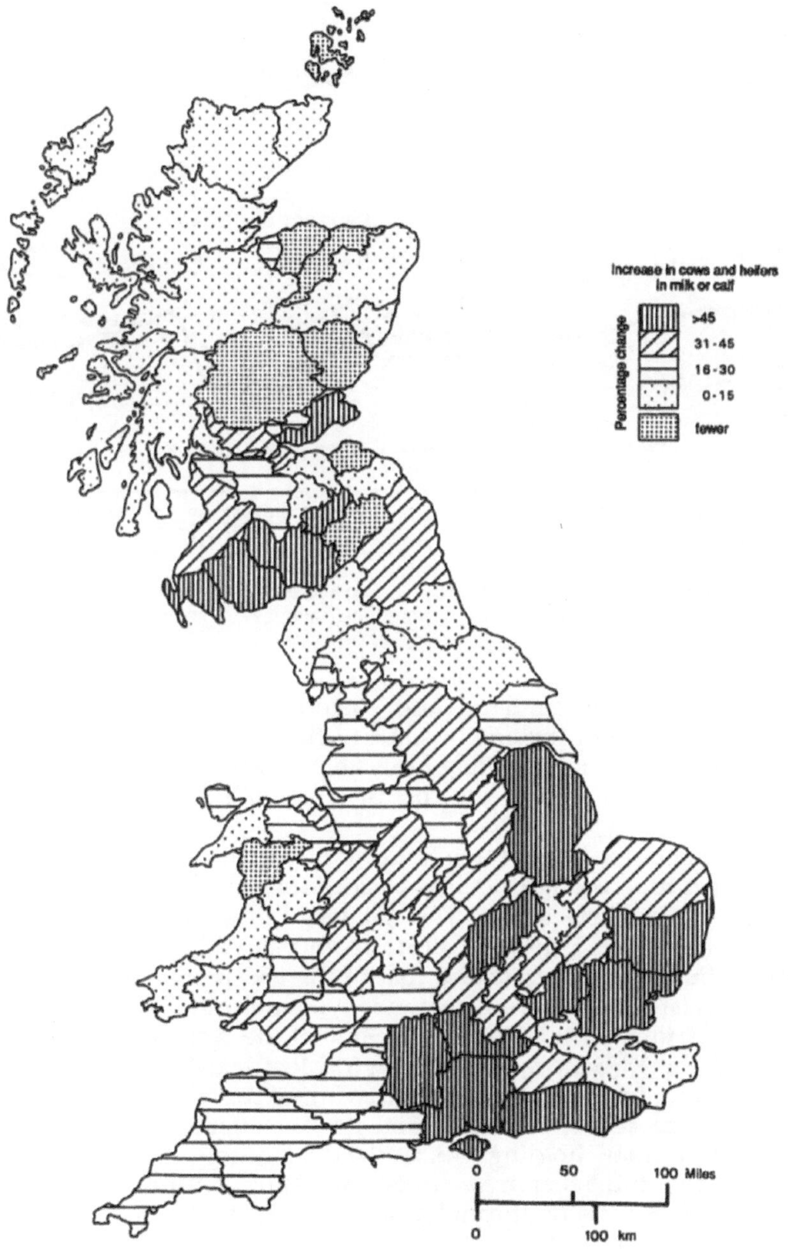

Figure 9.5 Changes in milking herds in Great Britain, 1870–1914.
Source: Pope (1989) p. 10.

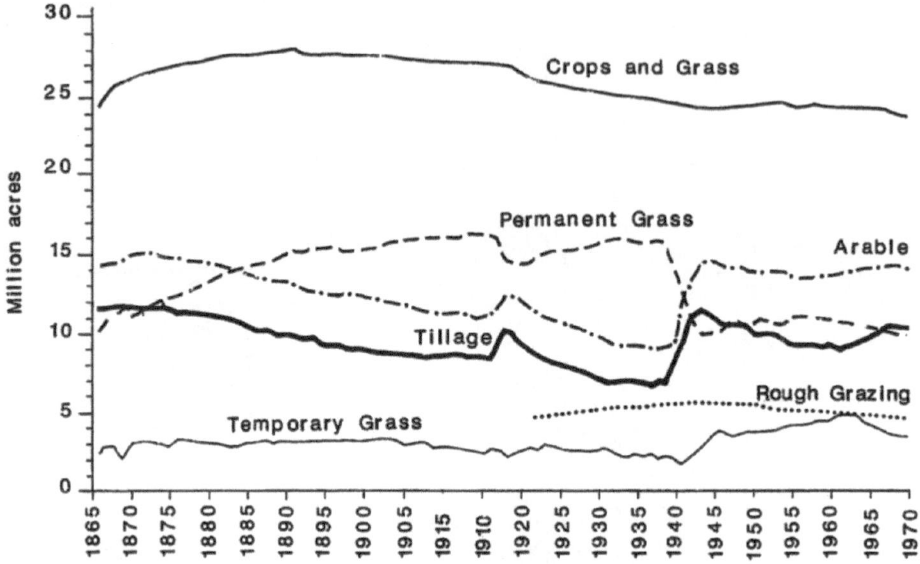

Figure 9.6 Major land-use changes in England and Wales, 1866–1970.
Source: Ministry of Agriculture, Agricultural Statistics, England and Wales.

ments on land rents (Figure 9.7) clearly demonstrate the contrasting fortunes of corn-growing midland and eastern areas and stock districts, though the low level of income in some of the latter, especially in Wales, must be taken into account. Falling rents helped farmers to some extent. So too did economies in labour, the cost of which fell by some 15 per cent, 1871–91.

Despite the endorsement of free trade after 1846, the problems of the late nineteenth century brought many basic issues concerning agricultural policy into the political arena. Though less influential than before the Reform Acts (Chapter 7), the agricultural lobby was still powerful, especially the aristocratic landed interest and that of the prosperous tenant farmers, many of them large cereal producers who were prominent witnesses to the Royal Commissions on the Depressed State of the Agricultural Interest of 1879–82 and the 1893 Royal Commission. Most legislation of the period favoured tenants (for example the outlawing of restrictive leases in Agricultural Holdings Acts of 1875, 1883 and 1906) or offered palliatives such as allotments and small holdings to landless labourers. But the plight of farming failed to shift belief in free trade and state aid came only in the First World War and, especially, through price subsidies and protection from the 1930s (Chapters 12 and 14).

Conditions of life in the countryside

The tendency for studies of nineteenth-century Britain to concentrate on urban life and neglect the countryside reflects a time of unparalleled industrialization, urbanization and and unprecedented urban problems.

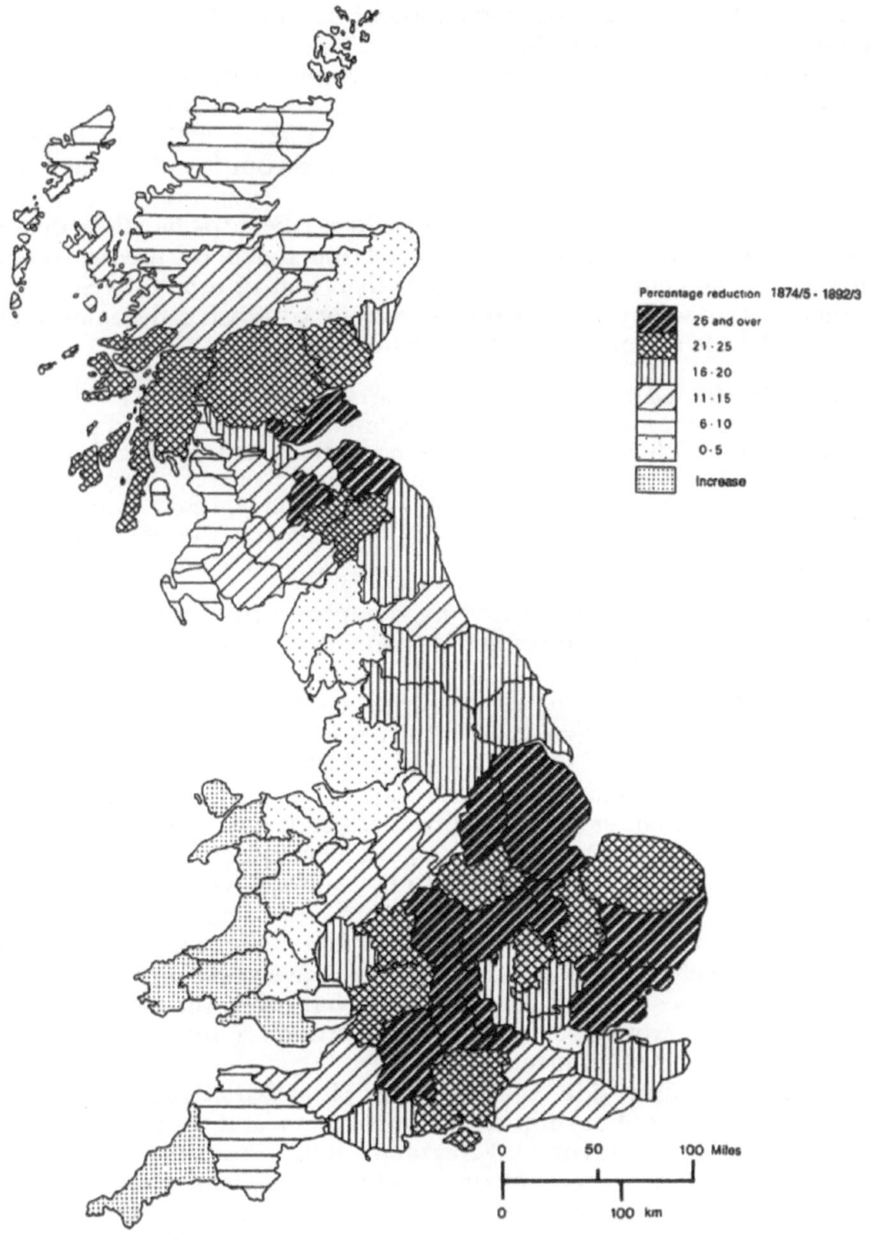

Figure 9.7 Changes in income tax (schedule B) assessments on land rents, 1874/5 to 1892/3.
Source: Pope (1989) p. 7.

Yet, in 1851 nearly half the population of Britain lived in rural areas and many more had been born in the countryside or had experienced rural life. Indeed, it can be argued that for most of the nineteenth century a rural view of the world continued to exert a significant influence in Britain. Although successive Reform Acts redistributed power after 1832, much political power and personal wealth remained in the countryside until the late nineteenth century.

Two further myths about rural life should also be dispelled: first, that rural life was in some way separate and distinct from that of the towns; secondly, that life in the countryside was easier than that of urban dwellers. Such commentators as Engels (1845) misleadingly contrasted the images of an idyllic rural life and the horrors of urban living:

> They did not need to overwork; they did no more than they chose to do and yet earned what they needed. They had leisure for healthful work in garden or field, work which, in itself, was recreation for them, and they could take part besides in the recreations and games of their neighbours . . . They were, for the most part, strong, well-built people . . . Their children grew up in the fresh country air, . . . while of eight or twelve hours work for them was no question'
>
> (F. Engels, 1972 edition p. 38).

Rural life in Britain had never been separate from the towns and, as nineteenth-century urbanization developed, the interconnectedness of countryside and town became stronger and more obvious. Even in the 1830s few well-settled areas were without urban contact and by mid-century few rural dwellers had no contact with the nearest market town: by the 1890s even upland Wales and the Highlands and Islands of Scotland were being integrated socially and economically into a regional system focused increasingly on the larger towns. Connections between countryside and town took many forms: most obviously expanding transport networks provided initially by turnpiked roads and then by railways which, by 1890, linked most villages into a complex and comprehensive transport system; through economic dependence of towns on rural labour; through rural to urban migration which could lead to family links between countryside and town and in some cases through the rural-based but urban-financed putting-out industries; and through social interaction between countryside and town at fairs, markets and other meeting places.

Life in the countryside could be every bit as harsh as that in towns: a combination of poor housing, lack of employment and poor social prospects frequently impelled townward migration rather than any specific urban attractions. Although undoubtedly built to lower densities, ameliorating the consequences of poor sanitation and associated disease, the density of occupation of rural housing was often as high or higher than that in towns. High natural increase in rural areas mostly offset migration losses up to the 1840s, rural population densities continued to increase and in many rural areas the housing supply expanded more slowly than population: indeed some large landowners demolished cottages and took less responsibility for housing their agricultural labour force. Many rural parents brought up eight or more children in tiny two-room cottages. Overcrowding and lack of privacy encouraged many young adults to migrate to towns in the nineteenth century where it was usually possible to rent a room; lack of housing had often forced young adults to delay marriage and remain in the parental home in the late eighteenth century.

Although the quality of rural housing varied greatly, for the very poor it was often worse than its urban counterpart. Increasingly, urban housing had proper foundations, solid walls and slate roofs; in some cases good quality housing filtered down to the poor. In contrast much rural housing was severely substandard when first built. Most landowners accepted little responsibility for the provision of decent housing and, even in more prosperous areas such as North West England, cottages were frequently small, cold and wet. In southern England, where there was more abject rural poverty, cottages often had mud walls, earth floors and neglected thatch which provided little protection from the elements. In remote districts of Wales and Scotland conditions were often worse: the traditional 'black house' of western Scotland remained a single room built of rough stones with a turf roof (Chapter 4).

Such conditions persisted until at least the 1850s but, during the later nineteenth century, housing gradually improved as out-migration lessened population pressure on the countryside and sanitary and housing reforms began to percolate into rural areas. On large estates and big arable farms there was considerable rebuilding from the 1860s, but only in richer areas, often close to large towns particularly in northern England, did significant improvement occur before the 1890s. Nevertheless, not all rural housing was bad: surviving nineteenth-century dwellings include not only good-quality homes of landowners, farmers and artisans, but well-built estate cottages and good-quality late eighteenth-century dwellings of rural factory workers.

For many rural families poor housing was combined with acute poverty. Although, by the 1830s, many rural areas were beginning to emerge from the worst rural distress of the agricultural depression, and rural protests, such as the 'Captain Swing' riots of the 1830s in southern England, were not substantially repeated, rural wages still remained low and highly variable from one area to another. James Caird's survey of agricultural wages in England in 1851 found variations from 13–14s per week in the West Riding, Lancashire and Cumberland to only 7–8s per week in such southern counties as Gloucestershire, Wiltshire, Dorset, Berkshire, Suffolk and Cambridgeshire. Northern wages were higher because of the greater prosperity of mixed and pastoral areas compared to wheat-growing counties and, particularly, because of competition for labour from industrial towns where wages were generally higher. In southern England counties close to London (Sussex, Essex, Hertfordshire) also had higher rural wages (9–10s per week). In the second half of the century agriculture around London also became more varied and prosperous because of the growth of market gardening, cash cropping and milk production for the urban market.

Rural industrial workers were usually rather better off. In such areas as the south Pennines survival of a dual farming-weaving economy gave some protection against poverty though, as the textile industry became increasingly dominated by factory processes, the distress of rural textile workers became acute and well-documented. Standards of living of both agricultural and factory workers in the countryside could also be affected by the continuation of the truck system of payment in kind. Although most agricultural labourers were, after 1830, independent in that they lived away from the landowner and paid rent, they were still highly dependent upon patronage. In northern England especially, agricultural labourers were usually hired for only short periods at hiring fairs and landowners were still liable to give them part of their payment in kind (i.e. farm produce). This lessened independence and reduced real incomes, already lower than those in industry.

Coalmining communities were particularly distinctive: the chapel; the workmen's club (or inn); later, the Colliery Welfare with its hall and sports teams; pigeon racing; allotments (and prize leeks); colliery bands and choral singing (especially male voice choirs). All were part of a male-dominant society. Mining villages were isolated and limited in facilities, and had a distinctive pattern and structure. Following the boom and decay of the mining cycle, miners and their families moved on to new pits in adjacent or, sometimes, distant areas. There were thus specialist intra- and inter-regional flows of labour, although much of the workforce was recruited from the miners' own large families.

The effects of rural poverty was also seen in malnutrition and associated ill-health. A survey of 1863 showed that most English rural labourers relied heavily on a diet of bread and potatoes, with meat consumption varying from season to season and area to area, though men were generally better fed than the rest of the family. Even so, the food supply in the countryside was rather better than that available to the urban poor: it was fresher and there were more opportunities to supplement it informally or illegally from gleaning, fishing or poaching, or from the cottage garden.

The social composition of rural areas also changed during the period 1830 to 1890: selective rural out-migration removed many younger and more active members of the community, but areas close to towns began to experience urban to rural movement of rich families seeking a house in the countryside. Commuter villages grew around such cities as Leeds, Manchester and especially London in the nineteenth century, particularly where there were good rail connections. Rural resort areas also began to be exploited. Thus from the late nineteenth century, Windermere became a centre of the invasion of the Lake District for recreation. This was especially true of Manchester merchants who could afford to establish second homes around the fringes of Lake Windermere and elsewhere. These trends, though more dominant in the period after 1890, represent a new phase in the development of perceptions of the countryside. While the reality of rural life in the nineteenth century was, for many, harsh and unpleasant, the image of rural idyll had, by the 1890s, become firmly implanted as a middle-class vision of the countryside which was increasingly imprinted on rural areas through residence, landownership and conservation movements.

Bibliography and further reading for Chapter 9

Many of the references for Chapter 4 also cover the whole nineteenth century. Only the more important references are repeated.

On agricultural change see:

ATKINS, P. J. 'The retail milk trade in London, 1790–1914' *Economic History Review* **33** (1980), pp. 522–37.

BRYDEN, J. and HOUSTON, G. *Agrarian change in the Scottish Highlands* (London: Martin Robertson, 1976).

CAIRD, J. *English agriculture in 1850–51* (1854, London: Cass, 2nd edition, 1968).

BROWN, J. *Agriculture in England: a survey of farming, 1870–1947* (Manchester: Manchester University Press, 1987).

COLLINS, E. J. T (ed.) *Agrarian history of England*, vol. vii, *1850–1914* (Cambridge: Cambridge University Press, forthcoming).

ERNLE, LORD, *English farming past and present* (London: Heinemann, 6th edition, 1961).

GRIGG, D. *English agriculture: an historical perspective* (Oxford: Blackwell, 1989).

JONES, E. L. *The development of English agriculture, 1815–1873* (London: Macmillan, 1968).

KAIN, R. J. P. and PRINCE, H. C. *The Tithe Surveys of England and Wales* (Cambridge: Cambridge University Press, 1985).

ORWIN, C. S. and WHETHAM, E. H. *History of British agriculture, 1846–1914* (Newton Abbot: David and Charles, 2nd edition, 1971).

PHILLIPS, A. D. M. *The under-draining of farmland in England during the nineteenth century* (Cambridge: Cambridge University Press, 1989).

SYMM, J. A. *Scottish farming past and present* (Edinburgh, Oliver and Boyd, 1959).

TROW-SMITH, R. *A history of British livestock husbandry, 1700–1900* (London: Routledge, 1959).

WALTON, J. R. 'Mechanization in agriculture' in Fox, H. S. A. and Butlin, R. A. (eds) *Change in the countryside* (Oxford: Institute of British Geographers, Special Publication **10**, 1979).

On life in the countryside see:

ARMSTRONG, A. *Farmworkers: a social and economic history 1770–1980* (London: Batsford, 1988).

BRIGDEN, R. *Victorian farms* (Marlborough: Crowood, 1986).

CHARLESWORTH, A (ed.) *An atlas of rural protest in Britain* (London: Croom Helm, 1982).

EVERITT, A. *The pattern of rural dissent in the nineteenth century* (Leicester: Leicester University Press, 1972).

GAULDIE, E. *Cruel habitations: a history of working-class housing 1780–1918* (London: Unwin, 1979).

HORN, P. *Labouring life in the Victorian countryside* (Dublin: Gill and MacMillan, 1976).

MARSHALL, J. D. and WALTON, J. *The Lake Counties from 1830 to the mid-twentieth century* (Manchester: Manchester University Press, 1981).

MILLS, D (ed.) *English rural communities* (London: Macmillan, 1973).

MILLS, D. *Lord and peasant in nineteenth-century Britain* (London: Croom Helm, 1980).

MINGAY, G. *Rural life in Victorian England* (London: Heinemann, 1976).

MINGAY, G. (ed.) *The Victorian countryside* (London: Routledge, 1981).

THOMPSON, F. M. L. *English landed society in the nineteenth century* (London: Routledge, 1963).

WOODFORDE, J. *Farm buildings in England and Wales* (London: Routledge, 1983).

10

Industry and industrialization from the 1830s to the 1890s

The structure and growth of the national economy

The Victorian era was an age of coal and iron. Mechanization and steam power transformed transport through the development of the railway system and, from mid-Victorian times, of marine transport. While many goods were still largely hand-made, steam-powered machines with standardized and eventually, in some cases, semi-automatic systems of production achieved substantial improvements in productivity and lowering of unit costs in the textiles industries, heavy industry and engineering. By late-Victorian times they began to permeate some sectors of consumer industry (for example clothing and footwear, furniture-making and food-processing).

Britain's industrial leadership was reflected in its domination of a progressively extending world economy: international trade expanded at over three per cent per annum throughout the greater part of this period, averaging 4.6 per cent between the 1840s and 1870s, and falling back only in the Great Depression of the 1880s. Despite an adverse trade balance throughout, earnings from overseas and other 'invisible' trade (in services and shipping, for example) generally kept the annual balance of payments in surplus, notably so from the mid-1850s. The value of manufactured exports (Table 5.2) rose rapidly during early- and mid-Victorian times, more than compensating for substantial increases in imports of food and raw materials (Table 5.1). Nevertheless, as the competitiveness of British industry was challenged from the late 1880s, especially by the USA and Germany, the value and proportion of imported manufactures grew substantially, especially of luxury goods and products of some newer industries. Britain continued to depend largely on exports of textiles, iron and steel products, machinery and, increasingly, coal, but by the 1890s foreign competitors were making greater progress in trade in chemicals, newer types of machinery and electrical goods.

These trends are reflected in changes in the basis of the nation's wealth and the structure of its labour force. The value of land and farm buildings fell to less than one-quarter of the national capital by the late 1880s, 40 per cent

DOI: 10.4324/9781003174783-10

Table 10.1 Structure of the National Product of Great Britain, 1801–1955

Sector	Percentage of the total national income in year							
	1801	1831	1851	1871	1891	1901	1924	1955
Agriculture, forestry and fisheries	32.5	23.4	20.3	14.2	8.6	6.1	4.2	4.7
Mining, manufacturing and building	23.4	34.4	34.3	38.1	38.4	40.2	40.0	48.1
Transport, trade and income abroad	17.4	17.4	18.7	26.3	29.8	29.8	35.0	24.9
Government, domestic and other services	21.3	18.4	18.6	13.9	15.1	15.5	14.4	19.2
Housing	5.3	6.5	8.1	7.6	8.1	8.2	6.4	3.2

Figures for 1924 and 1955 include Northern Ireland
Sources: Dean and Cole (1967), Table 76; Mitchell and Deane (1962), p. 366

less than half a century before, and the agricultural labour force fell from one-quarter in 1831 to only one-tenth by 1891. By then manufacturing, industry, trade and transport employed over two-thirds of the workforce and accounted for rather more of the national product (Table 10.1).

In a workforce that nearly doubled between 1841 and the end of the century (Table 10.2), industry continued to be dominated by textiles, metal manufacture and engineering, and mining. By 1901 nearly 1.5 million worked in textiles, mainly in fully mechanized mills, well over one million in metal manufacture, machine-tool making and vehicle manufacture, and almost a million in mining and quarrying (Table 10.3). Between them, these sectors provided one-quarter of the workforce, one-half of all industrial production and three-quarters of Britain's exports.

Continuous growth of 2.2–3.3 per cent in the national product of Victorian Britain was accounted for principally by sustained growth of between 2.7 and 3.5 per cent per annum in manufacturing, mining and building, and in trade and transport (2.8–3.7 per cent), though by the late nineteenth century it was outstripped by the USA (4.5 per cent) and Germany (2.8 per cent). Moreover, the steadily increasing output per head of the industrial workforce (1–1.3 per cent p.a.) was far behind that achieved in the USA (1.9–3.2 per cent p.a.) and in the 1880s and '90s in Germany (1.7 and 2.1 per cent p.a).

Much of Britain's increasing industrial output was achieved through greater mechanization of a traditionally trained workforce in established industries, rather than by application of science and technology to new industries. Nevertheless, notable innovations in engineering gave Britain a strong and competitive international trade in shipbuilding, locomotives and railways, heavy engineering (e.g. bridge building), and manufacture of agricultural and industrial machinery.

Technological advance and progressive mechanization had considerable implications for the organization of many industries: new skills of machine-

Table 10.2 Labour force participation rates in Great Britain 1841–1951

	MALES Labour Force (millions)	MALES Working-age population (millions)	MALES Participation rate	FEMALES Labour Force (millions)	FEMALES Working-age population (millions)	FEMALES Participation rate	TOTAL Labour Force (millions)	TOTAL Working-age population (millions)	TOTAL Participation rate
1841	5.09	6.69	76.1	1.82	7.08	25.7	6.91	13.77	50.2
1871	8.22	9.33	88.1	3.65	10.09	36.2	11.87	19.42	61.1
1891	10.01	12.03	83.2	4.49	13.06	34.4	14.50	25.09	57.8
1911	12.58	13.45	91.8	5.22	14.79	35.3	17.81	28.24	63.0
1931	14.79	16.34	90.5	6.27	18.32	34.2	21.05	34.66	60.7
1951	15.65	17.86	87.6	6.96	20.05	34.7	22.61	37.91	59.6

The Labour Force (in millions) for 1841–91 is the total occupied population excluding all 'unoccupied' (from whatever cause); for 1911–51 it is the 'economically active population' including the unemployed.
The total working-age population (in millions) is that over 10 years for 1841–1911, over 14 years in 1931 and over 15 years in 1951.

$$\text{Participation rates} = \frac{\text{total labour force}}{\text{population of working age}} \times 100$$

Source: Mitchell and Deane (1962), Table II.1

minding rather than individual craftsmanship; of engineers rather than wrights; of process workers (often on specific parts of a product) rather than sole creators of finished articles; bigger units of manufacture, whether factory or workshop, and larger firms. A newly structured workforce with a hierarchy from a skilled 'aristocracy of labour', maybe no more than one-sixth of the whole, with a mass of semi-skilled machine operatives and casual unskilled labourers brought new problems in industrial and social relationships between workers and employees and within the working classes. In 1851 factory industries employed 1.75 million (half a million in textiles) as against an estimated 2.5 million in traditional craft industries. By 1871 two million (half the industrial workforce) were employed in 23,346 factories as against a little over half a million in 106,988 workshops. However, handicraft workers were found not only in traditional rural industries (stockings and knitwear, gloving or straw hat-making) and individual crafts (tailors and dressmakers, smiths, bakers, building workers, etc) but also, increasingly, in the workshops and 'putting-out' systems of the urban 'sweated' trades – clothing, furnishing, box-making, toy-making. These trades continued to employ many (especially juvenile and women workers), but used little power, underlining the continuing importance of small-scale, unmechanized industry before the widespread adoption of electric power in the twentieth century (Chapter 15).

Nevertheless the scale of organization of both the productive unit and the firm increased in late-Victorian Britain. Most pre-Victorian firms were individual, often family, enterprises and with some exceptions – for example in smelting and processing (e.g. brewing) – production units were small (Chapter 5). Even large factories seldom employed more than a few hundred workers by early-Victorian times. While some big companies employed thousands of workers their workforces were often dispersed among hundreds of small workshops: at mid-century the Dowlais company employed some 7,000 workers at 18 blast furnaces; but the metal trades of Birmingham and cutlery industry of Sheffield largely operated in small, simply-equipped workshops and even the machine-tool trade was largely small-scale.

As mechanization and standardized production spread into the hosiery, knitwear, shoemaking and clothing trades, and into engineering, iron manufacture and shipbuilding, and as transport – especially railways – improved, so the size of companies and their productive units grew. Late-Victorian competition for markets eliminated many of the smaller, less competitive firms, producing notable concentration in the textiles, coal, chemicals and some processing industries. Integration of productive processes within industries and between large firms with complementary interests (e.g. coal and steel; related branches of chemicals; general engineering; food refining and processing) saw the emergence of the first modern industrial giants, though neither their number nor size of workforces should be exaggerated. The average workplace of the 1890s was still small, lightly mechanized and used little motive power, though the situation was beginning to change rapidly. For example, in footwear manufacture one of the most widespread of early-Victorian handicraft industries, large-scale factory production was firmly established by the 1880s: by 1895, 70 per cent of

Table 10.3 Occupational structure for selected industries by region, Great Britain 1851 and 1891

REGION	1851									1891								
	Mining	Metals	Engineering	Ship building	Metal manufacturing	Textiles	Clothing	Vehicles	ALL selected	Mining	Metals	Engineering	Ship building	Metal manufacturing	Textiles	Clothing	Vehicles	ALL selected
ENGLAND AND WALES South East (T) Percentage of GB	10.5 2.7	40.0 13.8	22.1 22.3	8.5 27.2	14.9 15.5	81.5 6.3	259.4 29.0	15.0 32.1	2278.7 24.3	17.4 2.3	63.9 12.2	54.4 18.1	15.7 16.7	26.4 15.1	77.4 5.6	357.8 31.3	24.3 28.2	3966.8 27.4
East Anglia	1.7 0.4	7.8 2.7	1.8 1.8	0.7 2.2	1.1 1.1	15.0 1.2	42.1 4.7	3.3 7.1	426.2 4.5	1.9 0.2	8.2 1.6	3.7 1.2	1.1 1.2	1.5 0.9	10.3 0.7	39.1 3.4	3.4 3.9	447.1 3.1
South West	53.6 14.0	21.7 7.5	4.6 4.7	3.6 11.5	3.8 4.0	56.1 4.3	114.2 12.7	3.8 8.2	991.3 10.6	34.8 4.5	19.6 3.7	10.6 3.5	4.5 4.8	4.4 2.5	34.0 2.5	105.1 9.2	6.9 8.0	1050.0 7.2
West Midlands	48.3 12.6	62.0 21.4	14.5 14.7	0.5 1.6	36.6 38.1	42.7 3.3	71.2 7.9	5.6 12.0	780.6 8.3	67.4 8.8	99.1 18.9	29.0 9.7	0.6 9.7	62.9 35.7	32.2 2.3	76.5 6.7	17.0 19.7	1171.7 8.1
East Midlands	18.8 4.9	12.7 4.4	4.0 4.0	0.3 1.0	3.5 3.6	125.2 9.7	73.3 8.1	4.3 9.2	659.6 7.0	69.1 9.0	27.1 5.2	15.1 5.0	0.9 1.0	3.9 2.2	105.8 7.6	119.2 10.4	7.8 9.1	930.4 6.4
North West	40.6 10.6	27.5 9.5	23.9 24.2	3.9 12.5	8.2 8.5	430.4 33.2	92.0 10.3	5.2 11.1	1226.6 13.1	101.3 13.2	76.4 14.5	57.0 19.0	10.3 11.0	13.0 7.5	588.6 42.4	139.7 12.2	10.3 12.0	2222.8 15.3
Yorkshire and Humberside	31.6 8.3	30.3 10.5	9.7 9.8	1.0 3.2	18.1 18.8	252.3 19.5	58.1 6.5	2.9 6.2	753.9 8.0	91.4 11.9	55.2 10.5	39.3 13.1	2.8 3.0	44.0 25.2	300.6 21.7	90.1 7.9	7.0 8.1	1312.0 9.1
North	56.1 14.7	19.5 6.7	5.2 5.3	6.8 21.8	2.8 2.9	22.1 1.7	42.6 4.8	1.9 4.1	481.3 5.1	136.5 17.8	54.2 10.3	33.5 11.2	30.1 32.0	4.4 2.5	16.2 1.2	51.4 4.5	2.3 2.7	882.3 6.1
South Wales	39.3 10.3	22.5 7.8	1.3 1.3	0.5 1.6	0.9 0.9	2.2 0.2	12.9 1.4	0.4 0.9	182.3 1.9	108.4 14.1	35.0 6.7	7.2 2.4	2.6 2.7	1.8 1.0	4.5 0.3	23.1 2.0	1.5 1.7	420.9 2.9
Mid and North Wales	27.9 7.3	9.0 3.1	0.9 0.9	1.1 3.5	0.9 0.9	10.6 0.8	24.1 2.7	1.2 2.6	315.4 3.4	36.7 4.8	13.8 2.6	2.0 0.7	2.0 2.1	1.0 0.6	8.2 0.6	23.5 2.1	1.1 1.3	347.0 2.4
SCOTLAND Strathclyde	34.4 9.0	18.6 6.4	6.5 6.6	2.0 6.4	2.4 2.5	135.1 10.4	44.2 4.9	1.0 2.1	479.8 5.1	61.6 8.0	49.4 9.4	33.7 11.2	18.5 19.7	6.8 3.9	84.9 6.1	56.0 4.9	2.2 2.6	768.4 5.3
East Central	16.0 4.2	11.5 4.0	3.7 3.7	1.2 3.8	2.4 2.5	94.2 7.3	34.5 3.9	1.0 2.1	415.7 4.4	34.7 4.5	17.7 3.4	10.2 3.4	3.7 3.9	3.2 1.8	90.5 6.5	36.8 3.2	1.4 1.6	564.7 3.9

Highlands	1.7 0.4	4.1 1.4	1.1 1.1	1.1 3.5	0.4 0.4	17.0 1.3	17.1 1.9	1.0 2.1	262.7 2.8	3.5 0.5	4.6 0.9	2.5 0.8	1.3 1.4	0.8 0.5	15.9 1.1	16.5 1.4	0.7 0.8	296.3 2.0
Southern	1.6 0.4	2.0 0.7	0.5 0.5	0.1 0.3	0.3 0.3	10.8 0.8	10.1 1.1	0.2 0.4	113.0 1.2	1.6 0.2	1.6 0.3	0.8 0.3	0.0 0.0	0.4 0.2	18.7 1.3	7.7 0.7	0.3 0.3	118.3 0.8
GREAT BRITAIN Percentage of all occupations	382.0 4.1	289.3 3.1	98.9 1.1	31.2 0.3	96.1 1.0	1295.5 13.8	895.8 9.6	46.8 0.5	9367.1 (38.4)	766.6 5.3	525.7 3.6	299.8 2.1	94.0 0.7	174.4 1.2	1387.6 9.5	1142.6 7.9	86.1 0.6	14499.7 (30.9)

Totals (T) are in 1,000s; regional percentages in each column are for each occupation as a proportion of those in Great Britain as a whole: the last line gives selected occupations as a percentage of all occupied in GB.

Occupations are selected for the following from the Registrar General's (Census) industrial orders: 2 – mining and quarrying; 6 – metal manufacture – iron, steel and non-ferrous smelting, founding and finishing, blacksmiths; 6 and 7 – mechanical and instrument engineering; 10 – shipbuilding and marine engineering; 12 – galvanised sheet, ferrous and non-ferrous objects, gold and silver smiths; 13 – all textiles; 15 – clothing and footwear; 11 – vehicles (rail, tram, cycle, coaches and carriages).

Regions comprise of the following counties:

ENGLAND AND WALES
South East: London, Middlesex, Kent, Surrey, Sussex, Hants, Berks., Oxford, Bucks., Bedford, Hertford, Essex
East Anglia: Cambridge, Huntingdon, Norfolk, Suffolk
South West: Cornwall, Devon, Somerset, Gloucester, Wiltshire, Dorset
West Midlands: Hereford, Shropshire, Stafford, Warwick, Worcester
East Midlands: Derby, Nottingham, Leicester, Lincoln, Northampton, Rutland
North West: Lancashire, Cheshire
Yorkshire and Humberside: West and East Ridings
North: North Yorks, Durham, Northumberland, Cumberland, Westmorland
South Wales: Glamorgan, Monmouth
Mid and North Wales: rest of Wales
SCOTLAND
Strathclyde: Argyll, Ayr, Bute, Dumbarton, Lanark, Renfrew
East-Central: East, West and Mid-Lothian, Clackmannan, Fife, Kinross, Perth
Highlands: Aberdeen, Banff, Kincardine, Moray, Caithness, Inverness, Nairn, Ross and Cromarty, Sutherland, Orkney and Shetland
Southern: Dumfries, Kirkcudbright, Wigtown, Berwick, Peebles, Roxburgh, Selkirk

Source: C. H. Lee, *Regional statistics of Great Britain 1841–1971* (1979)

England's 123,000 footwear workers were employed in factories, some of them very large, and by 1905, 82.5 per cent were factory workers.

Such concentration of production was accompanied by considerable geographical concentration of manufacturing with some notable specialization within particular industries. Reconstruction of the economy and population initiated in the early industrial revolution (chapters 3 and 5) transformed the regional map of industrial Britain. Comparative advantages in raw materials (as in the case of coal and heavy industry) and of inherited skills – as in the metal and engineering trades of the West Midlands, shipbuilding on Tyne and Wear, textiles in Lancashire and West Yorkshire, or pottery and ceramics in north Staffordshire – encouraged specialization (Table 10.3). By 1851 the North West was dominated by textiles (one-third of its workers) and engineering (one-quarter); the West Midlands had nearly two-fifths of Britain's metal workers and one-third of those in metal working and engineering. These emphases remained in 1891 and were strengthened in South Wales, northern England and western Scotland (Strathclyde) by growing dependence on mining, heavy industry and shipbuilding.

Concentration of individual industries or processes was even more striking, producing a dangerous dependency on a limited industrial base in many places; for example specialist Lancashire cotton towns; shipbuilding at Barrow, Sunderland or Greenock; above all in coal-mining communities. That dependency could be socially claustrophobic and eventually, as in the inter-war slump, economically disastrous (Chapter 15). In contrast, large markets progressively attracted a wide range of industries to the many major cities particularly of midland and southern Britain, including a substantial share of the new light industries of the late nineteenth century (Chapter 15).

The losers, to both the staple industries of the first industrial revolution and to the later-mechanized trades and new industries of the late nineteenth century, were the small-town and village craftsmen. Except where local specialization succeeded – as in Lincoln's agricultural and general engineering, or Kendal's and mid-Northamptonshire's boot-and-shoe, clothing and knitwear trades – a general and, from the 1870s, rapid loss of rural craftsmen and industry from the countryside and its small towns progressively impoverished rural life.

Coal mining and energy use

By the early 1830s coal production in Britain averaged nearly 32 million tons per annum. All surface coalfields were actively mined, but some smaller fields – such as Shropshire, North Wales and Cumberland – were in decline. Coal production was progressively focused on the larger coalfields as the development of deep mining enabled the concealed coal measures to be exploited. Large inland coalfields benefited as railway companies competed for the coal trade to distant markets such as London.

Technological innovation enabled pits to become bigger and more productive through improvements in mining techniques, particularly the long wall method of extraction. A key to such methods lay in improved ventilation by air pumps from the early nineteenth century and, from the 1830s, by steam-powered fans. Detection of and safeguarding against fire-damp, a major cause of explosions, through the improved Davy lamp (1816) made mines safer. Improved drainage by more effective steam pumps and developments in shaft-sinking and lining facilitated deep mining so that by the late 1830s coal was being raised in east Durham, for example, from depths of over 2,000 feet.

Underground transport by wagon way and pony, then by stationary steam engine linked to continuous rope-haulage (introduced at Wallsend on Tyne in 1844) greatly extended the range of underground haulage to the shaft. From the 1840s, better engines and winding gear brought coal to the surface in bulk. Such innovations lowered production costs and helped promote wider use of coal in industry, transport and domestically. As the price of coal dropped in real terms, that of wood fuel rose, so that coal became virtually the sole supplier of energy in most places from the 1850s (Table 10.4).

Steam horsepower increased by 37 per cent (from 1.67 to 2.28 million h.p.) between 1870 and 1907 in factories and workshops covered by the Factory Inspectorate and by two-thirds or more in mines, water- and gasworks, and grain and other mills. The greatest steam capacity was in textile mills, though their share fell from 36 to 19 per cent between 1870 and 1907; mining, the biggest user of steam engines in the early industrial revolution, had around one-quarter; smelting, metal working and engineering used a little over one-fifth. Indeed the biggest single increase in coal consumption was by demands of such public utilities as water, gas and then electricity systems developed in the growing cities, pushing up power capacity from a mere 2.6 per cent in 1870 to over one-sixth at the turn of the century.

Following William Murdock's work on gas lighting at the Boulton and Watt Soho Works in 1798, commercial exploitation for mill illumination, initially at Salford Mill in 1806–7, and in street lighting, pioneered in Manchester in 1805 and along London's Pall Mall in 1807, spread rapidly. By 1830 there were 200 gas companies in Britain which had increased to 500 by 1882. Domestic demand, fuelled particularly by the constant increase in population and housing, increased rapidly for home heating and, through the development of the gas industry, in cooking and lighting.

Through by-products of coke, coal-tar and its distillates coal became increasingly significant as an industrial raw material. High quality coking coals of south west Durham, South Wales and the South Yorkshire-Derby-Nottingham fields were a key locating factor in iron and steel works of early and mid-Victorian times. One of coke's by-products, coal tar, previously used only in caulking, provided a coal tar derivative, nitrobenzene, for synthetic aniline dyes developed from 1856 by W. H. Perkin, and one of the first-fruits of the industrial application of organic chemistry.

Finally, coal exports to European and more distant markets and in the bunkers of foreign ships became increasingly significant. By the late nineteenth century exports sustained thousands of jobs in coalmining and transport and contributed substantially to British trade.

The differential impact of changing utilization and supply of coal and of new technology led to substantial changes in regional patterns of coal

Table 10.4 Coal output and consumption in the United Kingdom, 1869–1950

	1869[1]		1887[1]		1913[2]		1929[2]		1932[2]		1937[2]		1950[3]	
	T(m.t.)	(%)	T(m.t.)	(%)	T(m.t.)	(%)	T(m.t.)	(%)	T(m.t.)	(%)	T(m.t.)	(%)	T(m.t.)	(%)
Output	107.4		162.1		287.3		257.9		208.7		240.4		216.3	
Export and foreign bankers	10.2	(9.5)	23.3	(14.4)	94.4	(32.9)	76.7	(29.8)	53.1	(25.4)	52.0	(21.6)	13.6	(6.3)
HOME USES														
Gasworks	6.3	(6.5)	9.5	(6.8)	16.7	(9.1)	16.8	(9.7)	16.4	(10.9)	18.2	(10.0)	26.2	(13.3)
Electricity	–	–	–	–	4.9	(2.7)	9.8	(5.7)	9.8	(6.6)	14.7	(8.1)	32.9	(16.7)
Railways	2.0	(2.1)	6.2	(4.5)	13.2	(6.8)	13.4	(7.7)	11.7	(7.8)	13.1	(7.2)	14.2	(7.2)
Mines	6.7	(6.9)	10.9	(7.9)	18.0	(9.3)	13.7	(7.9)	12.0	(8.1)	12.2	(6.7)	10.7	(5.4)
Blast Furnaces	16.3	(16.7)	14.3	(10.3)	21.2	(11.5)	14.5	(8.4)	6.5	(4.4)	14.7	(8.1)	9.9	(5.0)
Other Industrial and domestic uses	65.9	(67.8)	97.9	(70.5)	109.8	(56.9)	105.3	(60.6)	93.1	(62.2)	109.9	(59.9)	104.9	(53.3)
Total home consumption	97.2		138.8		183.8		173.5		149.5		181.8		198.8	
Surplus over home consumption	N.A.		N.A.		+ 9.1	(+4.7)	+ 7.7	(+4.2)	+ 6.1	(+3.9)	+ 6.6	(+3.5)	+ 3.9	(+2.9)

Column 2 of home uses gives percentages of: total available for home consumption in 1869 and 1887; actual consumption from 1913, less than the coal available for home use, giving surpluses shown in the last line

N.A. = not available

T(m.t.) = Total (million tons)

Sources: [1]Mitchell and Deane (1962), Tables 5–7; [2]Smith (1949), Table XXXIII; [3]Mitchell and Jones (1971), Tables 3–5

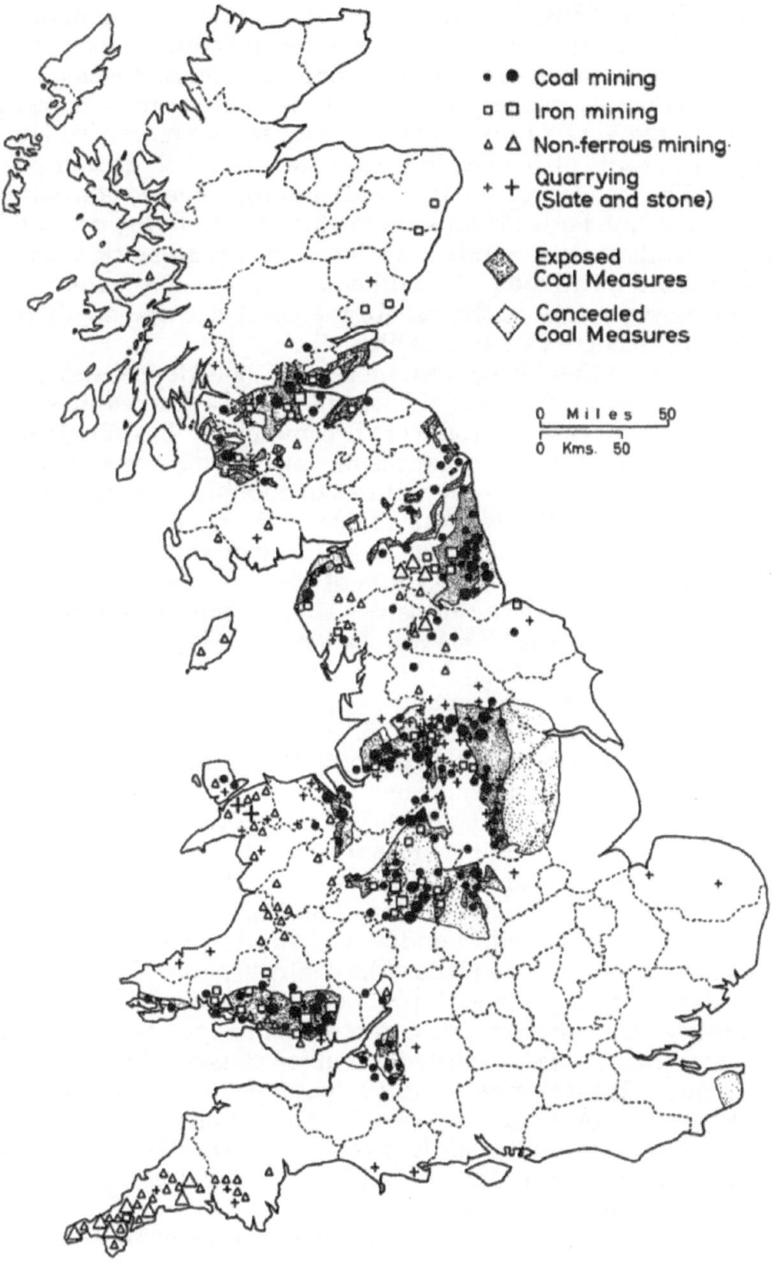

Figure 10.1 Employment in mining industries in Great Britain, 1851. Symbols are placed in districts where numbers occupied in specific industries were important; large symbols signify districts of special significance. Based on A. Petermann's map of occupations, Census of Great Britain, 1851.

production. By the 1850s all British coalfields, except the as yet undiscovered Kent field, were being mined using both open and concealed coal measures (Figure 10.1). New mines were bigger, more heavily capitalized and more productive. Whereas in the early nineteenth century few pits employed more than 100 workers, from the 1850s the bigger ones had several hundred, with annual outputs of 100,000 tons or more and much greater per capita production. The trend towards large-scale mining involved both landed coal owners, such as the Earls of Durham, the Marquis of Bute (in South Wales) and the Fitzwilliams (in South Yorkshire), and joint-stock colliery companies which often had interests in other industries such as the Wigan Iron and Coal Company and, in the case of the Quaker Pease family of south-west Durham, in chemicals and textiles.

As production shifted away from more accessible deposits to those with the greatest resources, the most efficient pits and best access to markets by rail and water, the leading producers became North East England, South Wales, Scotland, Yorkshire and Lancashire, all with more than 10 per cent of a rapidly increasing total output that reached its historic peak in the years just before the First World War (Table 5.4). Such producers as east Shropshire, north-east Wales, and west Cumberland waned as their reserves declined. Indeed the exhaustion of coal in early-exploited coalfields prompted the anxiety over the industry's abilty to meet foreseeable demand expressed in W. S. Jevons' *The Coal Question* published in 1865. The 1871 Royal Commission on the Coal Industry aired similar concerns, not least the difficulty and cost of mining in some areas reflected in falling output per miner between 1871 and 1891 (from 381 to 310 tons, nationally): geological problems led to a fall in production in South Wales from 425 tons per man in 1868–72 to 275 tons twenty years later.

By the turn of the century coalfields with larger reserves, deeper and less fragmented seams that were more amenable to mechanical cutting and uninterrupted working dominated production. The Yorkshire-Derbyshire-Nottinghamshire field which developed strong regional markets and increasing railborne trade with London and south eastern England experienced greatly increased production. In contrast parts of the Scottish and South Wales fields experienced production difficulties, although South Wales was in a strong market position as a major exporter of coal. Cardiff, a veritable 'coal metropolis', exported 30 million tons of steam coal in 1913 as against a mere 214,000 tons in 1850: and the quality of South Wales's anthracite gave it world-wide markets. North East England's varied coking, gas, steam and household coals had retained strong European as well as coastal markets, which it shared with Fife's steam coals, and supplied a growing range of regional industries.

In contrast, the decline of Lancashire reflected the exhaustion of coal in west Lancashire and Rossendale, and the difficulties of extraction from steep-dipping faulted seams along its southern margin. Staffordshire's famous 'thick clod' coals – the basis of the old Black Country coal-iron industry – were largely worked out by the 1870s: and by the end of the century its principal mines were in Cannock Chase and the Potteries.

The balance of production within individual coalfields shifted firmly to the more productive seams which were followed to increasing depth beneath the post-carboniferous rock cover. By the later nineteenth century, the larger collieries employed upwards of 1000 men, were much more highly capitalized and equipped, and supported larger settlements with distinctive long brick terraces and a wider infrastructure and range of services than older pit villages.

Textiles

By 1840 many sectors of the British textiles industry had reached maturity. East Lancashire's cotton factories and the Bradford area's worsted mills were to dominate their respective regional economies until their decline after the Second World War (Chapter 15). In other branches of the industry the transition to mechanized factory production was gradual and incomplete. While employment in textiles rose to a peak of 1.07 million men and 1.7 million women by 1911, the proportions in the workforce fell, after 1851, to under one-twentieth and one-sixth, respectively, reflecting greater productivity and a more diversified national economy.

Rapid take-up of steam power from the 1830s progressively concentrated woollen and worsted manufacturing on West Yorkshire. By mid-century 95 per cent of England's worsted looms and 87 per cent of its spindles were in factories in the Bradford – Halifax – Keighley area. Bradford, an archetypal specialist mill town which took up water-frame spinning only in 1794, had 67 steam-powered mills by 1841 and a population of 103,778 in 1851. Mechanization of wool combing from the 1840s – using improved Lister machines from 1851 – completed the industry's concentration largely at the expense of East Anglia. Norwich was left with a small and declining fine mixed cloth production using silk, a declining proportion of which was spun in the area. Other new materials, for example alpaca, were developed in the Bradford area where Titus Salt established not only a new industry but a new model community around his mill at Saltaire (1851); Samuel Lister specialized in velvet production at his Manningham mills (1873).

In contrast, woollens remained more diversified in character and location. The changeover to steam power and factory organization was more gradual: in 1851, there were as many factory workers as handloom weavers (c. 120,000). Two-fifths were in West Yorkshire while the West Country still accounted for one-eighth of woollen manufactures, much of it fine broadcloth. In the Scottish borders, specialization in tweed cloth and knitwear from the late eighteenth century expanded with new technology and, switching rapidly to factory production from the 1840s, it largely accounted for Scotland's one-sixth share of British machine spinning and one-fifth of weaving. By the end of the century, despite a relative decline in weaving, the area remained important for stockings and knitwear and its worsted yarn production was second only to that of West Yorkshire. The Scottish Highlands retained a scatter of small water-powered woollen mills, their most distinctive products from the 1870s being Harris tweed and hand-knit woollens.

The heavy woollen districts of West Yorkshire had 70 per cent of Britain's factory spindles and 75 per cent of its power looms, and were involved in a wide range of cloth manufactures, including the processing of rags in the mungo and shoddy trades of the Batley-Dewsbury area and carpet manufacture. Deserting previous upland sites, the industry was increasingly focused on the coalfield, especially in small towns of the Aire and Calder valleys in which vertically-integrated firms carried out all processes apart from dyeing. Leeds, originally a major focus of the industry and the location of the first large mills (for example Benjamin Gott's Bean Ing mill adopted power spinning in the early 1820s), became a more diverse industrial city and the commercial capital of the region, though Bradford dominated the wool trade.

Despite the growing dominance of West Yorkshire, woollen manufacture persisted in a number of areas with such distinctive products as Welsh flannel or Whitney blankets. In the 1890s the West Country's Stroudwater valley retained one-fifth of the industry, making fine broadcloth (for example for military uniforms) and baize in water-powered mills.

Knitwear and yarns – worsted, cotton and silk – continued to be made in small workshops and by cottage handloom weavers. In 1844 90 per cent of Britain's stocking frames were in the East Midlands where, in addition to the 56,000 operatives recorded in the 1851 census, there were perhaps as many part-time workers. The cotton famine during the American Civil War helped to boost production of worsted yarn in the area during a phase of mechanization of the hosiery and knitwear trades. Although the East Midland's first power-operated factory (Pagets of Loughborough) opened only in 1839, the switch to William Cotton's rotary power-frames (1864) led to rapid change. Hand spinning declined sharply from the 1870s with a mere 5,000 framework knitters by 1890 when 95 per cent of production came from the large power-operated mills of big firms such as Corahs of Leicester and Mundella of Nottingham.

Cotton, the dominant export industry of the early nineteenth century, was increasingly concentrated in east Lancashire and adjacent Pennine valleys where, by the 1890s, 78 per cent of Britain's spindles, 88 per cent of its looms and three-quarters of the cotton workers were located (Table 10.5). Elsewhere specialist types of production remained: cotton for lacemaking and stockings in the East Midlands, chiefly Nottingham; fine thread and distinctive patterned cloths in Paisley. Lancashire's growing range of finishing and ancillary trades from clothing to industrial belting, added diversity to the economies of the larger towns. The regional economy, including much of the commercial activity of Manchester and Liverpool, was closely related to cotton; dangerously so as the fall in international markets in the twentieth century were to show (Chapter 15). In large cotton towns over half the occupied population was employed in the mills: 60 per cent of Blackburn's workers in the 1880s, 57 per cent of Burnley's, 50 per cent of Preston's and Oldham's; over three-quarters of working women were commonly employed in the cotton mills. Thus, in recession unemployment was rife, as in 1862 and the slumps of the late 1870s and 1880s.

Emerging specialization and horizontal integration of different branches of the industry within different firms led to growing geographical segregation of major branches of cotton manufacture. Whereas in mid-century over half the country's spindles and four-fifths of the power looms operated

Table 10.5 Distribution of cotton operatives in the British Isles in 1838 and 1898–9

Region	1838 Number of operatives	percentages of British Isles	percentages of England and Wales	1898–99 Number	percentages of British Isles	percentages of England and Wales
Cheshire	36400	14.0	16.6	34300	6.5	6.9
Cumberland	2000	0.8	0.9	700	0.1	0.2
Derbyshire	10500	4.0	4.8	10500	2.0	2.1
Lancashire	152200	58.8	69.5	398100	75.7	80.2
Nottinghamshire	1500	0.6	0.7	1600	0.3	0.3
Staffordshire	2100	0.8	0.9	2300	0.4	0.5
Yorkshire	12400	4.8	5.7	35200	6.7	7.1
Rest of England and Wales	2300	0.9	1.0	13500	2.6	2.7
England and Wales	219200	84.5	100.0	496200	94.3	100.0
Scotland	35600	13.7	–	29000	5.5	–
Ireland	4600	1.8	–	800	0.2	–
British Isles	259400	100.00	–	526000	100.00	–

Source: S. J. Chapman, *The Lancashire Cotton Industry* (1904), p.149

in combined mills, by 1890 these proportions had fallen to 30 and 42 per cent, respectively. The rapid spread of power-loom weaving from the 1840s was focused on Rossendale which had 62 per cent of the region's looms by 1884, often in relatively small weaving sheds; south-east Lancashire had 78 per cent of the spindles, principally in large multi-storeyed mills. As transmission of power from mill engines changed from shaft to flywheel rope drive in the 1860s, and ring-spinning and fire sprinkler-systems became more common, a new wave of large multi-storey mills in red Accrington brick came to dominate the landscapes of the northern mill towns. Many (as in Preston) were along the railways rather than, as in the first phase of mill development, along the canals. Similar specialization focused bleaching, printing and dyeing on stream-side locations largely outside the major urban areas. Thus, by 1890 a distinctive internal geography of specialization had emerged within cotton Lancashire (Figure 10.2).

Formerly a widespread domestic activity, the move to factories saw considerable concentration of linen yarn production on large spinning mills. Leeds accounted for half English production by 1856, but the major concentration was increasingly in Angus and Fife where urban-centred power spinning then, from the 1850s, power looms progressively dominated British production. In a remarkable example of specialization, Dundee's switch to jute after 1832 gave it a virtual monopoly of fully mechanized jute production by the 1850s and promoted associated industries, such as linoleum manufacture, using jute backing.

Silk manufacture remained widely scattered, but older producers, notably London and East Anglia, declined as machine spinning developed in West Yorkshire and Lancashire, further concentrating silk production in the south Pennines around originally water-powered mills in Macclesfield, Congleton and Leek. However, the abolition of duties on silk imports in

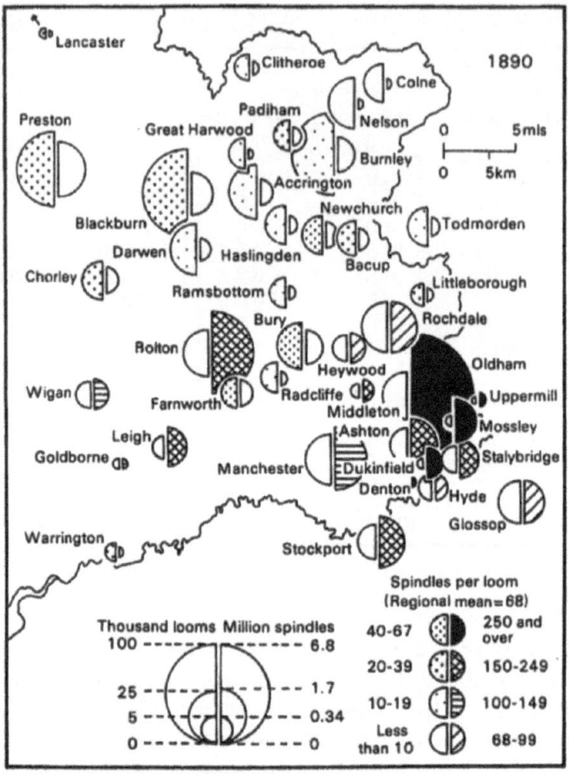

Figure 10.2 Regional specialization in the cotton industry in Lancashire, 1890.
Source: Langton and Morris (1986) p. 113.

1860 prevented further growth and killed off such specialist branches as the Coventry ribbon-weaving trade. In all, employment in the silk industry fell from its 1856 peak (56,000) to 36,000 by the mid 1890s.

Iron and steel manufacture

Iron and, later, steel provided the frame and sinews of the machine age and were central to the growth of mass transportation by land and sea, to a range of construction for railways, bridges, industrial and commercial buildings, and for equipment in all aspects of domestic life from kitchens to middle-class bathrooms and gardens. Iron and steel machinery and tools helped make late-Victorian Britain the workshop of the world.

Despite substantial improvements in fuel efficiency achieved by the hot-blast process, coalfield locations continued to dominate iron smelting up to mid-Victorian times. Blast furnaces consumed one-sixth of British coal production (Table 10.4) and Coal Measure ores remained the main source of raw material until, from the 1850s, cheaper Jurassic ores (Figure 5.3) began to make an impact on iron and, from the mid-1870s, steel production.

From the mid-1830s to 1870 pig iron production increased six-fold (Table

5.5). While castings remained important in heavy engineering most pig was converted to wrought iron for plates, rails and bars for a wide range of engineering products. The industry peaked in the 1860s as world markets expanded with the growth of railways, iron ships and a variety of constructional and mechanical engineering trades; but the bubble of prosperity burst in the 1873 slump, a prelude to the depressed years of the late 1870s and 1880s.

Some of the reasons for Britain's relative decline in iron manufacture were technological; some were economic. As Coal Measures ores approached exhaustion in old-established iron-making areas such as the West Midlands and the north eastern rim of the South Wales coalfield, increased costs of raw material assemblage disadvantaged them with Scottish and newer producers in Teeside and Furness. The newly worked 'lean' Jurassic ores, which ranged from 22 to 33 per cent iron content, were initially unsuitable for steel-making. Although British manufacturers lost some of their initial advantages in pig and wrought iron production, as steel technology developed they offset this in a number of ways. Better blast furnace technology through mechanical loading, gas recovery to superheat furnaces with consequent fuel economies, and less wasteful techniques of tapping the furnace contributed to lower unit costs. Skilled work-forces in old-established districts were tapped for new areas of production (Chapter 8). Changes in the structure of the industry made for greater efficiency in larger, integrated firms with involvements in coal and engineering.

The future lay with steel-making. A hitherto slow and expensive process was transformed by a series of innovations initiated by Henry Bessemer in 1856 and Siemens and Martin (1856–64), and completed by adaptation to all types of ore through the Gilchrist-Thomas process of 1879. Yet Britain's transition from iron to steel was protracted. Pig-iron output increased from 6.0 to 7.9 million tons between 1870 and 1890 and averaged 9.7 million tons per annum 1904–13. Wrought iron production, 2.8 million tons in 1882, was only exceeded by steel production in 1886. The United States' steel output, only half of Britain's in the early 1870s, was twice that of Britain (4.9 million tons) by the end of the century, mainly from big steel mills with fast blast furnaces.

The key to the new steelmaking technology was controlled decarburisation. Pig iron, with 2.5–4 per cent carbon content, is brittle and effectively restricted to castings. Wrought iron's very low carbon content (0.1 per cent) suits smith's work, machine parts, etc. Steel, with a carbon content of 0.1–2 per cent, combines workability with tensile strength and the carbon content can be controlled to produce steel for different needs, an asset enhanced by steel alloys from the 1870s: tungsten and manganese for edge tools, pioneered by R. F. Mushet from 1868; hard chrome steel developed in French armaments works from 1877; and nickel steel produced at Le Creusot from 1888.

The transition to a steel age between the 1850s and the 1890s was a period of both innovation and conservatism. The simple Bessemer process was based on hot air blown over molten pig iron in a 'converter' causing rapid decarburisation in which the heat of oxidation of the metal kept it molten. It was very economical in fuel: Bessemer entitled his paper on the invention, delivered at the British Association's meeting of 1856, 'On the manufacture of malleable iron and steel without fuel'. While it produced cheap mild steel suitable for rails and castings and quickly replaced wrought iron in rail manufacture, it was difficult to accurately control the carbon content and characteristics of the metal so that initially it was not suitable for many metallurgical industries.

Moreover, the Bessemer process was suitable for refining basic ores but not, initially, those with acidic impurities (chiefly phosphorus) like those of the extensive Jurassic orefields. That problem was not overcome until the late 1870s when Sidney Gilchrist-Thomas and his cousin Percy Gilchrist succeeded in drawing off acidic phosphorus by means of a limestone charge (or flux) in a furnace lined with dolomite instead of the siliceous brick lining hitherto used. This new 'basic steel' process could be used in both Bessemer furnaces and the more sophisticated open-hearth method perfected by Emil and Pierre Martin in France in 1861–63 and put into production in Birmingham and then at the Lodore Steel Company, Swansea in 1864 by the Siemens brothers. The Siemens-Martin method used both pig (or crude steel) and scrap in a furnace (or hearth) fuelled by low-grade coal, using waste gases to heat a regenerator in which the carbon content of the steel could be closely regulated and alloys added to produce the type and quality of steel required. Though slower than the Bessemer process, it gave better steels of more consistent quality.

As steel-making was perfected, the price ratio between steel and wrought iron narrowed: thus steel rails, 2.65 times as expensive as iron in 1867, were only 1.16:1 by 1875. However, many British iron and steel makers were slow to adopt basic steel-making. With capital locked up in the existing plants, many did not switch to newer methods of production and to integrated, automated plants until the end of the century. Hence they were out-priced in some markets (e.g. rails and certain kinds of rolled plate), particularly by the USA, Belgium and Germany.

New sources of ore, both Jurassic and, initially, imported high-grade basic ores (Table 10.6) modified the regional pattern of iron manufacture by 1870. In North East England, using abundant high-quality south-west Durham coking coal, Cleveland and imported basic ores, the boom town of Middlesbrough became a symbol of the age of iron and steel. At Barrow in Furness Schneider and Hannay, using local haematite for iron and steel production, produced one-fifth of British steel by 1880. Other regions experienced relative decline. The West Midlands, dependent on declining coal and Coal Measure ores, turned more to engineering, and the number and output of its furnaces fell sharply from the 1870s. Even the blacksheet and galvanized iron trade declined from the 1890s: increasingly adverse transport costs led to its migration to coastal plants in South Wales and North West England (especially at Ellesmere Port) and much of the British export market was lost to Belgian competitors. In South Wales the loss of the early dominance in the iron rail trade to Scotland was reflected in the peaking of its iron output in the early 1870s.

Steel production's location was increasingly dominated by coastal plants using imported ores (for Bessemer and open-hearth production alike), or specialist inland producers in central Scotland and, especially, in the

Table 10.6 Iron ore output in the United Kingdom 1855–1950

	Thousands of tons							
	UK	Cleveland	East Midlands		North West England	West Midlands	South Wales	Scotland
1855 (%)	9554	865 (9.1)	74 (0.8)		538 (5.6)	1221 (12.8)	1666 (17.4)	2400 (25.1)
1870 (%)	14371	4073 (28.3)	1104 (7.7)		2093 (14.0)	1698 (11.8)	560 (3.9)	1980 (13.8)
1880 (%)	18026	6487 (35.9)	2705 (15.0)		2758 (15.3)	1989 (11.0)	344 (1.9)	2659 (14.7)
1890 (%)	13781	5618 (40.8)	2940 (21.3)		2399 (17.4)	1283 (9.3)	50 (0.4)	999 (7.3)
1900 (%)	14028	5494 (39.2)	4536 (32.3)		1733 (12.4)	1118 (8.0)	20 (0.1)	849 (6.1)
1913 (%)	15997	5941 (37.2)	6556 (41.0)		1767 (11.1)	895 (5.6)	55 (0.3)	592 (3.7)
			Lincs.	Rest				
1938 (%)	11859	1514 (12.8)	3439 (29.0)	5774 (48.7)	795 (6.7)	127 (1.1)	188 (1.6)	17 (0.1)
1950 (%)	12963	1020 (7.9)	5012 (38.7)	6458 (49.8)	343 (2.6)	–	103 (0.8)	–

Sources: Mitchell and Deane (1962), pp. 129–30; for 1950, Mitchell and Jones (1971), p. 77.

Sheffield area (Table 10.7). With the development of basic steel making from the 1880s the inland Jurassic orefields of north Lincolnshire and the East Midlands began to be important. In South Wales production based on Coal Measure ores fell sharply. However, Bessemer steel-making persisted in the Merthyr area and open hearth plants in the Swansea area boosted Welsh steel production to one-third of the national total by 1880. Despite new integrated coastal steel works at Cardiff, Port Talbot and Newport that share fell to around one-fifth by the end of the century. Scottish steel-making fared better, not least because of close links with the heavy and marine engineering and ship-building industries of Clydeside. Proximity to the coast facilitated ore and coke imports and despite falling pig iron manufacture its share of national steel production increased to one-fifth by the end of the century.

Teeside epitomized the radical transformation in steelmaking. Cleveland ironstone came into its own with the Gilchrist-Thomas process and by 1895 North East England produced nearly one-third of British steel, 97 per cent of it from local ore. In the North West, Furness gave an early lead, but south Lancashire's role in steel production was confined to smelting of imported ores for the local rail and engineering trades at coalfield sites until new plants were drawn to waterside locations along the Manchester Ship Canal at the end of the century.

Sheffield, the remaining major steel producer, had a long-established reputation as a maker of high grade steel for its edge-tool and cutlery trades, and developed Bessemer steelmaking along the Don valley from 1860. By the late 1870s it produced nearly one-third of British steel though that declined by the 1890s when, like the West Midlands, it used a good deal of scrap in its furnaces.

Engineering

The important association between ferrous metal manufactures and engineering became increasingly significant during the Victorian era. The cast-iron bridge across the Severn gorge near Coalbrookdale is a potent symbol of the first industrial revolution: Fowler and Baker's Forth Bridge – that 'brontosaurus of technology' as Sir Kenneth Clark describes it – is testimony to its 54,000 tons of open-hearth steel and the ingenuity, utility and functional beauty of design and construction of the enormous tubular steel supports and girders, held together by 6.5 million rivets, that bear the railway over its one-and-a-half mile length.

In some cases engineering companies' interests spanned all stages from iron and steel to the finished product: Dorman Long of Middlesbrough, steelmakers and constructional engineers, focused on bridge-building; Schneider and Hannay's and Vickers' interests spanned steel, shipbuilding and armaments; the Peases began in textiles but diversified into coal and iron on Teeside. However, large-scale engineering developed mainly in the late nineteenth and twentieth centuries. Many Victorian firms were small and produced for local markets, though some branches of the industry experienced substantial growth in their scale of manufacture, degree of

Table 10.7 Steel Production in Great Britain by regions, 1880–1950

	South Wales		Scotland		Sheffield Area		North East Coast		North West Coast		Others		Total, GB
Year	Production	(%)	Production	(%)	Production	(%)	Production	(%)	Production	(%)	Production	(%)	Production
1880	424.2	32.8	84.5	6.5	296.9	22.9	147.2	11.4	328.2	25.3	14.3	1.1	1295.4
1890	873.7	24.4	485.2	13.6	428.4	12.0	897.4	25.1	731.3	20.4	163.0	4.5	3579.0
1900	990.4	20.2	963.3	19.7	586.2	12.0	1335.8	27.2	659.0	13.4	366.5	7.5	4901.0
1913	1807.0	23.6	1431.0	18.6	879.0	11.4	2031.0	26.5	398.0	5.2	1118.0	14.7	7664.0
1950	3450.0	22.1	2020.0	12.9	2130.0	13.6	3250.0	20.8	400.0	2.6	4400.0	28.1	15640.0

Production figures are given in thousands of tons and percentages of the total national production
Others include the West Midlands, Lincolnshire, East Midlands and Lancashire/West Yorkshire

Source: H. Roepke (1956), Table 17

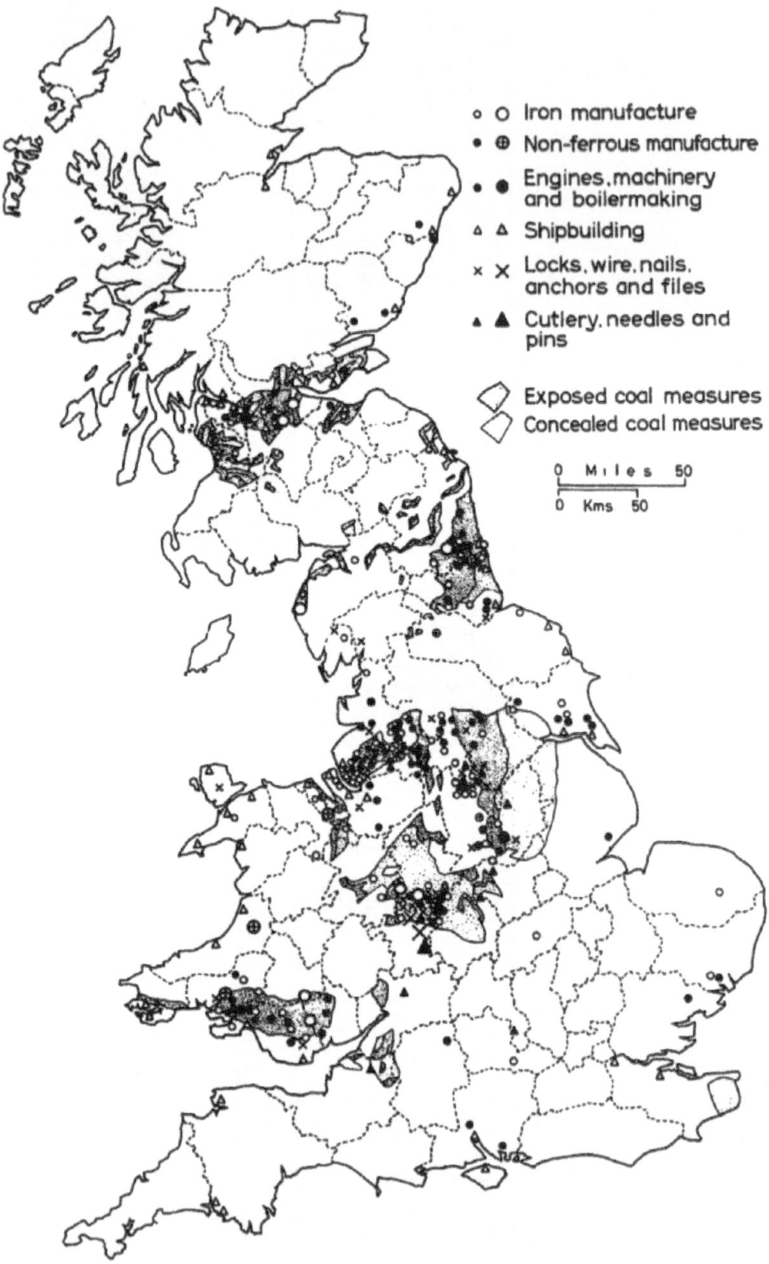

Figure 10.3 Employment in metallurgical and engineering industries in Great Britain, 1951. Symbols are placed in districts where numbers occupied in specific industries were important; large symbols signify districts of special significance. Based on A. Petermann's map, Census of Great Britain, 1851.

regional specialization and contribution to exports and the national economy (Tables 5.2 and 10.3).

Key innovations central to mechanization of industry and transport had been patented for steam engines and textile machinery before 1800 (Chapter 5). Precision instruments and machine tools, the key to further progress in civil mechanical engineering, were progressively developed from the early nineteenth-century: screw-cutting lathes at the Bramah and Maudsley London works from 1797; planing machines, advanced lathes and standard machine parts made by Whitworths of Manchester from the 1830s; drop-forging of standard beams and plates for large machines by James Nasmyth's steam hammer from 1839. Yet standardized interchangeable machine parts and uniformity of equipment were mainly the work of the late nineteenth century. Better quality steel alloys, precision milling and grinding machines, and the development of automation from the 1870s helped the machine tool trade to reach some 5 per cent of British engineering output by the end of the century though that lagged behind the achievements of the USA.

Victorian engineering firms were diverse in character and location (Figure 10.3). Most were workshop industries producing bespoke items, yet some large concerns employed thousands and had substantial national and international markets. Local and regional demand ensured a relatively wide spread of metal finishing and machine-making, in village smithies, small-town foundries or as producers of textile or agricultural machinery. Early technological and entrepreneurial skills gave some comparative advantage in world-wide markets. Cornwall, a pioneeer in beam-engines for mine drainage, specialized in heavy mining machinery. Metal-working and steam-engine manufacture by the Boulton and Watt partnership gave the West Midlands a leading role in general engineering in which its importance in brass-founding (critical for the manufacture of steam valves) was an additional advantage. Moreover, British firms continued to improve the effectiveness of steam power. High speed double- then triple-expansion engines were developed between 1884 and 1900, significantly advancing British dominance of steamship and locomotive manufacture. That dominance was enhanced by the parallel development of steam-turbines by C. A. Parsons which laid the foundations for modern high-powered engines.

Iron and steel were the bulkiest and costliest components of heavy engineering products. Shipbuilding, a hitherto widely-dispersed industry, became more concentrated as it gradually adopted steam power and iron construction. In 1838 the 'Sirius' narrowly beat I.K. Brunel's 'Great Western' (built in 1837) to the first transatlantic crossing by an iron ship. The first major vessel to use screw propulsion, the 'Great Britain' was launched in 1843. Such events put the iron steamship on the map, and the launch of the 'Great Eastern' at the Isle of Dogs in 1858 ushered in a new era of larger ships. Greater efficiency of high-compression boilers opened the way to full economic utilization of their potential, as demonstrated by Alfred Holt's use of compound engines in his ships after 1856. Whereas in the 1840s 90 per cent of the world's merchant shipping was wooden, half of it built in North America, between 1850 and the 1880s iron steamships shifted the balance to

British shipyards. By the 1890s 82 per cent of the world's merchant tonnage was British-built, mostly from two areas with advantage in iron- and steel-plate production and the design and manufacture of boilers and marine engines: Clydeside, which built iron ships from 1839 and paddle-steamers (for the Cunard Line) from 1840, made around one-third; North East England (an old-established shipbuilding area) around one-half. In contrast the output of the Thames yards declined sharply from the 1860s. In North West England shipbuilding was concentrated on the Mersey, chiefly at Laird's Birkenhead Works, and in the Vickers naval yards at Barrow in Furness leaving these towns dangerously exposed in recession (Chapter 15).

Locomotive engineering, in contrast, came to focus on three types of location: first, areas of early railway development and locomotive power, such as the North East; secondly, the railway workshops in key centres of the emerging rail network such as Crewe, Swindon and Ashford that had three-quarters of the industry's workforce by 1900; thirdly, in large city engineering complexes, such as Glasgow's Springfield, the Beyer and Peacock plant at Gorton, Manchester, or the Hawthorn and Kitson works in Newcastle and Leeds.

Other types of engineering developed near their initial customers, though many achieved much wider markets. Textile machinery was developed principally in east Lancashire, West Yorkshire and, for hosiery machines, Nottingham and the East Midlands. Older producers – such as Lees of Oldham or Dobson and Barlow of Bolton – were late eighteenth-century pioneers. Others developed later in association with machine tool production. Similarly as agricultural machinery was improved and more widely adopted from early Victorian times specialist production, especially of steampowered harvesting and threshing machines, created a substantial agricultural engineering industry in some of the major arable areas of eastern England (Chapter 9).

The growth of general engineering is difficult to trace. Up to mid-Victorian times it was widely scattered in numerous small workshops. Few large late nineteenth-century firms existed before 1850. Nevertheless metal working, engineering, ship and vehicle building made up 9 per cent of the workforce by 1891 (Table 10.3) and engineering products were of growing importance in exports (Table 5.2). Heavy engineering was closely linked to the iron and steel industry, but light engineering was progressively drawn to specialist skills and markets. The West Midlands' wide range of constructional, tube, hardware, engine and machine part manufactures continued to dominate metal and engineering production. By the 1890s however engineering in London and the South East was growing in range and importance, using the technology, structure, organization and working methods characteristic of twentieth-century industry (Chapter 15).

Chemicals

The stimulus to innovation in chemical manufactures in the early industrial revolution came largely from textiles. The key raw materials – salt, coal, soda, limestone and sulphur – combined to dominate the location of the industry until mid-Victorian times, though by-products of coal and metal manufacture were increasingly important. The substantial changes in the scale and location of an increasingly varied range of late nineteenth-century chemicals came from three sets of factors: scientific and technological innovation using new processes and/or new raw materials; concentration of production into bigger and more diverse firms; and the differential impact of foreign competition, especially from Germany, in different branches of the industry.

The manufacture of bleaching powder (chloride of lime), for the textiles industries, remained one of the most basic and widely distributed of the chemicals industries, being drawn to both coal and raw material at sites on Tyneside and markets in pioneer areas such as central Scotland. However, gradual adoption of the Leblanc process of soda-making based on salt and sulphuric acid replaced Scottish kelp and imported soda ash and transformed the pattern of alkali manufacture from the 1840s. Production became focused largely on the Cheshire saltfield and the mid-Mersey, especially Widnes the new 'chemical town' of the 1840s. Sulphuric acid, made from the processing of copper and iron pyrites, was widely scattered in production until the recovery of sulphur from the alkaline waste of soda manufacture utilized that hitherto noxious pollutant, especially after the Alkali Act of 1863, and concentrated heavy chemicals on the mid-Mersey area by 1890 (Figure 10.4).

The cheaper Solvay method of ammonia soda manufacture (developed in Belgium in 1856–66) at the Brunner-Mond works at Northwich (Cheshire) completed the dominance of the mid-Cheshire – Mersey area in alkali manufacture, a position confirmed by two business mergers, the Cheshire Salt Union of 1884 and the 51 producers in the United Alkali Co. of 1890. By then new methods based on the electrolysis of brine to make chlorine and caustic soda were being developed, inaugurating a new phase in the industry.

The agricultural fertilizer industry, a major market for sulphuric acid, was much more widespread. The scientific work of Justus Liebig and J. B. Lawes inaugurated an 'inorganic' phase in farming in the 1840s (Chapter 14). Natural nitrates – Chile *caliche* or home-produced coprolite – could not be replaced until chemical then electric fixation of atmospheric nitrogen after the First World War. But superphosphates made from bone were replaced by phosphate of ammonia from gas and later coke by-products and by cheap basic slag, a by-product of the Gilchrist-Thomas process. Many fertilizer producers were small firms scattered in country towns, though large manufacturers in the main chemical areas and at the ports were of increasing significance.

New synthetic materials were of increasing importance for a range of chemicals by the late nineteenth century: dyestuffs, from coal-tar derivatives; soap from imported vegetable oils which, with soda, were behind the building of the massive Lever works from 1888 and its model village at Port Sunlight on the south bank of the Mersey estuary. New processes based on cellulose made from rags or wood pulp treated with sulphites provided a whole range of products from paper to artificial fibres and explosives. Paper is, indeed, a notable example of the growing contribution of industrial

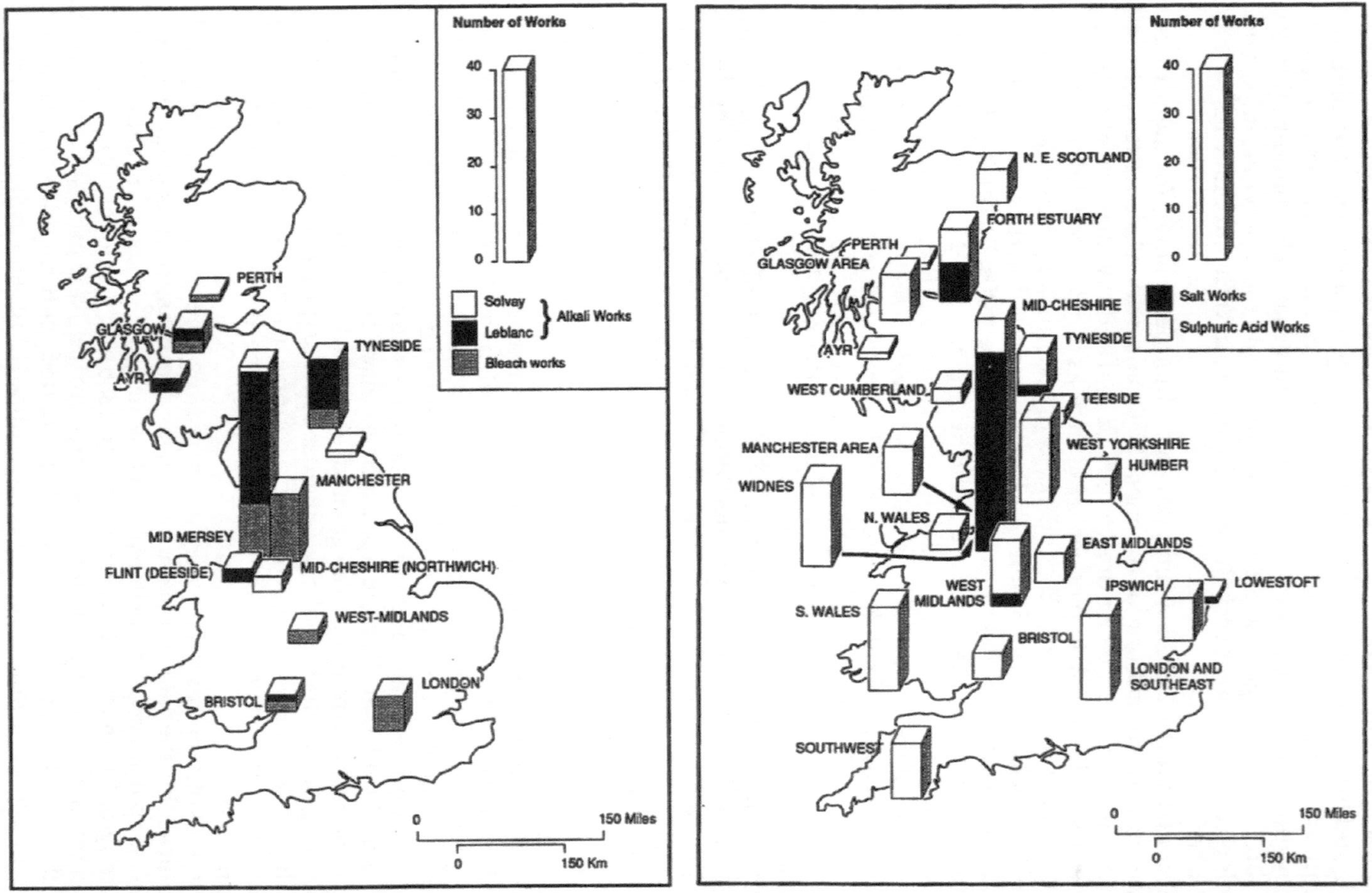

Figure 10.4 Chemical industries in Great Britain *c.* 1882–1890. Based on Langton and Morris (1986) pp. 114–18 and Pope (1989) p. 40.

chemistry to everyday life. The Victorian 'paper world' was the basis of commercial transactions; of newspapers in an era of growing mass literacy; of literary and musical entertainment; of advertising and packaging of goods; and, indeed, of a growing range of papier-mâché objects from japanned trays to decorative ornaments.

Chemicals were thus one of the symbols of the 'second industrial revolution' of the late nineteenth century. While Britain continued to dominate certain sections of the international heavy chemicals industries, it was progressively eclipsed – especially by Germany – in organic and electrochemistry, in newer applications of dyestuffs, in the manufacture of nitrocellulose and the growing range of pharmaceutical products important in twentieth-century light chemicals industries. Before the First World War Britain's share of world production of chemicals fell to one-quarter of that in 1876 (from 46 to 11 per cent), far behind the USA (34 per cent) and Germany (24 per cent).

Conditions of work and standards of living in industrial towns

Variations in standards of living, wages and working conditions were at least as great in towns as in the countryside. Average wages in towns (Table 10.8) were certainly higher but rent and food were also more costly so that urban dwellers were not necessarily better off than their rural counterparts. Women's wages were invariably well below those of men, and families dependent upon a sole female wage earner were often amongst the poorest

Table 10.8 Indices of average weekly wages in selected occupations (United Kingdom), 1770–1910

Year*	English agricultural workers	Scottish agricultural workers	Compositors	Building workers	Shipbuilding and engineering workers	Cotton factory workers
1770	51	18	–	–	–	–
1780	51	–	62	–	–	–
1790	53	31	62	44	40	–
1800	83	41	74	63	56	–
1810	105	58	88	88	71	62
1820	95	46	78	78	71	61
1830	76	38	78	78	71	56
1840	82	54	83	83	75	55
1850	72	61	83	83	73	54
1860	89	75	83	76	79	68
1870	96	85	85	86	84	81
1880	100		96	96	90	87
1890	100		98	100	100	96
1900	109		101	111	108	107
1910	110		–	111	–	–

1891 = 100

*Where figures were not available for selected year the nearest estimate is given.

Source: B. R. Mitchell, (1988) p. 156

of the urban population. Apart from life-cycle variations in income and expenditure so clearly demonstrated by Booth and Rowntree in the 1890s, the most important factor governing urban standards of living was regularity of employment. A job guaranteeing a regular weekly wage, with little cyclical unemployment, was rare, highly prized and jealously guarded. The majority of workers were affected by cyclical fluctuations in trade and short-term lay-offs due to the weather and local conditions so that they never knew when they would be without wages for a day or two or even much longer. Such fluctuations in employment were major factors in the urban labour market and, in turn, had a significant impact on the standard of living, the quality of housing and the residential areas to which people could aspire. Both constraints of income and access to work influenced housing choice.

At the bottom of the nineteenth-century urban labour hierarchy the genuinely casual workers formed a residual labour force which was often entered on initial migration to a town when no other work was available. Such work included hawking and street trading, scavenging, street entertainment, prostitution and some casual labouring and domestic work. Below these were begging and poor relief which, after the implementation of the new Poor Law Act of 1834, often meant the workhouse, the very nadir of Victorian living. Casual trades were mostly concentrated in large cities, especially London, and the numbers involved fluctuated considerably. Although some made a regular living from them, many others resorted to such jobs only when no other work was available. Very low and irregular incomes condemned families dependent on casual work to one or two rooms in the worst slums, but in London they would emerge from the rookeries of St Giles to sell their wares in the City or in middle-class residential districts. Their 'rounds' took them considerable distances between different social areas. Large numbers of street traders in prosperous middle-class districts caused antagonism and sometimes fear, so that the police were often called to control street trading activities. Such contacts helped reinforce middle-class stereotypes of a dirty and dangerous sub-class that should be confined to the slums (Chapter 11).

One step above casual street trading was a whole range of unskilled mainly casual occupations in which workers were frequently hired for a few hours at a time and could be laid off for long periods without notice. These include labourers in the building trade, in sugar houses and other factories, carters, shipyard labourers and especially dockers. All towns had such casual workers, but they were particularly significant in port cities such as Glasgow, London, Liverpool and Bristol and in industries such as coal mining or clothing which had a partly seasonal market thus influencing standards of living.

Due to imprecise census occupational classifications and frequent changes of job, the precise numbers involved in casual work in Victorian Britain are impossible to determine. In Liverpool over 22 per cent of the employed population in 1871 were general, dock or warehouse labourers of some kind, many casual. Liverpool dockers were hired for half-a-day at a time at one of 60 hiring stands developed along the dock system by the end of the century. Short-term fluctuations in trade depended on a ship's movements and the stevedore's men, who actually loaded and unloaded boats, were the most affected. Porter's work on the quayside was more spread and gave rather more consistent employment. When in work, some Liverpool dockers earned high wages, ranging from 27s for a six-day week for quay porters to 42s for a stevedore in 1871, but few maintained such incomes for any length of time and in a bad week many earned only a few shillings.

Conditions changed little between the mid-nineteenth century and the early twentieth century. A diary of income and expenditure kept in 1910 by a Liverpool dock worker with a family of three young children showed that his earnings fluctuated between 2s and 20s per week with an average of only 10–12s. The wife's occasional cleaning work brought in about 2s a week and they sublet part of their house to a lodger for a further 3s 6d a week for board and lodgings, a total average income of 15s 6d to 17s 6d per week. They spent 6s 6d a week on rent for a five-room house and 6–10s per week on food depending on income. They were frequently in debt and regularly pawned clothes. Although in good times they would eat meat or fish, frequently their diet consisted mainly of bread, margarine and tea. This budget was typical of many urban workers, and illness or industrial accident (common in dangerous dockland working conditions) would have led to financial disaster.

Casual workers needed to live close to their workplace since employment was usually allocated on a first-come, first-served basis necessitating knowledge of work availability and quick access to it. Liverpool dockers mostly lived close to the docks in which they usually worked, and where they were most likely to gain employment (Figure 10.5). Their housing choice was thus restricted, often to old, insanitary and, above all, affordable accommodation.

Factories provided more regular employment in the nineteenth century, as did such public services as railway companies and many commercial organizations. Skilled manual labour was relatively privileged: a factory cotton spinner in Lancashire in the 1830s earned 27–30s per week, and a skilled iron foundry worker up to 40s. In coal mining too skilled underground workers earned good wages and in key jobs such as shot-firing, putting, hewing and shaft sinking usually had regular employment although this often meant moving from colliery to colliery and between coalfields. Such textile towns as Manchester, Bradford and Leeds and metal and engineering areas such as Sheffield and the Black Country tended to suffer less from poverty due to irregular earnings than cities like Glasgow, Cardiff, Liverpool or London. Although conditions of work in factories and mines were often poor and hazardous, legislation gradually controlled conditions of work more quickly than in the casual trades. By 1853 a ten-hour day was normal for most textile workers, although lower-paid factory operatives did not always welcome the reductions in hours as it could also reduce their income.

Skilled engineering trades were amongst the earliest to unionize, along with artisans and craftsmen, particularly in London and northern industrial towns. Part of Hobsbawm's 'labour aristocracy', they tended to protect their interests jealously. Entry to many skilled trades was restricted both by long apprenticeships and union control, and new migrants often found it difficult

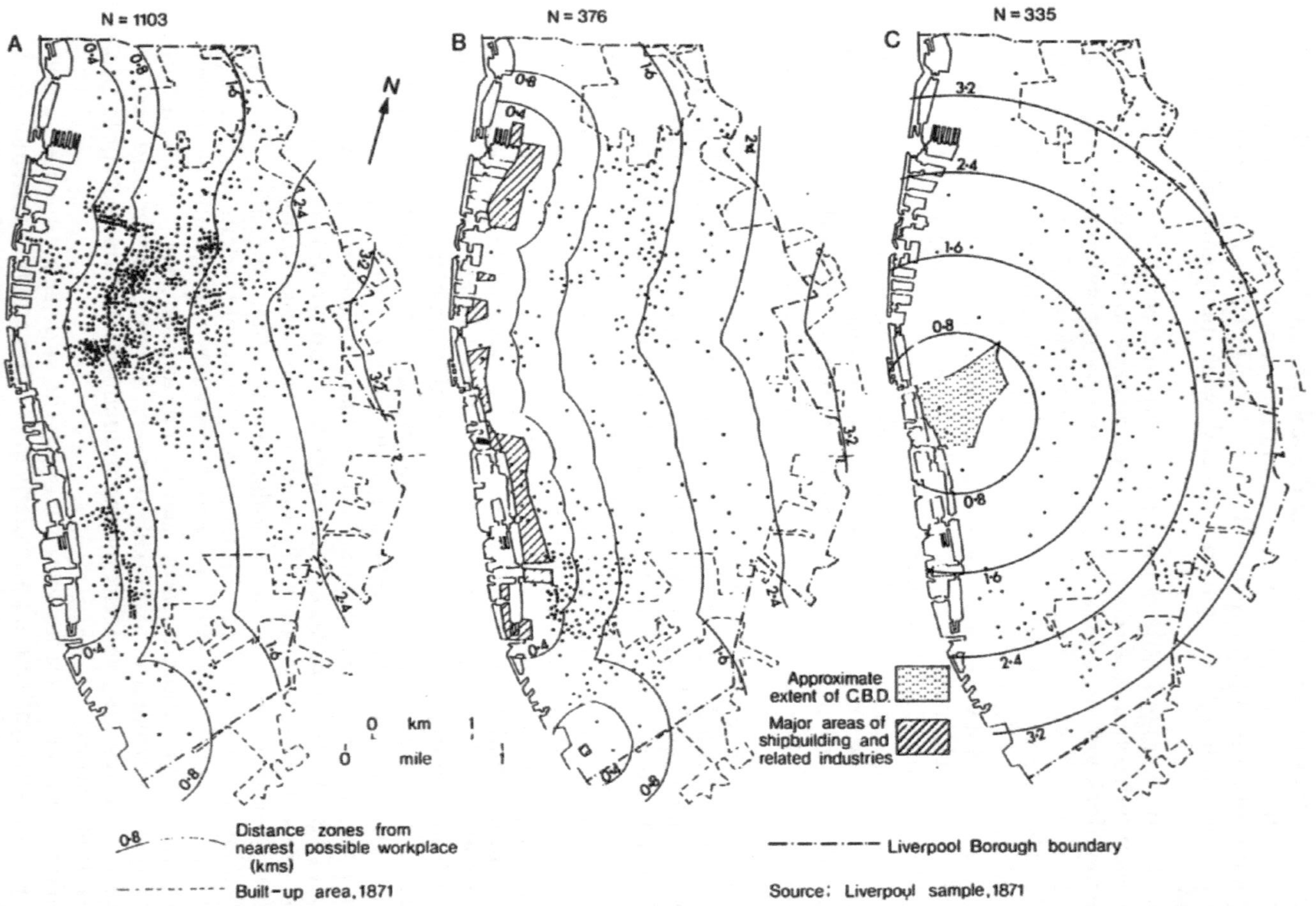

Figure 10.5 Residential location of dock workers (A), those employed in shipbuilding and repair (B), and office workers (C) in Liverpool, 1871. (Approximate distances to nearest possible places of work are shown.)
Source: Census enumerators' books and Gore's directories, Liverpool, 1871.

to enter certain trades. Such workers' higher wages and regular employment conferred many advantages: most skilled workers could afford to rent a decent terrace house in the suburbs and thus avoid the squalor of Victorian slums (Figure 10.5). Although their journey to work was usually on foot they were less dependent on direct access to their job than casual workers, and a two- or three-mile walk gave them a much wider range of residential choice.

In the later nineteenth century the number of workers in white-collar occupations increased and a 'lower-middle class' emerged amongst the petit-bourgeoisie of small shopkeepers and white-collar salaried occupations of clerks, commercial travellers and school teachers. White-collar employment increased from 2.5 per cent of the employed population in Britain in 1851 to 5.5 per cent by 1891. Although such employment was found in all towns it was particularly concentrated in London, Edinburgh and such commercial and financial centres as Glasgow, Bristol, Liverpool and Manchester. As with skilled manual workers, white-collar employees were a diverse group: insurance and bank clerks commanded the highest incomes (frequently over £3 per week) and the greatest prestige; in contrast railway clerks often earned little more than skilled manual workers but had greater security of employment. White-collar employees certainly perceived themselves (and were perceived by others) to be in a secure and privileged position. A survey of London clerks living in working-class districts in 1887 reported that 90 per cent of them considered their employment secure: consequently they spent more on rent and other long-term commitments than manual workers. White-collar workers could thus not only afford a decent terrace house, but by the late nineteenth century could also afford to commute over longer distances by public transport, especially from the 1890s when tramways and suburban railways were well-established (Figure 10.5). Despite long hours of work for clerks and shopkeepers, their occupations were less hazardous than most factory employment and, with more regular incomes and better housing, they were more likely to enjoy good health than most industrial workers.

Although women were employed in all categories of work, and in textile districts female factory employment was very significant, many women could not easily gain regular employment. Single women often entered domestic service but married women who needed to supplement a low male wage, or widows supporting several children, were severely limited in choice. Away from the textile districts most found jobs as domestic cleaners, laundry workers, in sewing, dressmaking, boot and shoemaking and other trades carried on in either the home or small workshops. Even in areas of female factory employment such as Lancashire and Yorkshire many women with responsibilities for young children could not commit themselves to the long and regular hours of factory work and were forced into domestic or sweated trades. Wages in these trades were always low with relentless piece rates producing incomes ranging from 5s to 20s per week. However, few women could put in the long hours necessary to earn 20s in a week, and cramped conditions and poor light at home or in a workshop were severely detrimental to health. Moreover, most married women in the workforce had two full-time jobs: one in the factory or workshop and one as housewife and mother in the home. The great physical and mental strain placed on many

Victorian women was reflected in premature ageing and ill-health.

At the other end of the labour market wealthy businessmen, bankers, merchants and professionals to a large extent controlled their own working conditions and usually had the resources to live where and how they wished. However, there were few 'idle rich' within this middle and professional class. Most businessmen put in long hours, often travelling widely between factories or shops and visiting suppliers. Those newly aspiring to middle-class lives often had especial difficulty in maintaining appearances and downward as well as upward mobility was not uncommon. Despite northern England's principal dominance in industrial enterprise, London – the financial and commercial capital of Britain – had a larger concentration of the very wealthy than any other town, although commerical and industrial cities had significant wealthy minorities. Even in landed wealth London was pre-eminent, despite notable concentration of land in a few hands in parts of Scotland. The lives of the rich, be they in London, Manchester or Edinburgh, were worlds away from the experiences of most residents of Victorian towns. Not only did they control most of the national wealth, but they dominated positions of power and influence.

Bibliography and further reading for Chapter 10

General industrial change:

General trends in the British economy are mainly covered in the bibliographies to chapters 1 and 2. In addition see:

ALDCROFT, D. H. (ed.) *The development of British industry and foreign competition, 1875–1914* (London: Allen and Unwin, 1968).

BRIGGS, A. *Victorian things* (London: Batsford, 1988).

CHURCH, R. A. (ed.) *The dynamics of Victorian business. Problems and perspectives to the 1870s* (London: Allen and Unwin, 1980).

SAUL, S. B. (ed.) *Technological change: the United States and Britain in the nineteenth century* (London: Methuen, 1970).

Contemporary accounts include:

GIBBONS, H. de B. *Industrial history of England* (London: Methuen, 1890).

McCULLOCH, J. R. *The literature of political economy* (London, 1845).

PORTER, G. R. *The progress of the Nation . . . from the beginning of the nineteenth century* (London, 1851).

Specific industries

Relevant references in Chapter 5 should be consulted in addition to those listed below.

Coal and power:

CHURCH, R. A. *The history of the British coal industry*, vol. iii, *1830–1913, Victorian preeminence* (Oxford: Clarendon, 1986).

DUCKHAM, B. F. *A history of the Scottish coal industry*, vol. ii, (Newton Abbot: David and Charles, 1970).

HIRSCH, B. T. and HAUSMANN, W. J. 'Labour productivity in the British and South Wales coal industry, 1874–1914. *Economica* **50** (1983), pp. 145–57.

MORRIS, J. H. and WILLIAMS, L. J. *The South Wales coal industry 1841–75* (Cardiff: University of Wales Press, 1958).

WILLIAMS, T. I. *A history of the British gas industry* (Oxford: Oxford University Press, 1981).

Iron and steel:

ALLEN, G. C. *The industrial development of Birmingham and the Black Country 1860–1927* (1929, reprinted London: Cass, 1966).

BURN, D. L. *The economic history of steel making, 1867–1940. A study in competition* (Cambridge: Cambridge University Press, 1940).

CARR, J. C. and TAPLIN, W. *History of the British steel industry* (Oxford: Blackwell, 1962).

PAYNE, P. L. *Colvilles and the Scottish steel industry* (Oxford: Oxford University Press, 1979).

WARREN, K. *The geography of British heavy industry since 1800* (Oxford: Oxford University Press, 1976).

WARREN, K. *The British iron and steel sheet industry since 1840* (London: Bell, 1972).

Engineering

CANTRELL, J. A. *James Nasmyth and the Bridgewater Foundry* (Manchester: Manchester University Press, 1984).

FLOUD, R. C. *The British machine tool industry 1850–1914* (Cambridge: Cambridge University Press, 1976).

LARKIN, E. J. and J. G. *The railway workshops of Britain, 1823–1985* (London: Macmillan, 1985).

MUSSON, A. E. with ROBINSON, E. *Science and technology in the industrial revolution* (Manchester: Manchester University Press, 1969).

ROLT, L. T. C. *Tools for the job. A short history of machine tools* (London: Batsford, 1965).

SMILES, S. *Industrial Biography: iron workers and tool makers* (1863, reprinted Newton Abbot: David and Charles, 1970).

WILSON, C. and READER, W. *Men and machines: a history of D. Napier and son, engineers, ltd. 1808–1958* (London: Weidenfeld and Nicolson, 1958).

Textiles.

COLEMAN, D. C. *Courtaulds: an economic and social history*, vol. 1, *silk and crepe* (Oxford: Clarendon Press, 1969).

ELLISON, T. *The cotton trade of Great Britain* (1886, reprinted London: Cass, 1968).

GATRELL, V. A. C. 'Labour power and the size of firms in Lancashire cotton in the second quarter of the nineteenth century.' *Economic History Review* **30** (1977), pp. 95–139.

Chemicals.

GITTENS, L. 'Soapmaking in Britain 1824–1851: a study in industrial location.' *Journal of Historical Geography* **8** (1982), pp. 12–28.

READER, W. J. *Imperial Chemical Industries: a history.* (2 vols.) (Oxford: Oxford University Press, 1970).

WARREN, K. *Chemical foundations. The alkali industry in Britain to 1926* (Oxford: Oxford University Press, 1980).

Transport and other service industries.

ANDERSON, G. *Victorian Clerks* (Manchester: Manchester University Press, 1976).

FREEMAN, M. J. and ALDCROFT, D. H (eds) *Transport in Victorian Britain* (Manchester: Manchester University Press, 1988).

HALL, P. G. *London's industries since 1861* (London: Hutchinson, 1962).

JEFFREYS, J. B. *Retail trading in Britain, 1850–1950* (Cambridge: Cambridge University Press, 1954).

LEE, C. H. 'The service sector: regional specialization and economic growth in the Victorian economy' *Journal of Historical Geography* **10** (1984), pp. 139–56.
SIMMONS, J. *The railways of Britain: an historical introduction* (London: Macmillan, 1968).

On working conditions and standards of living see:

BOOTH, C. *Life and labour of the people of London* (London: Macmillan, 1889–1903).
BYTHELL, D. *The sweated trades: outwork in nineteenth-century Britain* (London: Batsford, 1978).
CROSSICK, G. 'The labour aristocracy and its values: a study of mid-Victorian Kentish London'. *Victorian Studies.* **19** (1976), pp. 301–28.
CROSSICK, G. (ed.) *The lower-middle class in Britain* (London: Croom Helm, 1977).
ENGELS, F. *The condition of the working class in England* (1845, reprinted with an introduction, London: Granada, 1969).
FORSTER, J. *Class struggle and the industrial revolution. Early industrial capitalism in three English towns* (London: Methuen, 1974).
GRAY, R. Q. *The aristocracy of labour in nineteenth-century Britain, 1850–1914* (London: Macmillan, 1981).
GREEN, D. 'Street trading in London: a case study of casual labour 1830–1860' in Johnson, J. H. and Pooley, C. G. (eds) *The structure of nineteenth century cities* (London: Croom Helm, 1982), pp. 129–151.
HOBSBAWM, E. J. *Labouring men: studies in the history of labour* (London: Weidenfeld and Nicolson, 1964).
JORDAN, E. 'Female unemployment in England and Wales 1851–1911' *Social History* **13** (1988), pp. 175–190.
JOYCE, P. *Work, society and politics: the culture of the factory in later Victorian England* (London: Methuen, 1982).
LOVELL, J. *Stevedores and dockers: a study of trade unionism in the port of London, 1870–1914* (London: Macmillan, 1969).
MAYHEW, H. *London labour and the London poor* (London: Cass edition, 1967).
PENNINGTON, S. and WESTOVER, B. *A hidden workforce: homeworkers in England, 1850–1985* (London: Macmillan, 1989).
ROBERTS, E. *Women's work 1840–1940* (London: Macmillan, 1988).
ROSE, M. *The relief of poverty 1834–1914* (London: Macmillan, 1972).
RUBENSTEIN, W. D. 'The Victorian middle classes: wealth, occupation and geography' *Economic History Review, 2nd series* **30** (1977).
SAMUEL, R (ed.) *Miners, quarrymen and saltworkers* (London: Routledge, 1977).
SOUTHALL, H. 'Regional unemployment patterns among skilled engineers in Britain 1851–1914' *Journal of Historical Geography* **12** (1986), pp. 268–86.
STEDMAN-JONES, G. *Outcast London: a study in the relationship between classes in Victorian London* (Harmondsworth: Penguin, 1976).
VANCE, J. 'Housing the worker: Determinative and contingent ties in nineteenth-century Birmingham' *Economic Geography* **43** (1967), pp. 95–167.

11

Urbanization and urban life from the 1830s to the 1890s

The urban hierarchy

Transport played an essential role in the development of bigger, functionally more specialized towns by 1830 (Chapter 6). It was only with the coming of the railway and the establishing of a national rail network in the 1840s that a fully integrated urban system developed and the constraints of time and distance which kept all cities – London apart – tightly bounded in early-Victorian times were progressively reduced (Chapter 7).

This profound social revolution encompasses a period of great change in the structure of the urban system and the extent, characteristics, and internal and external relations of cities. First, urbanism became more pervasive and individual towns more populous. In 1831 some 44 per cent of the population of England and Wales and 32 per cent of Scotland's was urban-dwelling. By 1891 those proportions had increased to 75 and 65 per cent respectively (Table 6.1). Big towns grew at the expense of small. In the 1830s London was the only 'million city' but about one-sixth of Britain's population lived in large towns of over 100,000. By the 1890s nearly two-fifths did so and, in addition to London, there were, another four or five city-regions of over one million people: Glasgow, the 'second city of the Empire', Manchester, Birmingham, Liverpool and possibly Leeds. Except in Scotland and Wales, other regional capitals mostly exceeded a quarter of a million. As a result the break in the slope of the rank-size curve was less abrupt than in the early nineteenth century (Figure 6.1). Such provincial capitals were major centres of commerce and industrial services: Liverpool and Manchester for cotton imports and brokerage; Leeds for the woollen trade; and Birmingham for skilled engineering and metal manufactures. Major ports such as Liverpool and Glasgow rivalled, and in some activities surpassed, London. They dominated not only shipping services to parts of the world (Liverpool to Africa and South America, for example), but commodity trades (Liverpool for grain, cotton, sugar, oil-seeds and tobacco) and related insurance and banking. They were essentially multi-functional.

As striking as the rapid growth of great cities was the increase in both the numbers and size of manufacturing towns. Many were highly specialized

 DOI: 10.4324/9781003174783-11

but the progressive concentration of commerce and much of industry into larger units and fewer areas of production gave locational advantage at the risk of potentially dangerous over-dependence on one set of activities. The total number of towns of over 2,500 in England and Wales more than doubled between 1831 and the end of the century (from 412 to 895) (Figure 11.1). Up to mid-Victorian times the fastest-growing clusters, producing towns of up to 100,000, were in the major manufacturing areas of the industrial revolution – the West Midlands, the Potteries, south Lancashire, west Yorkshire, North East England and Clydeside. By 1871 some of the new industrial towns – such as Cardiff, Birkenhead, St Helens, Middlesbrough and Clydebank – had almost outstripped slow-growing historic cities such as Chester, York or Exeter. Such growth rates slackened by the late nineteenth century apart from in the coalfields of South Wales and North East England and in some heavy industrial areas (for example towns of Clydeside and the mid-Mersey chemicals area). Towards the end of the century renewed urban concentration of economic activity led to spill-over of great cities into surrounding residential and satellite towns. In parallel, some older cities were revitalized as new industries sought out skilled labour from declining crafts or as shifting locational values drew industries back to older towns such as Norwich, Coventry, Northampton, Leicester (an archetypal Victorian town), Derby and Edinburgh where 1860s expansion paralleled its late-Georgian surge.

The first phase of railway construction confirmed the new regional hierarchy of the nineteenth century in its focus on London, the provincial capitals and industrial areas (Figure 7.1). It also created new towns such as Crewe, Swindon, Ashford and Wolverton, workshops and company headquarters at strategic sites and junctions within their regional system. Rail companies also added new impetus to old-established towns such as Doncaster, Derby and Newton Abbot, while specialist suburbs or satellites focused on railway and engineering works for example at Springburn (Glasgow), Hunslet (Leeds), Gorton (Manchester) and Saltley (Birmingham).

Railways also played a key role in the growth of specialist resorts and residential towns. Georgian spas and seaside towns were exclusive and, apart from Bath, mostly small (Chapter 6). The railway widened access, provided a broader function and a wider regional spread. Although Bath's population remained around 53,000 throughout the Victorian period other towns such as Leamington Spa, Cheltenham, Harrogate, Llandrindod Wells (and other small Welsh spas), Pitlochry and Peebles developed as specialist resorts and, by the late nineteenth century, as residential towns.

The seaside town was a phenomenon of the age. Older resorts such as Brighton, Hastings, the Kent coastal resorts and Scarborough expanded greatly. In the south links to London and a growing middle-class clientèle promoted growth then, as cheap excursion trains and Bank Holidays (under an Act of 1871) gave working-class day trippers access to the coast, a new generation of seaside resorts came into being. Although legislation to provide a week's holiday with pay for all did not come until 1938, many northern textile towns had their 'wakes' (or holiday weeks) when factories closed and the towns emptied for the seaside. In some places the whole

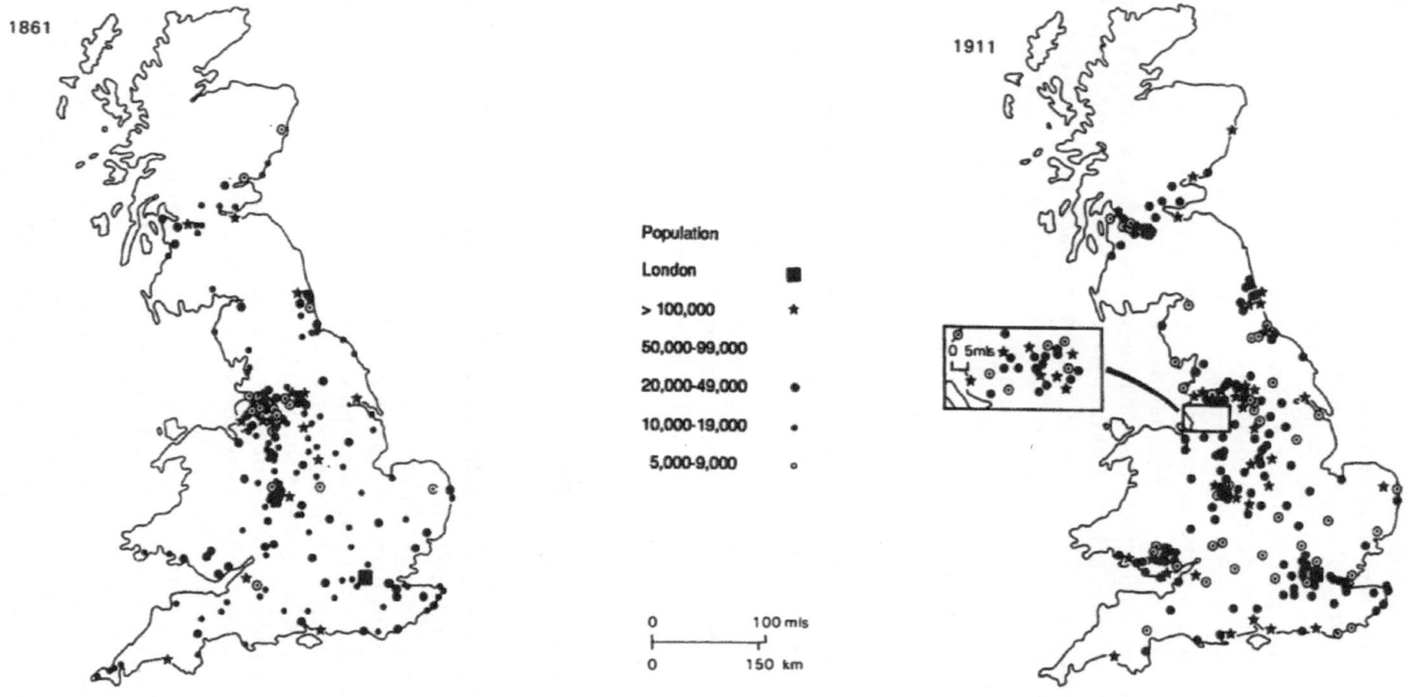

Figure 11.1 Principal urban populations of Great Britain, 1861 and 1911. Based on Langton and Morris (1986) p. 165.

town took a rest, as in the July Glasgow Fair when excursion steamers on the Clyde and the railways to the Ayrshire coastal resorts were packed. Many of the creations of the dawning of the era of mass holidays became substantial late-Victorian towns. Blackpool, Lancashire's playground (47,000 population in 1901), was one of the biggest. Its smaller Lancashire rivals were different in 'tone' and hinterland: Southport catered for Liverpool's and Manchester's middle classes; Morecambe for Bradford, the Midland Railway's link being of critical importance. On the south coast Brighton increasingly became a popular resort with Hove as its 'quality' end and the new resort of Eastbourne (a planned town on the Devonshire estates) the 'exclusive' resort: Southend was for London's eastenders, Broadstairs and Ramsgate were more 'select'. On the east coast Skegness provided for East Midlands workers; in Wales resorts ranged in character from Rhyl (close to and popular with working-class Merseysiders) to the more exclusive Llandudno, and from Barry (Cardiff's playground) to Tenby.

One level in the urban hierarchy – the small country town – lost ground and the percentage of Britain's population in towns under 10,000 had changed little by the 1890s. Rural depopulation reduced the demand for crafts and services in market and many county towns; cottage industries lost ground to factory production; and increased accessibility by rail to larger towns reduced the range of shopping and services, leading to a decline in many hitherto thriving little towns. Often the crucial factor in retaining their stock markets and wider functional base was a rail link. Those that kept out the railway (such as Stamford, Lincolnshire) or were by-passed (such as Newport Pagnell near Wolverton) declined: in those that retained local industry, such as many small towns of Northamptonshire and Leicestershire, growth often continued.

Urban planning and administration

By the 1830s, the administrative and electoral map of Britain was at odds with economic and demographic facts. Prior to the Municipal Corporations Act of 1835 in England and Wales and the Burgh Acts of 1833 and 1834 in Scotland the archaic legal structure of local government created three major sets of problems for urban government (Chapter 6). First, urban status was often unrelated to contemporary size and function: major cities – such as Manchester, Defoe's 'greatest mere village in England' and the east Lancashire cotton towns and the Black Country industrial centres – were without formal urban status; in contrast many decayed towns with Parliamentary representation (for example such rotten boroughs as Old Sarum) or a handful of inhabitants in the 'pocket' of aristocratic landowners retained borough status. London's metropolitan area of some eight miles radius from St Pauls had a population of 1.75 million in 1831 but lacked a coherent overall administrative structure.

This points to a second problem. Even where urban administrations were in place in large towns, as in the Scottish Royal Burghs of Edinburgh, Glasgow, Aberdeen and Dundee or in Incorporated Boroughs as at Liverpool, Bristol, Newcastle and Kingston upon Hull, their built-up areas were under-bounded and often tightly restricted in terms of continuing expansion. Herein lies a third problem: the effective control of a range of issues – physical, environmental, health, economic and social – which often affected areas outside existing corporation boundaries. Thus, although London's parish vestries sought to provide better sanitation and health their efforts lacked integration. Despite the work of Improvement Commissioners in most larger English cities and of the Dean of Guild Courts in Scotland, there were severe limitations to the range and integration of their activities (Chapter 6).

Between the 1830s and 1890s such difficulties led to two major restructurings of urban and local government and much state legislation on specific urban problems, together with a restructuring of the franchise and of parliamentary and civic representation (Chapter 7). The widening of the franchise created a more equal, though incomplete, accordance between parliamentary representation and property ownership and population size, and increased the urban voice in national affairs. The 1835 Act in England and Wales, and the Acts of 1833 and 1834 in Scotland, laid the basis for municipal planning and control over a wide range of issues and recognized the true administrative map of urban Britain by giving full urban status to many hitherto unincorporated towns. Some, such as Manchester, Birmingham and Sheffield were already very large; others, such as Bradford, Salford, Bolton, Huddersfield, Wolverhampton and Brighton were growing rapidly. They also permitted incorporation of adjacent townships over which urban development had spread, as reflected in the considerable 1835 boundary extensions of Liverpool and Leeds and of Glasgow in the 1830s.

The 1835 Act did not solve the problems of integrated urban government. Intervention through bye-laws in key issues – health and sanitation, housing, public amenities, poverty – was either conferred piecemeal or, as in the case of the Poor Law and the provision of compulsory state education (made over to local authorities in 1919 and 1902, respectively), was reserved for central government. Moreover, they were often out of tune with the times: thus the reformed Poor Law of 1834 created an administrative framework of 624 Unions focused on old market towns and regional centres, a pre-railway age pattern of functional regionalism which had to be constantly adjusted to meet the changing population distribution.

By the 1860s there was already a growing recognition of the need for more coherent urban administration to implement the legislation made since the 1840s on health, housing and sanitation. The first attempts to create an integrated government for a large urban area came in London with the Metropolitan Management Act of 1855 (following the Royal Commission of 1854) which reorganized the previously haphazard structure into a Metropolitan Board to control sewerage, highways, lighting and health in London's 36 Registration Districts with an 1861 population of 2.8 million.

Elsewhere, despite the addition of 554 new urban areas between 1848 and 1868 in England and Wales under Local Board procedures, confusion remained. A Royal Commission to investigate Local Government was set up in 1869 and its Second Report initiated the transition to the Acts of 1888 and 1894 which established the late nineteenth and early twentieth-century framework of local government. Meanwhile the Public Health Act of 1872 created a coherent administrative framework of Urban and Rural Sanitary Districts under a Local Government Board set up in 1871, and Acts of 1872 and 1875 developed both a code of urban administration and recognized the need for a coherent system of local government to replace the multiplicity of fragmented *ad hoc* administrative areas that had evolved over the previous 40 years.

The 1875 Local Government Act and the Municipal Corporations Act of 1882 defined the principles and functions of a new system of urban administration, but the Commissioners of the Board set up under the Local Government Boundaries Bill of 1887 and the decisions made under the local

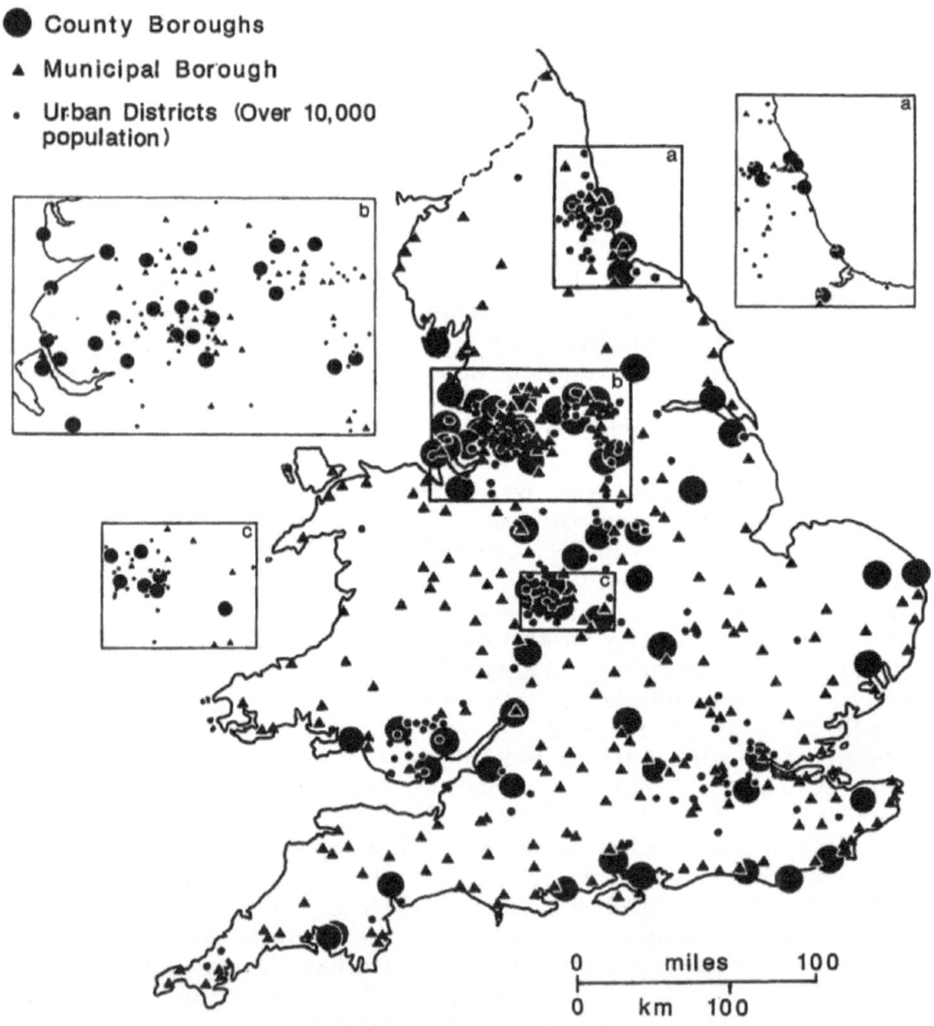

Figure 11.2 Urban administrative areas over 10,000 population in England and Wales, 1901.
Source: Census of England and Wales, 1911.

Government Acts of 1888 and 1894 determined its geography. Acts recognized that the needs of large towns could best be met by integrating all the functions of local government within comprehensive all-purpose administrations of 63 Counties and 61 County Boroughs. London became an Administrative County incorporating its 41 Metropolitan Board Areas. Borough Charters were given to those towns with a population of over 50,000 in 1881 together with all 'counties of cities' (such as Exeter or Canterbury) regardless of size. In 1894 the remaining urban areas were consolidated into Municipal Boroughs and Urban Districts each with a range of powers but subordinate to their Administrative Counties for Education, Police and Fire, and some other services (Figure 11.2).

In Scotland there were similar problems in achieving unified urban administration: for example, in satellite residential towns ('villa burghs') around Glasgow developed under the General Police and Improvement Act of 1862. Generally speaking burgh status in Scotland, however, implied urbanization (Figure 11.3) and, as in Glasgow's 1891, 1905 and 1912 incorporation of a number of such burghs, ultimately responded to the pressure of urban expansion.

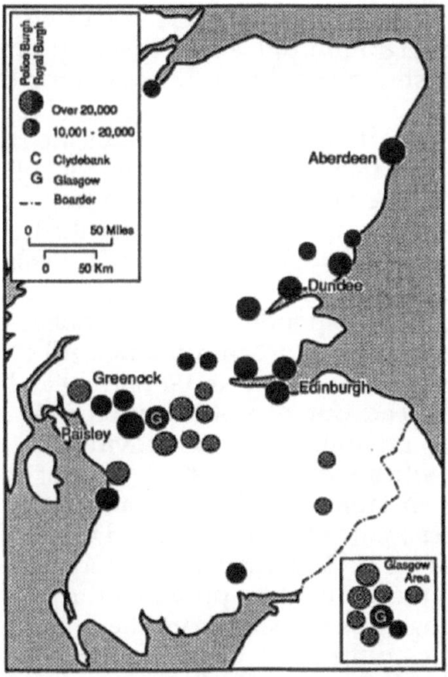

Figure 11.3 Royal Burghs and Police Burghs over 10,000 population in Scotland, 1891. Based on D. Turnock *The historical geography of Scotland.* (Cambridge: 1982) p. 147.

Yet such legislation delegated functions to towns but separated them from their county areas. Simplicity was achieved at the cost of reality at a time when, especially around large cities, town and country were becoming more interrelated. Flexibility for new boroughs and urban districts to be incorporated and for boundary extensions recognized a dynamic situation, not least the increasing suburbanization of Britain. But it was anachronistic in two ways. First, in a long list of nearly 1,000 towns many small historic centres, some of less than 2,000 people, retained their urban status. Secondly, major urban regions – for example Merseyside, the West Midlands, Greater Manchester, Leeds or Clydeside, to say nothing of Greater London – were still sub-divided into a complex of administrative areas with no overall powers in the urban region as a whole. The planning and administration of the city region remains a challenge for local government in the twentieth century (Chapter 16).

Population and housing

Demographic trends suggest that the social geography of Victorian cities should have been in a constant state of flux. Many residents in all large cities were migrants from a considerable distance (Chapter 8): in 1851 38.3 per cent of the population of London and 55.9 per cent of Glasgow's population had been born outside their respective city boundaries. Moreover, many of them did not stay long in one place: 45–55 per cent of urban populations either died or moved from a town within ten years. Victorian cities were constantly changing, although migration did not necessarily fundamentally change their social characteristics nor did high levels of mobility necessarily affect either the social geography or the processes of social control. Indeed, where social and geographical patterns were well established, and newcomers had similar characteristics to those who left, the impact of population mobility may have been slight. The following sections examine the extent to which a clearly defined social geography became established in urban areas by the 1850s, the degree to which change occurred by the 1890s, the extent to which these issues were perceived as problems within urban society, and the ways in which different groups responded to such problems.

Housing was essential for all urban residents. All cities had some homeless who slept rough in doorways or under railway arches, but the ability to get a roof over your head was one of the attractions of urban life to new migrants and young urban-born adults leaving the parental home. For established families with some resources the range of housing and variety of residential locations in large cities was an attraction. Most housing was rented and owner-occupancy rarely amounted to more than 10 per cent of the housing stock before 1918. Rented accommodation came in a vast array of types. In central areas, most was provided through the construction of purpose-built working-class housing from the mid eighteenth century, or was in large multiply-occupied dwellings filtered down from the middle classes who had moved to suburban villas or more spacious town houses. From the mid nineteenth century terraced suburbs increasingly housed the skilled working class. For those on low incomes, rent levels were crucial to housing availability. Although cheap working-class housing had been built in many cities in the early nineteenth century (Chapter 6), by the 1850s it became impossible to construct new housing to rent at much below 5s per week, well beyond the means of those on low and irregular incomes (Chapter 10). Such families had little alternative but to rent lodgings or take

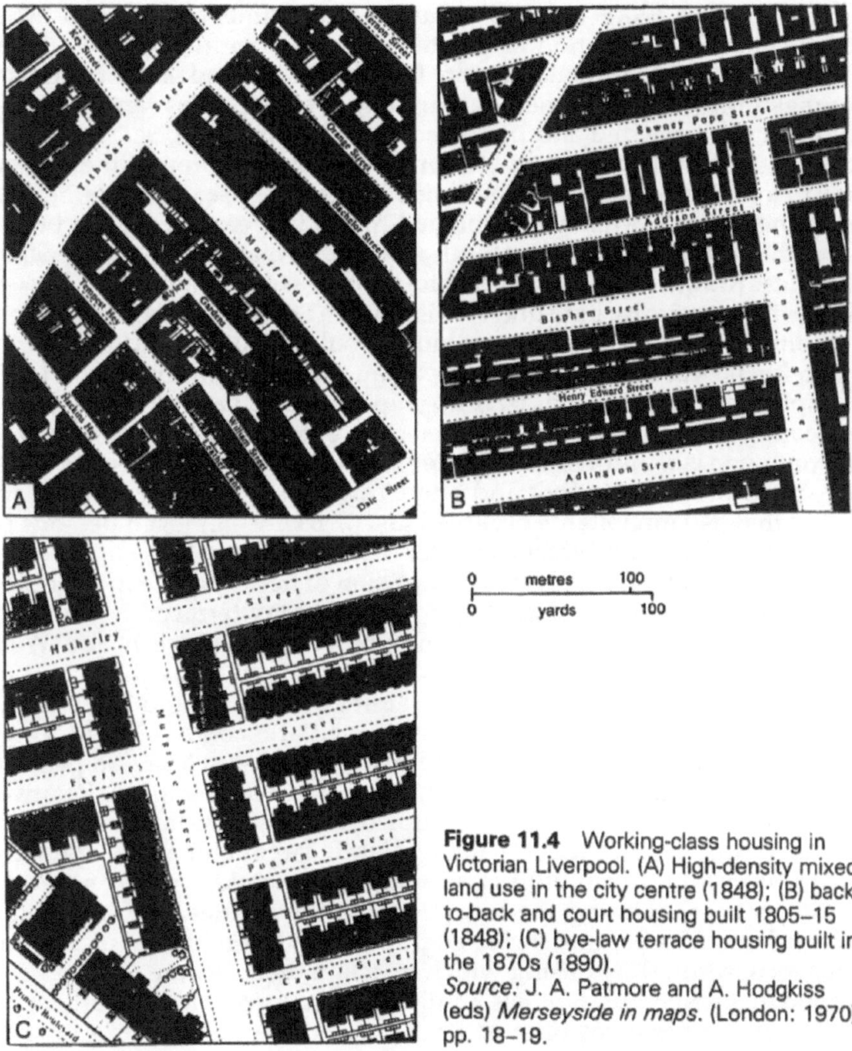

Figure 11.4 Working-class housing in Victorian Liverpool. (A) High-density mixed land use in the city centre (1848); (B) back-to-back and court housing built 1805–15 (1848); (C) bye-law terrace housing built in the 1870s (1890).
Source: J. A. Patmore and A. Hodgkiss (eds) *Merseyside in maps*. (London: 1970) pp. 18–19.

slum housing in the city centre. Thus income largely determined where you lived and construction costs controlled the type of housing that was built in different locations (Figure 11.4).

In such areas as Whitechapel or St Giles in London or dockside areas and commercial districts of Liverpool slum accommodation could be obtained with few questions asked and little required in the way of references or rent advance. Accommodation was confined and relatively expensive; for example a single room 12ft square could be rented for 1s 6d or more per week in a provincial town, and for rather more in London. It could be dirty, and facilities were shared with many other residents in the house. In Liverpool, Manchester and London cellar dwellings were not unusual: whilst providing privacy these were inevitably dark, dirty, damp and unhealthy. In Scottish cities a combination of greater poverty than in England and the 'feuing system', an annual levy on land sold, inflated land prices and encouraged high-density tenement building. Most families rented a single room in an overcrowded and poorly serviced tenement block.

A young family with greater resources would try to rent either several rooms in a multiply-occupied building or a small house: typically, in most English industrial towns, a two-roomed house in a crowded court built in the first thirty years of the nineteenth century. In many towns, particularly in the West Riding and east Lancashire, the house might be built back-to-back, further restricting light and ventilation. In Scotland, an apartment on a 'common stair' in a large tenement block would be rented. Although most Scottish families occupied fewer rooms than their English counterparts, the living space in each room was larger. A typical court in a northern industrial town could contain eight or ten houses, rented at 4s–5s 6d per week depending on size, and with one water standpipe and 1 or 2 privies for the entire court: thus in the Boot and Shoe Yard in Leeds in the 1840s it was reported that 340 inhabitants occupied just 57 rooms (almost 6 persons per room), with the nearest water a quarter of a mile away.

By the mid nineteenth century construction of new housing in the central area of towns had almost ceased, but lower-density terraced housing was expanding rapidly in new residential suburbs of all English and Welsh towns. In Scotland tenement construction continued to be the norm, but better-class tenement blocks were increasingly built around the edges of towns. An English or Welsh family with a regular income of 20s to 30s per week could probably afford to rent a new terrace house at 5s–6s 6d per week for four small rooms, its own privy and in-house water supply. Relatively few such properties were multiply-occupied, although the family might take in a lodger. Moving up the social scale, slightly larger and better-appointed terrace houses for lower-middle class white-collar workers were rented at 7s or more per week. During the second half of the nineteenth century such residential suburbs became increasingly socially differentiated as different groups focused on the housing which matched their income.

Although prosperous families continued to occupy large Georgian terraces in some inner areas of most British cities, the wealthy progressively fled most inner city areas for new suburban villas. Many of the very rich owned their own homes, and some prosperous and prudent working-class families also aspired to homeownership. This was made possible through terminating building societies in which a group of skilled workers would come together to raise finance and build their own homes, and when all members were housed and all loans repaid the society would be wound up. However, relatively high repayments of around 10s per month meant that working-class homeownership was feasible only for those with relatively stable incomes in prosperous areas: high levels were found in parts of north east Lancashire, County Durham, the West Riding and South Wales. Housing provided by employers or by philanthropic organizations (such as the Peabody Trust in London) was also locally significant, but never accommodated more than a few per cent of the population: moreover, managers tended to select tenants from among skilled workers.

The geography of housing essentially showed little fundamental change between 1830 and 1890. Although all towns spawned a succession of new residential suburbs, these were mainly for relatively affluent working and lower-middle class families who could leave the older parts of the city centre, and new skilled in-migrants. The poor remained trapped in low-cost, sub-standard housing, and the spatial segregation of social groups was clearly structured by the economic realities, reflected in income and occupation, that controlled access to different types of housing.

Cultural groups in the Victorian city

Among the many migrants into Victorian cities, some were set apart by easily identifiable social and cultural characteristics. Most obvious were the Irish, but European migrants, especially Jews, also formed cohesive and distinctive communities in many towns, whilst Welsh and Scottish migrants were sometimes perceived as distinctive cultural groups in England and vice-versa. Attracted to towns by perceived opportunities for work and housing, there were also strong push factors for many of them: famine in Ireland; religious persecution of Jews in Eastern Europe; clearances in the Scottish highlands.

Although there was considerable Irish migration to Britain before the 1840s, with well-established Irish communities in London and western port towns, it was the famine of 1846–47 which brought a flood of destitute Irish to mainland Britain and which helped create images which were to persist throughout the nineteenth century. While most towns had some Irish-born residents in 1851, recent migrants were heavily concentrated in such ports as London, Liverpool, Cardiff, Bristol and Glasgow. By 1891 the relative position had changed only slightly, although the Irish had diffused into a wider range of towns offering employment opportunities, especially textile towns of central Scotland, Yorkshire and Lancashire and industrial centres in the midlands, North East England and South Wales (Table 11.1).

Analysis of Irish communities in Victorian towns is hampered by a lack of data on the precise origins, time of migration and occupations of migrants, and by their diversity. Although the majority of famine Irish were poor Catholics, the Irish community in Britain also included wealthier Catholics and Protestant Irish from Ulster. These could – and did – form quite distinct communities, yet images of poverty, Catholicism and nationalism tend to dominate the picture. In most cities the Irish-born (which is all that census tabulations give) were highly concentrated in specific locations. Studies of London, Liverpool, Glasgow, Cardiff, Leeds, Bradford and many other towns all establish the same picture (Figure 11.5). Yet within the Irish community, there were distinctive residential patterns. Protestants and Catholics in adjacent but separate areas; and segregation between Irish of different social classes.

Most Catholic Irish famine migrants were poor and dominated the employment statistics in towns with a large Irish population. However, their concentration in poorly paid, low-status occupations shunned by other migrants obscures the fact that in all towns some Irish were found in most

Table 11.1 The Irish-born community in British towns* in 1851 and 1871

Number of Irish-born 1851		Percentage of Irish-born 1851		Number of Irish-born 1871		Percentage of Irish-born 1871	
Town	Number	Town	Per cent	Town	Number	Town	Per cent
London	108548	Liverpool	22.3	London	91171	Dumbarton	17.7
Liverpool	83813	Dundee	18.9	Liverpool	76761	Greenock	16.6
Glasgow	59801	Glasgow	18.2	Glasgow	68330	Liverpool	15.6
Manchester**	52504	Manchester**	13.1	Manchester	34066	Glasgow	14.3
Dundee	14889	Paisley	12.7	Dundee	14195	Aidrie	14.2
Edinburgh†	12514	Kilmarnock	12.1	Leeds	10128	Dundee	11.9
Birmingham	9341	Newport	10.7	Greenock	9462	Paisley	9.8
Bradford	9279	Stockport	10.6	Birmingham	9076	Middlesbrough	9.2
Leeds	8466	Bradford	8.9	Bradford	8381	Manchester	9.0
Newcastle	7124	Gateshead	8.6	Edinburgh	8031	Hamilton	8.6
Stockport	5701	Newcastle	8.1	Newcastle	6904	Newport	8.4
Preston	5122	Carlisle	8.0	Sheffield	6082	Stockport	7.5
Bristol	4761	Dumfries	7.4	Bolton	5383	Ayr	7.2
Sheffield	4477	Preston	7.4	Paisley	4703	Gateshead	6.8
Bolton	4453	Chester˙	7.3	Preston	4646	Kilmarnock	6.7
Paisley	4036	Bolton	7.3	Sunderland	4469	Merthyr Tydfil	6.6
Sunderland	3601	Wolverhampton	7.0	Plymouth[x]	4093	Bolton	6.5
Wolverhampton	3491	Edinburgh†	6.5	Stockport	3975	Stirling	6.0
Merthyr Tydfil	3051	Halifax	6.2	Bristol	3876	Chester	5.9
Hull	2983	Macclesfield	6.0	Middlesbrough	3621	Bradford	5.8

*The twenty British towns with the greatest number and percentage of Irish-born are listed.
**Manchester and Salford
†Edinburgh and Leith
[x]Plymouth and Devonport
Sources: Census of Great Britain, 1851; Census of England and Wales, 1871; Census of Scotland, 1871

Figure 11.5 Distribution of the Irish-born population in Liverpool, 1871.
Source: Census enumerators' books, Liverpool, 1871.

Table 11.2 Socio-economic structure of the Irish-born population in selected British towns in the nineteenth century

<table>
<tr><th>Town</th><th>Seg 1/2</th><th>Seg 3</th><th>Seg 4</th><th>Seg 5/6</th></tr>
<tr><td>London 1851[1]</td><td>3.0</td><td colspan="2">26.0</td><td>70.9</td></tr>
<tr><td>London 1861[1]</td><td>4.6</td><td colspan="2">21.3</td><td>74.0</td></tr>
<tr><td>Liverpool 1871[2]</td><td>6.6</td><td>9.5</td><td>24.0</td><td>57.4</td></tr>
<tr><td>York 1851[3]</td><td>8.0</td><td colspan="2">24.0</td><td>56.0</td></tr>
<tr><td>Cardiff 1851[4]</td><td>1.7</td><td colspan="2">15.1</td><td>83.0</td></tr>
<tr><td>Greenock 1851[5]</td><td>1.3</td><td colspan="2">37.5</td><td>61.2</td></tr>
<tr><td>Greenock 1891[5]</td><td>2.4</td><td colspan="2">35.9</td><td>61.7</td></tr>
<tr><td>Hull 1851[6]</td><td>6.0</td><td colspan="2">31.0</td><td>63.0</td></tr>
<tr><td>Lancaster 1871[7]</td><td>1.8</td><td>5.6</td><td>23.8</td><td>68.9</td></tr>
</table>

All figures are percentages of the total relevant population in each town
Categories are unlikely to be precisely the same in different studies because of the different classification systems used.
Seg 1/2 = Professional and intermediate occupations
Seg 3 = Skilled non-manual occupations
Seg 4 = Skilled manual occupations
Seg 5/6 = Semi-skilled and unskilled occupations

Sources: [1]L. Lees, *Exiles of Erin* (Manchester: Manchester University Press, 1979)
[2]C. G. Pooley, 'Migration, mobility and residential areas in nineteenth-century Liverpool' Unpublished Ph.D. thesis, University of Liverpool, 1978
[3]W. A. Armstrong, *Stability and change in an English county town: a social study of York 1801–51* (Cambridge: Cambridge University Press, 1974)
[4]C. R. Lewis, 'The Irish in Cardiff in the mid-nineteenth century' *Cambria* **7** (1980) pp.13–41
[5]R. D. Lobban, 'The Irish community in Greenock in the nineteenth century' *Irish Geography* **6** (1971) pp.270–81
[6]P. A. Tansey, 'Residential patterns in the nineteenth century: Kingston upon Hull 1851' Unpublished Ph.D. thesis, University of Hull, 1973.
[7]Authors' calculations

social groups. In Greenock, on Clydeside, the Irish worked mainly in sugar refining, textiles and paper mills; in Liverpool they were principally general labourers or employed in dock-related work (Table 11.2). But some Irish were found in most occupations and these were scattered among all residential areas and seemed to mix more freely with the rest of the population.

Many studies have attempted to assess the reasons for Irish residential concentration. As with other immigrants, clustering can be explained by a combination of structural (economic) and cultural reasons. Most obviously, poor Irish were constrained to live in low-cost housing areas in older central and industrial areas of cities. Within such districts many Irish seemed to cluster for cultural reasons: a grouping of Irish Catholics in a few courts provided mutual support for a religion that was still widely despised in England, or simply offered neighbourly help in an often hostile environment. While economic constraints were dominant, cultural factors – religion, a sense of common origin, or common experience of poverty and discrimination – were also significant.

As the volume of new Irish migrants decreased, Irish communities became gradually less prominent in Victorian towns although they still existed in 1890. However, many upwardly mobile Irish, both Catholic and Protestant, had moved out of the housing areas to which they had first gone.

Slum clearance also began to erode traditional Irish residential areas, although this did not begin to have a major impact until after the First World War. The Irish population in mainland Britain, however, remained a focus of both political and popular attention, fuelled by the growing political crisis in Ireland and the identification of Catholic Irish with its Nationalist cause. In the late nineteenth century Liverpool elected Irish Nationalist councillors in some wards, and a long-serving Member of Parliament in the Scotland Division, whilst sectarian violence continued to erupt on the streets of Glasgow and Liverpool, not least during Orange or St Patrick's Day parades.

The impact of other minorities was small in comparison to the Irish, although the Jews shared some of their characteristics and formed important communities in such towns as London, Manchester and Leeds. The total Jewish population in England in 1852 was some 28,500 (of which 20,000 were in London); and by 1880 this had expanded to around 65,000. The London Jewish community was by far the largest and contained both poor working-class Jews and those in higher status occupations, particularly dealing and commerce. London Jewish congregations were heavily concentrated in the East End and experienced many of the negative factors also suffered by the Irish. In contrast, Welsh and Scottish migrants to English cities, although having some cultural cohesion and residential separation (especially the Welsh), were in other respects easily assimilated into English society.

Disease in the Victorian city

Victorian cities were unhealthy places, though chances of illness and premature death varied considerably depending upon who you were, where you lived, how much you earned and how well you were fed. Not all towns had equally high mortality rates, and death rates in the countryside could match those in middle-class suburban areas of cities (Chapter 8). Contemporary opinion was most concerned about infectious disease, though as figures for Manchester show (Table 11.3) more people died from 'other causes' than from all infectious diseases together: from old age, heart attacks, cancers, bronchitis, pneumonia and most of the non-infectious diseases still important today. Nevertheless, malnutrition and the weakening effects of infectious disease increased susceptibility to death from non-infectious causes, and struck hardest at the inner city poor.

Infectious disease was spatially concentrated: deaths from tuberculosis, typhus, typhoid and cholera focused mainly on inner-city slum districts (Figure 11.6). However, big variations in the virulence and method of transmission of different infectious diseases – although not fully understood by contemporaries – affected both the geography of disease within towns and public health reforms.

Respiratory tuberculosis (phthisis or consumption) was the largest single adult killer in the nineteenth century. Few families were not touched by the effects of T.B. and even at the end of the century tuberculosis was responsible for around 10 per cent of all deaths nationally, despite a significant decline from 1850. Spread by a bacillus through droplet infection from coughs and saliva, tuberculosis is not highly contagious but its spread is

Table 11.3 Average annual standardized[1] mortality rates from selected infectious diseases: Manchester, Salford and Chorlton Registration Districts, 1851–60 to 1891–1900

Cause of death	1851–60	1861–70		1871–80		1881–90		1891–1900	
Smallpox	0.19		0.13		0.29		0.02		0.01
Measles	0.54		0.60		0.58		0.71		0.77
Scarlet fever }	1.46	1.30 }	1.46	0.98 }	1.08	0.51 }	0.73	0.27 }	0.49
Diphtheria }		0.16 }		0.10 }		0.22 }		0.22 }	
Whooping Cough	0.82		0.85		0.72		0.57		0.58
Typhus }	1.03		1.38	0.10 }	0.61	0.03 }	0.36	0.003 }	0.27
Typhoid fever }				0.37 }		0.31 }		0.25 }	
Simple continued fever }				0.14 }		0.02 }		0.005 }	
Diarrhoea, Dysentery and Cholera	2.07		2.04		1.48		1.07		1.35
Pulmonary tuberculosis	3.26		3.10		2.97		2.40		2.02
Non-respiratory tuberculosis	0.89		0.89		0.83		0.94		0.93
Total from above	10.26		10.45		8.56		6.80		6.42
All other causes	17.11		17.24		17.57		16.98		16.58
All causes	27.37		27.69		26.13		23.78		23.00
Mean population	479349	555566		650341		754429		845519	

[1]Rates per thousand are standardized to age structure of area 1891–1900

Source: Annual reports of Registrar General
See also M. E. Pooley and C. G. Pooley, 'Health, society and environment in nineteenth-century Manchester' in R. Woods and J. Woodward (eds) *Urban disease and mortality in nineteenth-century England* (London: Batsford, 1984) pp.148–75

encouraged by prolonged contact in crowded conditions, especially under poor nutrition. Although not immune, the middle classes were better able to withstand tuberculosis than the poor, malnourished working class.

Although typhus and typhoid fever were not separately diagnosed until 1869, they have completely different methods of transmission. Typhus, spread by body lice mainly to adults, is encouraged by filthy living conditions, overcrowding and malnutrition. Endemic in the nineteenth century, it became epidemic during economic depressions and poverty crises and was strongly associated with poor residential areas. In contrast, typhoid fever is spread by a bacillus contained in sewage-contaminated water, milk or food and is directly related to poor sanitation and hygiene. While it could spread through the water supply to all parts of a town, inner-city districts

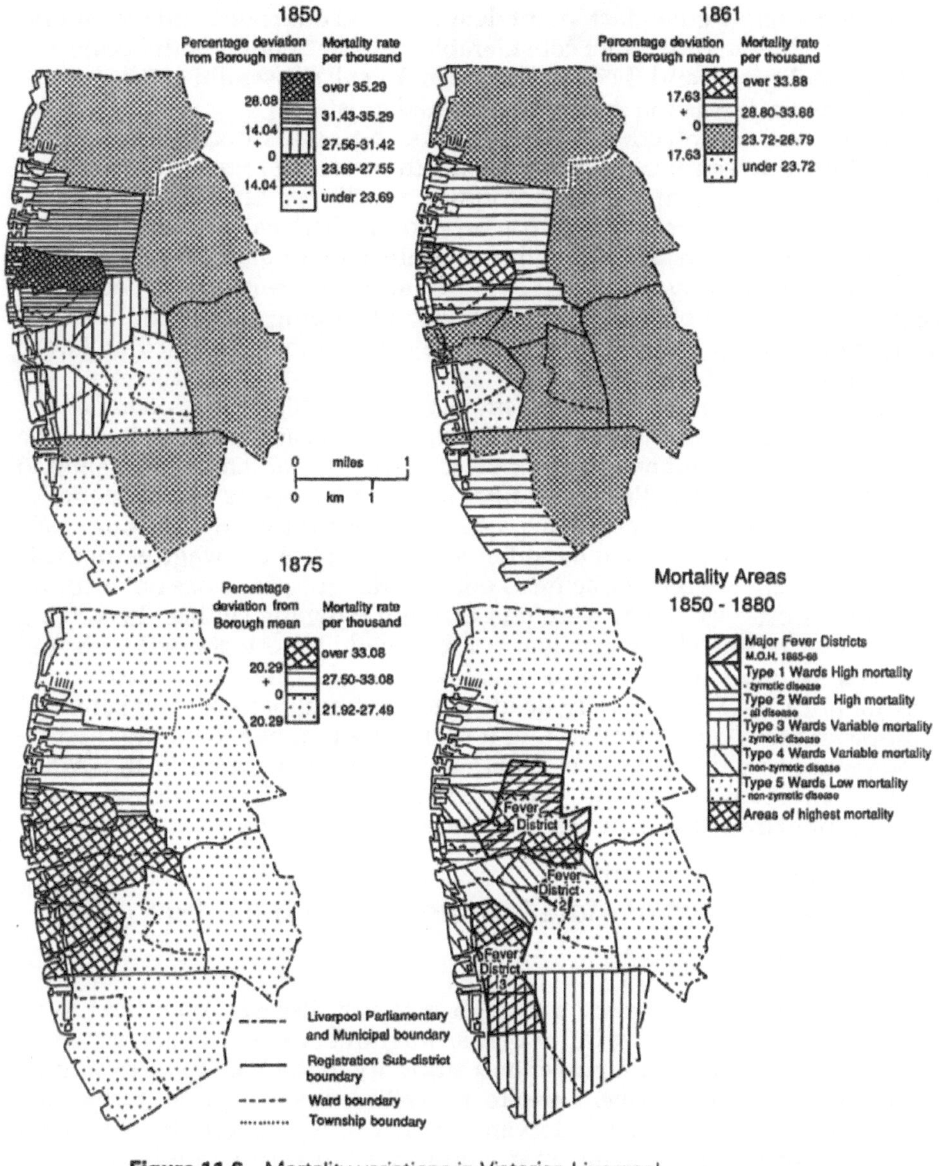

Figure 11.6 Mortality variations in Victorian Liverpool.
Source: Annual Reports of the Medical Officer of Health for Liverpool.

were most likely to have an inadequate or contaminated water supply and the poor were hit the hardest. Cholera is also spread by contaminated water and food but, unlike typhoid, it occurred only in specific epidemics introduced from Europe (1831–32, 1848–49, 1853–54 and 1866) and was not otherwise present in Britain. Epidemic mortality could be high in affected communities, but in general it was much less important than other infectious diseases: many deaths in the Registrar General's category which includes cholera were in fact infant deaths due to diarrhoea and dysentery. However, cholera did attract considerable public attention both because of its high fatality rate and the fact that it struck at all classes although, as with typhoid fever, it was the poor who suffered most.

Although most infectious diseases affecting adults declined gradually during the nineteenth century, infant death remained high and by the end of the nineteenth century still accounted for 25 per cent of all mortality (Chapter 8). As suggested earlier the very young died mainly from diarrhoea and dysentery contracted through contaminated water, milk or food and aided by filthy and overcrowded living conditions. Deaths from childhood diseases such as diphtheria, scarlet fever, whooping cough and measles were most widespread and fatal among weak and malnourished children. Both most afflicted the poor.

It is pertinent to ask what the impact of such high rates of infectious disease was on those affected. Death was only one, and not necessarily the most important, of many effects of disease. For a poor family struggling to pay rent and buy food, illness (whether fatal or not) created a whole series of additional strains: medical bills to pay; medicines to buy; extra heating costs; and problems of childcare if a mother was taken ill. If the wage earner was off work the crisis would be acute as not only did outgoings rise but incomes also fell. Short-term crises were met through pawning clothes, borrowing from kin or raising short-term loans. Prolonged illness increased costs and reduced incomes to such an extent that it could cause or increase malnutrition for the whole family, leading to further illness or to eviction for non-payment of rent. Families might then have to move to inferior accommodation or, *in extremis*, to be separated from one another in the Workhouse. The high level and concentration of infectious disease was a significant extra burden for working-class families in the Victorian city.

The development of state intervention in the urban environment

Poor housing, overcrowding and high levels of disease, often held to be exacerbated by the massive influx of Irish migrants, were certainly perceived as problems by those with power and authority in the Victorian city and by politicians at Westminster. Despite prevailing *laissez faire* attitudes, the development of municipal intervention in various aspects of the urban environment reveals a genuine structural crisis in urban living conditions with an increasing gap between public expectations and the realities of urban life. Whether this gap was due to increasing aspirations or deteriorating conditions is, however, open to question and certainly varied over space and time.

Much as they might have wished to, neither local nor national politicians could ignore urban living conditions. First, the increasing amount of statistical and other information was discussed and publicized by local societies and used as propaganda by medical men and others with first-hand experience of life in the slums. Edwin Chadwick was the best known propagandist, but at the local level many influential people became increasingly aware and concerned about conditions of urban life. Such evidence was unlikely to have been enough on its own to persuade ratepayers and their elected representatives to pass legislation and spend money improving housing and sanitation for the working classes. Self-interest was at the heart of political action. Concerned about events elsewhere in Europe, politicians genuinely feared that poor living conditions could lead to mass disturbances and urban violence. Closer to home, the impact of cholera in 1832 and 1848 brought home the fact that disease could affect all classes. The poor were blamed for the disease, but it was in the interests of the middle classes to improve conditions and prevent it recurring. Intervention was also rationalized through economic self-interest since a reduction in disease and improvement in housing would produce a more efficient workforce and therefore benefit industrialists and entrepreneurs.

Whatever the reasons, the second half of the nineteenth century saw unprecedented activity in the passing of both local bye-laws and national legislation affecting urban living conditions (Table 11.4). Local legislation was in practice more important than that passed by the national Parliament: national acts often incorporated what had previously occurred at a local level. The changing framework of local government from the 1830s (see above) made such legislation possible.

Although the 1848 Public Health Act did not effect any major changes in urban areas, it was the culmination of a concerted public health campaign in England and Wales marking acceptance of the fact that public health was an

Table 11.4 Principal public health and housing legislation in Britain, 1848–1900

1848	Public Health Act (England and Wales)
1851	Lodging Houses Act (Shaftesbury Act) (England and Wales)
1855	Dwelling Houses Act (England and Wales; Scotland)
1858	Public Health Amendment Act (England and Wales)
1866	Sanitary Act (England and Wales)
1866	Labouring Classes' Dwelling Houses Act (England and Wales)
1867	Public Health Act (Scotland)
1868	Artisans' and Labourers' Dwellings Act (Torrens Act) (England and Wales; Scotland)
1872	Public Health Act (England and Wales)
1874	Working Men's Dwellings Act (England and Wales)
1875	Public Health Act (England and Wales)
1875	Artisans' and Labourers' Dwellings Improvements Act (Cross Act) England and Wales; Scotland)
1879/80	Artisans' and Labourers' Dwellings Improvements Act (England and Wales; Scotland)
1882	Artisans' Dwellings Act (England and Wales; Scotland)
1885	Housing of the Working Classes Act (England and Wales; Scotland)
1890	Housing of the Working Classes Act (England and Wales; Scotland)
1890	Public Health Act (England and Wales)
1897	Public Health Act (Scotland)
1899	Small Dwellings Acquisitions Act (England and Wales; Scotland)
1900	Housing of the Working Classes Act (England and Wales; Scotland)

For further details see *English Historical Documents* Vol. XII(1) and XII(2) (eds G. M. Young and W. D. Hancock)

Sources: E. Gauldie, *Cruel Habitations* (London: Allen and Unwin, 1974): I. Adams, *The making of Urban Scotland* (London: Croom Helm, 1978)

issue of national concern. Not until the 1866 Sanitary Act were local Authorities obliged to provide a proper water supply, drainage and sewerage system, and even this Act lacked teeth to enforce its powers. Many towns acted independently: Manchester, for instance, took control of the city's water supply in 1851. But the power to force Local Authorities to act to improve water supply and sanitation did not become effective until the 1875 and, especially, 1890 Public Health Acts.

The effectiveness of such legislation in reducing disease can be called into question. Clean water and improved sewage disposal certainly reduced the impact of water-borne diseases like typhoid and cholera, but it can be argued that the impact of infectious disease was related more to poverty and levels of nutrition than it was to sanitation. McKeown argues that 50 per cent of the decline in infectious disease over the nineteenth century was due to general improvements in living standards and diet, 25 per cent due to autonomous changes in disease prevalence and 25 per cent due to sanitary improvements after 1850. Szreter (1988) and others place greater emphasis on public and environmental health measures, but it is clear that sanitary improvements alone must have had a limited impact on many nineteenth-century diseases.

Action to improve housing generally came after legislation on water supply and drainage, and not until the 1858 Public Health Amendment Act was there any significant national legislation on housing. In theory this banned back-to-back housing (something which Manchester for example had done in 1844), but in practice the effects of both national and local legislation were limited. The Act lacked teeth: most local legislation applied to old administrative areas and ignored expanding suburbs where back-to-backs could still be built. The Torrens Act of 1868 and the Cross Act of 1875 both explicitly tackled the problem of slum housing, but both were mainly destructive. Although Local Authorities could rebuild in areas of slum clearance, in practice few did and the effects of the Acts were to remove cheap housing stock without providing additional low-cost accommodation. Even those authorities which did provide Corporation housing (for example Liverpool) made little impact on the housing problem because they demolished more than they built and provided new housing that was mostly too expensive for the dispossessed poor. In the ten years following the 1875 Act only 12 towns applied for improvement orders under the legislation (although some acted under local legislation) and in most places relatively little was done until the Housing Act of 1890. For most of the urban poor, housing conditions were scarcely improved during the nineteenth century. In Scottish cities health and housing was, if anything, worse than in England due both to greater poverty and the unworkability of much British public health legislation north of the border. In public health Scotland probably did not catch up with England until the 1890s.

The spatial structure of Victorian cities

The social geography of Victorian cities was structured by realities of income, availability of jobs and housing, the effects of discrimination against particular groups and the impact of disease on insanitary urban areas. Their

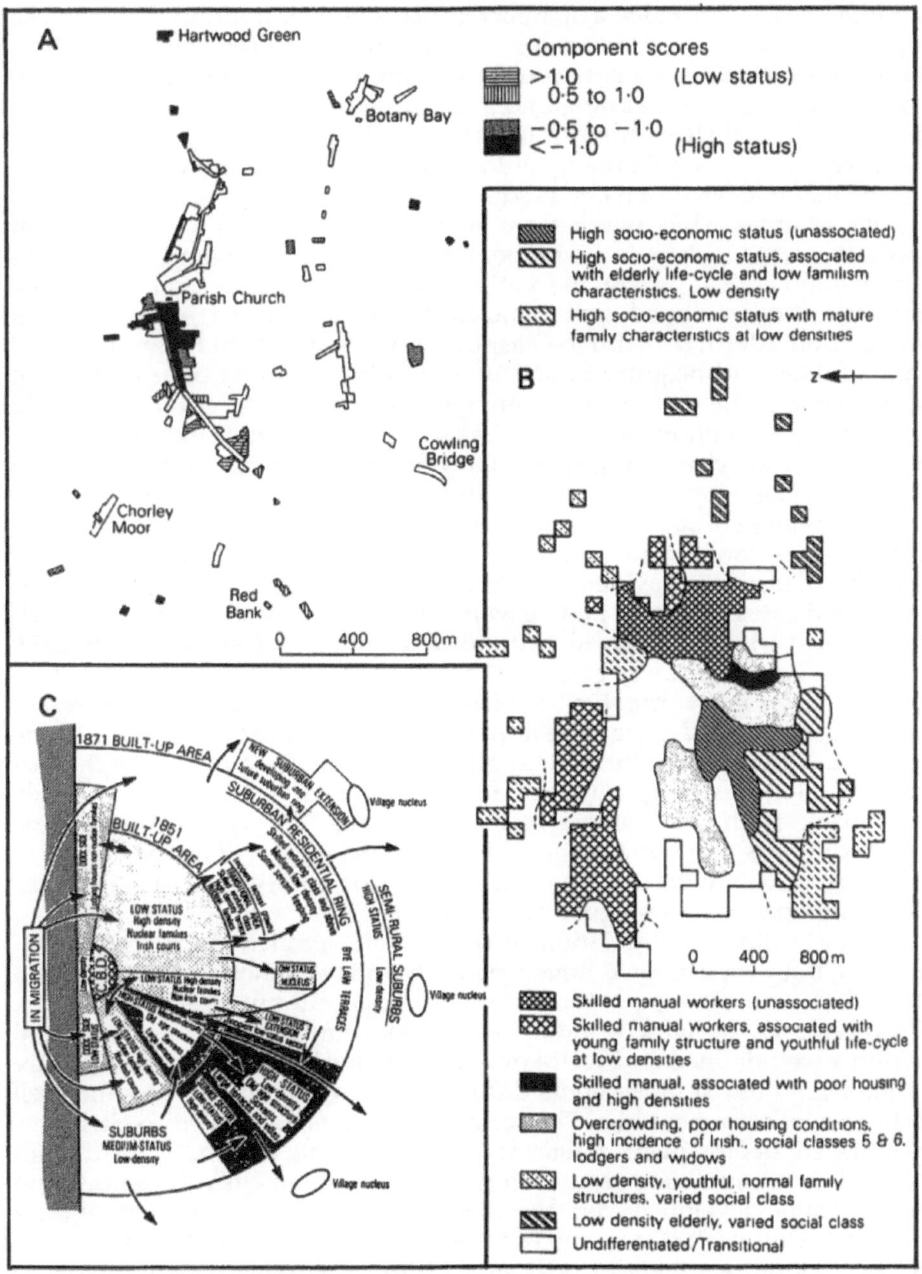

Figure 11.7 The spatial structure of Victorian cities. (A) High and low status areas in Chorley (Lancs), 1851; (B) social areas in Preston (Lancs), 1851; (C) simplified model of social areas in Liverpool, 1871.
Source: Lawton and Pooley (1988) p. 164.

net effect was to produce a spatially structured city, not only through the separation of rich and poor but also, more subtly, between particular ethnic and cultural groups and between those with different aspirations and life-styles. Although the mapping of census evidence clearly demonstrates the broad residential structuring of urban areas (Figure 11.7), such techniques rarely capture the subtle reality of local communities. Ward (1980) demonstrated that mid-Victorian Leeds contained areas which were socially mixed as well as areas which were more segregated depending on the scale of analysis used, much as in Liverpool and other cities (Pooley 1984). As the nineteenth century progressed patterns of segregation became more pronounced and developed earlier and most obviously in large towns. In all towns, from Merthyr Tydfil to Glasgow and from Leeds to Plymouth, local communities and neighbourhood units can be identified on a number of spatial scales. The patterns that emerge depend upon the criteria and scale used, but for individuals and families such local communities and neighbourhoods were clearly important. They contained people with sufficient in common to develop a sense of identity.

Measurement of small communities in urban areas is difficult, and there is a tendency to concentrate on those (like the Irish) with obvious characteristics. But descriptive evidence relating to individuals and small groups clearly indicates that most people were attached to a locality where they felt secure, had friends, received support and carried out most of their daily activities. This is certainly borne out by the diary of David Brindley, a dock railway porter in Liverpool in the 1880s, who lived at 12 different addresses in Liverpool over an eight-year period but remained within the same residential area with quite finite spatial limits. For Brindley his neighbourhood fulfilled specific functions and had real meaning, but it was not easily identifiable from, for instance, census evidence. The view of Cannadine (1982) supports that there is a need to match the shapes of residential areas on the ground with the functions that those areas serve within society, a task that geographers are only just beginning to tackle.

Not only did the population of individual Victorian cities change frequently, but people moved house regularly and over short distances within the urban area. The poor moved most frequently but over the shortest distances and, as with Brindley, much movement was circulatory and within a well-defined and well-known neighbourhood community (Figure 11.8): such internal movements did not weaken community ties and well defined spatial frameworks. The social geography of most British cities was established before mass public transport became important. Although horse-drawn omnibuses were common in the late nineteenth century and electric trams developed from the 1890s, these mostly served middle-class suburbs. Well-paid artisans and office workers might use public transport to get to work or for recreation, but most ordinary working people walked to work, shopping and social events.

In contrast, the railway undoubtedly had a great impact on the internal structure of cities as railway companies vied with each other to buy land and establish termini close to the city centre. Kellett (1969) estimates that railway companies owned some 5–9 per cent of central city land and up to 20 per cent of urban land overall. Cities such as Manchester quickly developed a

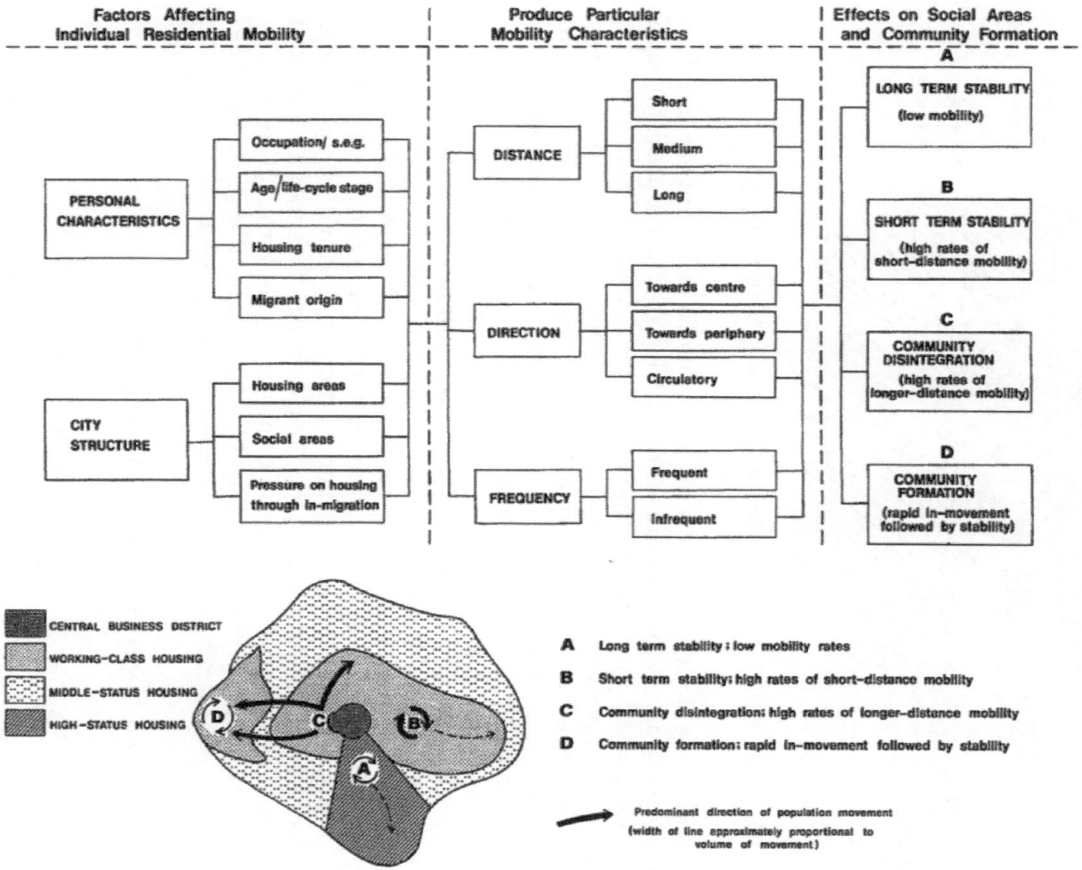

Figure 11.8 Summary of the process of urban residential mobility in Victorian cities.
Source: Pooley (1979) p. 274.

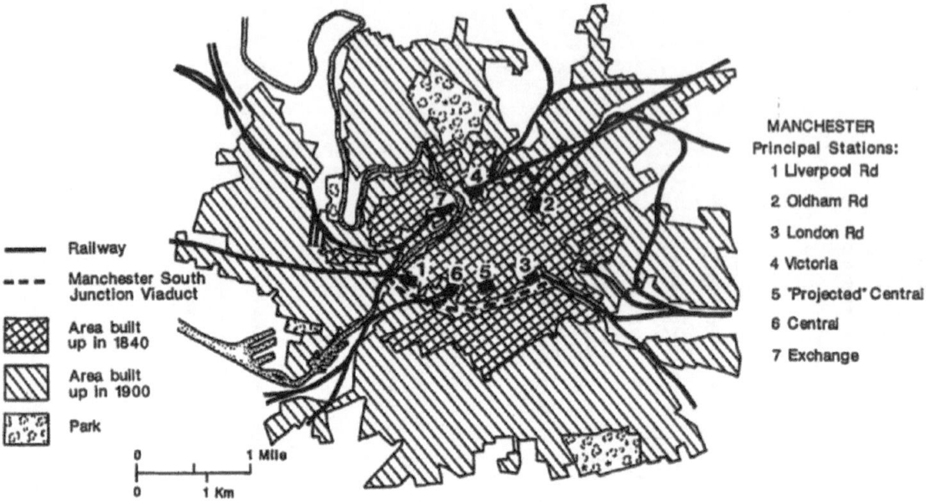

Figure 11.9 The impact of railway development on Victorian Manchester. Based on Kellett (1979) p. xxiii.

ring of competing stations (Figure 11.9) which had a major impact on urban life. Noisy and dirty, railways were detrimental to residential areas through which they passed; termini generated additional traffic which added to congestion and pollution; whilst the construction of track and railway buildings removed considerable quantities of working-class housing. In most cities railway companies were a more effective agent of slum clearance than the local authorities, but demolished housing was not replaced and this increased pressure on the rest of the housing stock.

Life in a Victorian city thus varied depending on income and location of work and residence, though for most people everyday life was acted out within a well-defined social area or neighbourhood community. A social geography which, in most cities, was emerging from the 1840s changed little by the 1890s, although neighbourhood boundaries could become more firmly defined and levels of segregation between rich and poor accentuated and new suburbs acquire character. For most people, however, it was the local environment which was important, and small-scale differences between courts and streets could have considerable significance.

Bibliography and further reading for Chapter 11

For overviews see:

ADAMS, I. *The making of urban Scotland* (London: Croom Helm, 1978).

BRIGGS, A. *Victorian cities* (Harmondsworth: Penguin, 1968).

CARTER, H. and LEWIS, R. *An urban geography of England and Wales in the nineteenth century* (London: Arnold, 1990).

CHERRY, G. *Cities and plans. The shaping of urban Britain in the nineteenth and twentieth centuries* (London: Arnold, 1988).

DENNIS, R. *English industrial cities of the nineteenth century* (Cambridge: Cambridge University Press, 1984).

GORDON, G. and DICKS, B (eds) *Scottish urban history* (Aberdeen: Aberdeen University Press, 1983), Chapters 4–7 and 9.

KELLETT, J. *The impact of railways on Victorian cities* (London: Routledge, 1969).
LAWTON, R. 'An age of great cities' *Town Planning Review* **43** (1972), pp. 199–224.
LAWTON, R. 'Peopling the past' *Transactions of the Institute of British Geographers NS***12** (1987), pp. 259–83.
WALLER, P. J. *Town, city and nation. England 1850–1914* (Oxford: Oxford University Press, 1983).

On urban growth see:
LAW, C. M. 'The growth of urban population in England and Wales, 1801–1911' *Transactions of the Institute of British Geographers* **41** (1967), pp. 125–43.
LIPMAN, V. D. *Local government areas, 1834–1945* (Oxford: Blackwell, 1949).
ROBSON, B. *Urban growth: an approach* (London: Methuen, 1973).
WEBER, A. F. *The growth of cities in the nineteenth century* (London: Greenwood edition, 1969).

On urban population mobility see:
ANDERSON, M. 'Indicators of population change and stability in nineteenth-century cities: some sceptical comments' in Johnson, J. H. and Pooley, C. G. (eds) *The structure of nineteenth-century cities* (London: Croom Helm, 1982), pp. 283–98.
DENNIS, R. 'Intercensal mobility in a Victorian city' *Transactions of the Institute of British Geographers NS***2** (1977), pp. 349–63.
POOLEY, C. G. 'Residential mobility in the Victorian city' *Transactions of the Institute of British Geographers NS***4** (1979), pp. 258–77.

On housing and social conditions see:
BURNETT, J. *A social history of housing 1815–1970* (Newton Abbot: David and Charles, 1978).
CHAPMAN, S. D. (ed.) *The history of working-class housing* (Newton Abbot: David and Charles, 1971).
DAUNTON, M. *House and home in the Victorian city: working-class housing 1850–1914* (London: Arnold, 1983).
DENNIS, R. 'The geography of Victorian values: philanthropic housing in London 1840–1900' *Journal of Historical Geography* **15** (1989), pp. 40–54.
DOUGHTY, M. (ed.) *Building the industrial city* (Leicester: Leicester University Press, 1986).
DYOS, H. J. *Victorian suburb. A study of the growth of Camberwell* (Leicester: Leicester University Press, 1961).
ENGLANDER, D. *Landlord and tenant in urban Britain 1838–1918* (Oxford: Oxford University Press, 1983).
GAULDIE, E. *Cruel habitations: a history of working-class housing 1780–1918* (London: Unwin, 1974).
RODGER, R. *Housing in urban Britain 1780–1914* (London: Macmillan, 1989).
SPRINGETT, J. 'Building development on the Ramsden Estate, Huddersfield.' *Journal of Historical Geography* **8** (1982), pp. 129–44.
TREBLE, J. H. *Urban poverty in Britain, 1830–1914* (London: Methuen, 1983).

On cultural groups and social structure in Victorian cities see:
LEES, L. *Exiles of Erin* (Manchester: Manchester University Press, 1979).
McLEOD, H. *Religion and class in the Victorian city* (London: Croom Helm, 1974).
MORRIS, R. J. (ed.) *Class, power and social structure in British nineteenth-century towns* (Leicester: Leicester University Press, 1986).
NEALE, F. *Sectarian violence: the Liverpool experience 1818–1914* (Manchester: Manchester University Press, 1988).
POOLEY, C. G. 'The residential segregation of migrant communities in mid-Victorian Liverpool' *Transactions of the Institute of British Geographers NS***2** (1977), pp. 363–82.

SWIFT, R. and GILLEY, S (eds) *The Irish in the Victorian City* (London: Croom Helm, 1985).
SWIFT, R. and GILLEY, S (eds) *The Irish in Britain, 1815–1939* (London: Pinter, 1989).
WILLIAMS, B. *The making of Manchester Jewry, 1740–1875* (Manchester: Manchester University Press, 1976).

On health and municipal intervention see:

CHERRY, G. 'The town planning movement and the late-Victorian city' *Transactions of the Institute of British Geographers NS*4 (1979), pp. 306–319.
EYLER, J. *Victorian social medicine* (Baltimore: Johns Hopkins, 1979).
FLINN, M. W. (ed.) *Edwin Chadwick's report on the sanitary condition of the labouring population of Great Britain, 1842* (Edinburgh: Edinburgh University Press, 1965).
FRASER, D. *Power and authority in the Victorian city* (Oxford: Blackwell, 1979).
KELLETT, J. 'Municipal socialism, enterprise and trading in the Victorian city' *Urban history yearbook* (Leicester: Leicester University Press, 1978).
ODDY, D. J. 'Urban famine in nineteenth century Britain: the effect of the Lancashire cotton famine on working-class diet and health.' *Economic History Review* **37** (1983), pp. 68–86.
ROSEN, G. 'Social variables and health in an urban environment: the case of the Victorian city' *Clio Medica* **8** (1971), pp. 1–7.
SUTCLIFFE, A. 'The growth of public intervention in the British urban environment during the nineteenth century: a structural approach' in Johnson, J. H. and Pooley, C. G. (eds) *The structure of nineteenth century cities* (London, Croom Helm, 1982), pp. 107–124.
SZRETER, S. 'The importance of social intervention in Britains mortality decline 1850–1914' *Social History of Medicine* **1** (1988), pp. 1–38.
WOHL, A. *Endangered lives: public health in Victorian Britain* (London: Methuen, 1984).
WOODS, R. and WOODWARD, J (eds) *Urban disease and mortality in nineteenth-century England* (London: Batsford, 1984).
YELLING, J. *Slums and slum clearance in Victorian London* (London: Allen and Unwin, 1986).

On the spatial structure of Victorian cities:

CANNADINE, D. 'Victorian cities: how different?' *Social History* **4** (1977), pp. 457–82.
CARTER, H. and WHEATLEY, S. *Merthyr Tydfil in 1851* (Cardiff: University of Wales Press, 1982).
HARRIS, R. 'Residential segregation and class formation in the capitalist city' *Progress in Human Geography* **8** (1984), pp. 26–49.
JOHNSON, J. H. and POOLEY, C. G (eds) *The structure of nineteenth century cities* (Part 4), (London: Croom Helm, 1982).
LAWTON, R. and POOLEY, C. G. 'The social geography of nineteenth-century cities' in Denecke, D. and Shaw, G. (eds) *Urban historical geography: recent progress in Britain and Germany* (Cambridge: Cambridge University Press, 1988) pp. 159–174.
POOLEY, C. G. 'Residential differentiation in Victorian cities: a reassessment' *Transactions of the Institute of British Geographers. NS*9 (1984), pp. 131–44.
WARD, D. 'Victorian cities: how modern?' *Journal of Historical Geography* **1** (1975), pp. 135–52.
WARD, D. 'Environs and neighbours in the 'Two Nations': Residential differentiation in mid-nineteenth century Leeds' *Journal of Historical Geography* **6** (1980), pp. 133–62.
See also other essays in *Transactions of the Institute of British Geographers NS*2 (1977), and *NS*4 (1979).

General studies of specific towns include:

BRIGGS, A. *History of Birmingham* (Oxford: Oxford University Press, 1952).
DAUNTON, M. *Coal metropolis: Cardiff 1870–1914* (Leicester: Leicester University Press, 1977).
FRASER, A. *A history of modern Leeds* (Manchester: Manchester University Press, 1980).
GIBB, A. *Glasgow: the making of a city* (London: Croom Helm, 1983).
HILL, F. *Victorian Lincoln* (Cambridge: Cambridge University Press, 1975).
OLSEN, D. J. *The growth of Victorian London* (London: Batsford, 1976).

SECTION III:

Britain from the 1890s to the 1940s

12

The political, economic and social context, 1890–1950

In comparison with the previous 60 years, the period from the 1890s to the 1940s was characterized by instability. At home traditional political values were challenged by the rise of the labour movement; social and cultural values were questioned through the rise of secularism; society was transformed by progressive involvement of women in work and politics and the development of state welfare; and the economy was rocked by the effects of recession, two World Wars and fundamental restructuring. Internationally, Britain's position as a major imperial power became progressively weaker. Yet, despite these shocks, many nineteenth-century trends continued: heavy industry remained important in the economy; the Empire was still significant economically and politically; and a whole series of political and related institutions, either explicitly or implicitly, worked to maintain traditional values and class structures. While the seeds of fundamental economic and social change were being sown, they did not mature until the second half of the twentieth century.

Political initiatives for social reform during the nineteenth century and influences from the rest of Europe led to the formation of the Independent Labour party in 1893. Although socialism remained politically marginalized for most of the following 30 years its growth not only led to the eventual replacement of the Liberals by the Labour Party as the main opposition, but was also reflected in changing attitudes and policies of the Conservative Party. In 1906 the newly-formed Labour Party had 29 seats in Parliament: by 1919 its 59 members of Parliament represented some 2,375,000 voters. Soon the effective opposition, the Labour Party thereafter played a significant role in a mostly Conservative-dominated Parliament. Following brief and largely unsuccessful periods of Labour government in 1923–1924 and 1929–31, the economic crisis which led to the formation of the National government in 1931 and a wartime Coalition government between 1939 and 1945 increased

 DOI: 10.4324/9781003174783-12

the influence of socialist politicians. This, and the collapse of the pre-1914 social and class system, paved the way for the election of Clement Attlee's Labour government in 1945. Over a full five-year term this government introduced genuinely socialist policies, marking the culmination of a long period of tentative state intervention in social affairs which had begun in the nineteenth century, and extending wartime controls in economic policy which eventually led to fully fledged regional planning.

For a significant part of this period domestic politics were overshadowed by international events, especially the effects of two World Wars. Although the Boer War (1899–1902) had more direct impact on the British public than the conflict in the Crimea, highlighting the poor medical condition of many recruits, the First World War (1914–18) had more fundamental effects on British society. Some 744,000 men from the United Kingdom were killed during the war, possibly three times that number seriously injured and by 1918 some nine million men and women were either serving in the armed forces or working in munitions factories. The removal of such large numbers, temporarily or permanently, from British society and economy had a massive effect on those left behind (particularly on women). The war was certainly partially responsible for the initiation of some of the major social and political changes after 1918.

The social effects of the Second World War (1939–45) were in some respects less dramatic, although it hastened the completion of trends begun earlier in the century. Although British casualties were considerably fewer than in World War I some 303,000 members of the armed forces died and the effects of air raids on most major British cities brought the horrors of war much closer to home. Some 60,000 civilians were killed in air raids and the destruction of large parts of British towns had a major impact on post-war planning and redevelopment. The financial cost of World War II was also substantial, marking a major turning point in Britain's position as a world power.

Victorian social values were challenged on a number of fronts after 1890 and in some cases led to significant legislation. Most importantly the role of women in society changed. Although economic opportunities for women were beginning to broaden in the late nineteenth century, in 1914 there were still only 212,000 women employed in engineering and munitions industries, 18,000 in transport and 33,000 in clerical work. With most of the male population sucked into the war, women had to fill the jobs left behind in industry, commerce and agriculture. Female participation in the labour force increased more than four fold between 1914–18 when over one million women took on paid work for the first time.

The effect of this was to strengthen claims for female emancipation begun before 1914 and, although many women lost their jobs when men returned from the war, it was not possible for women to return completely to Victorian domestic slavery in 1918. Attitudes to women gradually changed and, following energetic campaigning by Emmeline Pankhurst and others, the franchise was extended to include some women. The Representation of the People Act (1918) extended the vote to all men over 21 and women over 30 provided that she or her husband owned or occupied property with a value of at least £5. Some two million men and 8.5 million women were added to the electorate in 1918, although not until 1928 did all women achieve equal voting rights with men. Despite significant challenges to male-dominated Victorian values, most women did not achieve full domestic or economic emancipation in the inter-war years.

The rise of the labour movement and the spread of strikes and civil disturbances linked to calls for improved housing and employment conditions all had a major impact in the twentieth century. Although union membership was increasing before 1914, the effects of war accelerated its growth and employers made concessions to union leaders to ensure uninterrupted wartime production. The Bolshevik revolution in Russia (1917) sent shock-waves through Britain, strengthening labour protest which after the war was also directed at dissatisfaction with demobilization procedures and appalling living conditions in towns. In 1919 rioting and civil unrest affected a number of towns including east London, Cardiff, Glasgow and Liverpool and, unhappily for race relations, West Indian seamen were attacked in several ports. In 1921 over 85 million working days were lost through strikes as industry slumped, and 'Red Clydeside' became the focus for potential socialist revolution. Periodic strikes and unrest had occurred on Clydeside from 1915 and in 1919 troops and tanks were used to strike-break in Glasgow.

Industrial unrest flared up again in 1925 and, especially, in the 1926 General Strike. Nevertheless the strike was short-lived in most sectors: many rallied to the support of public services; only the miners held out until forced back to work after a bitter six months which cast a long shadow over relations in the industry. The capitulation of labour effectively softened it up for wage reductions and ensured that the 1930s slump passed without political or economic disruption. Thus, although most of these events were uncoordinated and posed little real threat to the stability of Britain they, together with other factors, cumulatively helped to create a political climate in which a whole series of state interventions, including social and economic welfare legislation, were made. While much of the legislation on housing, unemployment, pensions and health had precursors before 1914, and most was flawed and limited in scope, it cumulatively helped to create a climate of opinion favourable to welfare development, including the Beveridge Report in 1942 and the establishment of the 'Welfare State' by 1948 (Chapter 16).

The terms of international trade were also turning against Britain. From the 1880s imports of both foodstuffs and manufactured goods increased rapidly (Table 5.1) and the balance of trade worsened, despite the upturn in economic growth between the 1890s and the First World War and growth in coal exports which helped to bridge the gap (Table 5.2) in the short term. Earnings from overseas investments were vital: first, property income which paid for 28 per cent of Britain's imports by 1913; secondly, through the export of services from shipping, finance, insurance, and professional and cultural expertise (Table 10.1). Whether such investment was at the expense of home production is still debated.

Competition for markets in textiles was inevitable once major importers such as North America, India and the Far East began to supply their own home market and to compete internationally. By the First World War Britain was second best to the USA and Germany in steelmaking and some of the new industries – electrical and precision engineering, many branches of the metals and chemicals industries – and trailed the USA in the assembly industries, particularly automobiles. Despite a temporary wartime boost to many basic industries (as to agriculture) Britain's weakened trading situation, unbalanced industrial structure (over-dependent on declining industries) and the sale of many overseas investments to pay for the war created major problems. These were exacerbated by the breakup of free trade and the multilateral system of the nineteenth century in favour of protectionist bilateral trade systems. In the financial crises of the late 1920s and early 1930s Britain, the holder of a key currency but weak in gold reserves, was highly vulnerable. The result was high unemployment and stagnation in many hitherto prosperous industries and industrial regions which persisted into the post Second World War years. While the Bretton Woods (1944) and post-war monetary agreements which led to the International Monetary Fund and the establishment of the World Bank helped international investment and monetary stability, attempts to liberalize world trade and regulate trading agreements have been less successful. Moreover, the UK's unwillingness to commit itself to the EEC in the 1950s may have cost it dear.

The consequences of the economic effects of war and, especially, of the inter-war recession posed other major challenges to stability. Although under- and unemployment had been a major problem before 1914, it became a much more important political and social issue between the wars. The immediate post-war economic boom gave way between 1921 and 1940 to a series of lengthy depressions in which there were never less than one million people unemployed in Britain (some 10 per cent of the work force): at their peak in the winter of 1932–33 almost 3 million workers (25 per cent of the insured working population) were unemployed. Moreover, as official figures exclude some categories of workers, the actual figures were probably considerably higher. The effects of unemployment on society were considerable, although not everyone or every area was equally affected. Although international conditions contributed to the recession of 1929–33, the British economy was also undergoing restructuring involving the decline of its traditional staple industries – heavy manufacturing, coal, cotton and marine transport – and the expansion of new light engineering, food, drink and tobacco industries, and commercial and service occupations. Thus whilst some sectors and areas experienced poverty and hardship on a scale comparable with recessions of the 1880s, others – with new and expanding industries – were better off than ever before (Chapter 15). Economic change in the inter-war period thus led to a polarization of classes within society and an underclass of long-term unemployed which was only removed by the onset of the Second World War.

The geographical impact of these changes is clearly seen in the changing fortunes of the British regions. The inter-war years created a new geography of regional inequality which, although related to previous trends, established a North-South divide clearly seen in unemployment. In 1932 North East England, Wales and Scotland all had unemployment rates of over 28 per cent of the insured workforce whilst in London and South East England the rate was only 13.7 per cent (Table 15.10). Locally, conditions were much worse: over 60 per cent unemployed in Jarrow and Merthyr Tydfil in 1934, as against 5 per cent or less in such places as Oxford, Coventry and St Albans. The mid-Victorian concept of 'Two Nations' became more marked and left a much sharper imprint on the industrial population in the inter-war years.

Economic conditions were also reflected regionally in other forces for change. Support for the Labour Party was strongly concentrated in northern and Scottish industrial towns and other large cities (including parts of London), and it was no accident that cities like Glasgow and Liverpool had reputations for radical politicians – albeit in a minority until 1945 in Liverpool – and militant labour forces, and at the same time were ahead of most other cities in the provision of council housing. The extent to which the social and economic problems of the inter-war years were regional was recognised by several government investigative reports and the Special Areas Act of 1934. This was a precursor of post-war regional policy, which enabled government assistance to be given to areas thought to be hardest hit by the depression, although the regions selected (South Wales, Tyneside, West Cumberland and industrial Scotland) were but a small sub-set of those most severely affected. However, these policies did little to counter the inexorable trend which saw wealth, prosperity and economic power and expansion pass to central and southern England at the expense of Scotland, Wales and northern England.

Nevertheless, the general social well-being – the health and standards of living – of the British population undoubtedly improved in the period from 1890–1950. Despite recession, average *per capita* real incomes rose by one-third between 1920 and 1939 because of the effects of falling prices of food and raw materials in depressed world commodity markets and of cheap mass-produced manufactures, especially consumer goods. For many people the first half of the twentieth century offered opportunities for housing, material possessions, entertainment and travel undreamt of in the mid-nineteenth century. Personal well-being was enhanced; access to secondary and further education increased; and some social barriers began to be removed, not least as a result of wartime mixing.

One measure of these improvements was the death rate which fell, in England and Wales, from 16 per thousand in 1901–5 to 12.8 per thousand 1941–45, and from 17.0 per thousand to 14.1 per thousand over the same period in Scotland. A rapid improvement in infant mortality raised average life expectancy from 45 years in 1900 to almost 60 years in 1932 in the depths of the recession. However, national figures mask wide variations between regions, localities within regions, and different classes and social groups. In South East England, for example, not all groups benefited equally from economic growth. Likewise, not all women experienced an equal degree of freedom and emancipation, not all children had equal improvements in their education, and not all families had equal access to new council housing. Persistent and substantial class and regional differences continued to divide Britain and are reflected in the changes in its geography from the 1890s to the 1940s examined in subsequent sections.

Bibliography and further reading for Chapter 12

ALDCROFT, D. H. *The British economy*, vol. 1, *years of turmoil, 1920–1951* (Brighton: The Harvester Press, 1986).
ASHWORTH, W. *An economic history of England, 1870–1939* (London: Methuen, 1960).
BARKER, T. and DRAKE, M. (eds) *Population and Society in Britain 1850–1980* (London: Batsford, 1982).
FLOUD, R. and McCLOSKEY, D. (eds) *The economic history of Britain since 1700*, vol. 2 *1860–1970* (Cambridge: Cambridge University Press, 1981).
FRASER, D. *The evolution of the welfare state* (London: Macmillan, 1973).
GLYN, S. and OXBORROW, J. *Interwar Britain: a social and economic history* (London: Unwin 1976).
HALL, P. and PRESTON, P. *The carrier wave: new information technology and the geography of innovation, 1846–2003* (London: Unwin Hyman, 1988).
HALSEY, A. H. *Trends in British society since 1900* (London: Macmillan, 1972).
HARRISON, T. and MADGE, C. *Britain by mass observation* (London: Hutchinson, 1939; Crescent Library edition, 1986).
HOGGART, R. *The uses of literacy* (Harmondsworth: Penguin, 1957).
LAW, C. M. *British regional development since World War I* (Newton Abbot: David and Charles, 1980).
LEE, C. H. *British regional employment statistics, 1841–1971* (Cambridge: Cambridge University Press, 1979).
MARSHALL, M. *Long waves of regional development* (London: Macmillan, 1987).
MARWICK, A. *Britain in a century of total war: war, peace and social change 1900–1967* (London: Bodley Head, 1968).
MOWAT, C. L. *Britain between the wars 1919–1940* (London: Methuen, 1955).
PARSONS, W. *The political economy of British regional policy* (London: Routledge, 1989).
PRESTWICH, R. and TAYLOR, P. *Introduction to regional and urban policy in the UK* (London: Longman, 1990).
PERKIN, H. *The rise of professional society: England since 1880* (London: Routledge, 1989).
POLLARD, S. *The development of the British economy, 1914–80* (London: Arnold, 3rd edition, 1983).
POLLARD, S. *Britain's prime and Britain's decline, 1870–1914* (London: Arnold, 1990).
Report of the Royal Commission on the Distribution of the Industrial Population [The Barlow Report]. (London: HMSO, 1940).
ROBBINS, K. *The eclipse of a great power: modern Britain 1870–1975* (London: Longman, 1983).
ROBERTS, E. *A woman's place: an oral history of working-class women 1890–1940* (Oxford: Blackwell, 1984).
ROUTH, G. *Occupation and pay in Great Britain, 1906–79* (London: Macmillan, 1980).
STEVENSON, J. and COOK, C. *The slump: society and politics during the depression* (London: Cape, 1977).
STEVENSON, J. *British society, 1914–45* (Harmondsworth: Penguin, 1984).
THOMPSON, P. *The Edwardians* (London: Granada, 1977).
THOMSON, D. *England in the twentieth century, 1914–63* (Harmondsworth: Penguin, 1965).
WARD, S. *The geography of inter-war Britain. The state and uneven development* (London: Routledge, 1988).

13

Demographic change from the 1890s to the 1940s

Introduction

The 1890s were a turning point after which population trends reflected significant new demographic and distributional features by the mid-twentieth century. In the third phase of the demographic transition, from the late 1870s to the 1930s, mortality resumed its downward course. By 1938 the crude death rate was 11.6 per thousand in England and Wales and 12.6 in Scotland, little more than the present-day level, as compared with 19–20 per thousand in the early 1890s. The major contributor to this was the rapid fall in infant mortality from 163 per thousand live births in 1899 (134 in 1898 in Scotland) to 53 per thousand (70 in Scotland) in 1938, with a further substantial fall after the Second World War to 21 and 26, respectively, by 1961 (Figure 3.2). The epidemic instability which was a marked feature of mortality up to the 1890s largely disappeared after the European-wide 1919 influenza epidemic, a tribute to the social impact of improvements in living conditions and in preventive and curative medicine.

The fertility decline initiated among the middle classes of mid-Victorian Britain spread rapidly as wide-spread adoption of birth control led to a fall in crude birth rates from 36.3 per thousand in England and Wales (in Scotland 35.6) in 1876 to 28.7 (Scotland 29.6) in 1900 then rapidly to around 15 (Scotland 18) in the 1930s. Between the 1870s and 1930s the fertility rate (births per 1000 women aged 15–44) was halved in Scotland and reduced by 60 per cent in England. Hence, the fourth phase of the demographic transition, with controlled mortality and fertility, was largely attained by 1930. Since then fluctuations in natural growth have largely been guided by fluctuations in fertility, for example the trough of the 1930s' depression and the post-war baby booms.

Up to the First World War, however, a youthful structure (Table 3.2) still predisposed Britain's population to growth which, despite reduced rates of natural increase, maintained levels of total increment. Annual births remained over one million throughout the years 1890–1914 (1.08m in the peak year of 1903) and, with deaths falling from 670,000 per annum in 1891 to

 DOI: 10.4324/9781003174783-13

578,000 in 1913, it was only increased emigration that kept population growth in check.

From 1903 to 1914 considerable emigration to North America (increasingly to Canada) and Australasia produced an annual net loss from Britain to countries outside Europe reaching over one-quarter of a million between 1910 and 1913. There were further net losses after the 1914–18 war, but emigration fell and return migration rose as the deepening world depression of the late 1920s and 1930s hit the Americas and the primary producers even more harshly than the home labour market. Moreover, the influx of refugees from mounting oppression in Germany and Eastern Europe created a net inward migration of some half a million in the 1930s, the first such gain since the late 1840s.

Internal migration was a key factor in re-shaping regional patterns of population distribution from the late nineteenth century (Figure 13.1): continuing, though substantially reduced, rural to urban migration; a progressive focusing of jobs on London and the major conurbations; and increasing intra- and inter-urban mobility. Intra-urban mobility, assisted by

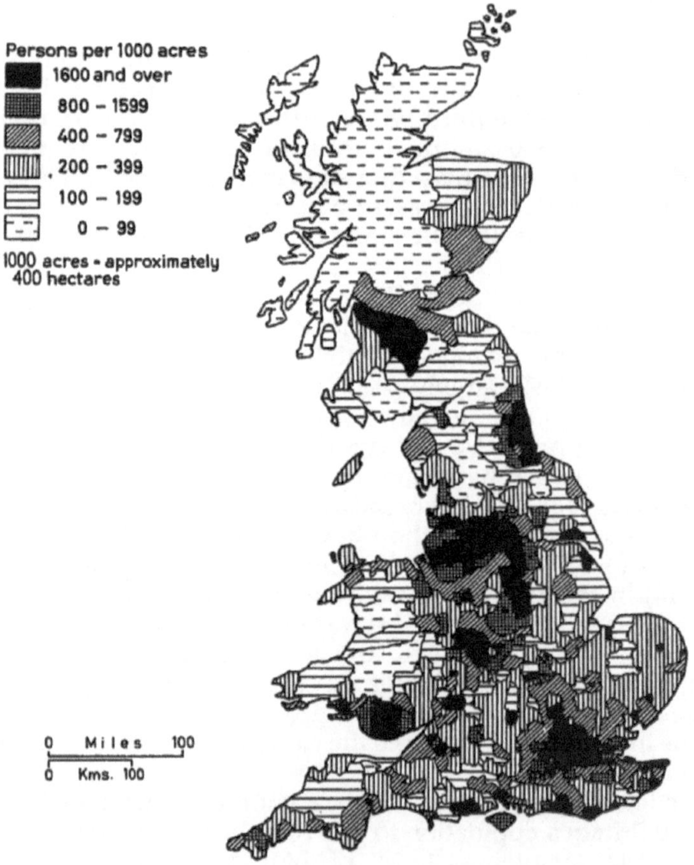

Figure 13.1 The density of population in Britain, 1911.
Sources: Census of England and Wales and of Scotland, 1911.

cheap mass transport, let to much increased commuting as suburbs sprawled after 1918 and as new technologies and economies of location tended to disperse much new urban industry to peripheral industrial estates (Chapter 15).

As the staple industries of the first industrial revolution declined so their labour markets weakened. Even before the First World War, there were net migration losses in older industrial areas, for example parts of the Black Country and worked-out areas of the major coalfields. These extended to virtually all heavy industrial areas and the textile regions of northern England by the 1920s. As the share of new manufacturing industries in Greater London and the midlands grew, so the 'drift south-east' became a flood. Between 1901 and 1951 the South East and West and East Midlands regions increased their share of the national population mainly through differential migration. In the post-Second World War years such net gains extended to East Anglia as London's influence spread (Table 3.4).

Fertility

One of the most remarkable demographic features of the period, rapid fertility decline, saw mean family size fall from 4.1, for women marrying in the 1890s, to 2.8 for the 1911 marriage cohort and to 2.1 and 2.0 respectively for women marrying in 1931 and 1941. The latter, representing below-replacement fertility, caused much concern in the 1930s and led to the setting up of a Royal Commission on Population in 1944. By the time it reported in 1949, the proportion of women marrying had increased, and a younger average age of marriage and rather larger families produced a recovery in the level of fertility in the immediate post-war years.

Nevertheless, substantial differences in the rates of fertility, both between different occupational groups and social classes and between one part of Britain and another, suggest continuing differential responses to the adoption of family limitation and varying patterns of nuptiality (Chapter 8). Effectively, the pattern established in the late nineteenth century continued into the twentieth. Despite higher marriage rates in industrial areas, the proportion of women over 20 who ever married was relatively low in areas of substantial female employment in factories (as in the textile districts) or domestic service (as in London and some residential towns), and substantially so in outer London and Sussex. Throughout Britain the highest levels and proportions of marriage, and the youngest mean age of marriage for both men and women continued to be in the mining and iron-manufacturing districts.

These were reflected in the first full survey of fertility in the Census of 1911. Gross reproduction rates (that is the number of female children born to women 15–49, the best indicator of future growth potential) were highest in the mining counties and on Clydeside, with relatively high levels in other heavy industrial areas. In contrast some of the most urbanized areas had moderate or low fertility, notably in North West England, London and the Home Counties. There were even larger differences between urban areas, cities such as Liverpool, Glasgow and Hull having high reproduction rates but Bristol, Leeds and Edinburgh low ones. Some of the major urban contrasts were internal: on Merseyside, for example Liverpool and Bootle were among those with the highest fertility while residential Wallasey was among the lowest. However, there were few rural areas of high fertility by 1911 and rural fertility generally fell sharply as reproductive potential was diminished by heavy out-migration.

The impact of the First World War on British fertility levels is difficult to assess. The loss of over 600,000 men in excess of normal mortality must have reduced marriage opportunities for women and produced fewer births. On the other hand, compared with net extra-European emigration of one million between 1910 and 1913 – largely cut off by the war – these losses seem less significant. Higher post-war marriage rates continued into the 1930s; the age of first marriage for women was slightly lower, and the proportion of women marrying increased. But the marked fall in births between the wars, after a short post-war baby bulge, was almost entirely attributable to birth control. As compared with mid-Victorian parents, only one in five of whom tried to limit their families, two-thirds did so by the mid-1930s producing a marked decline in marital fertility, a sharp increase in the number of one- and two-child families and an equally marked fall (to only one in ten) in families with five or more children.

A second feature of the post-war fertility decline was its rapid spread through most sections of society. Between 1911 and 1931 birthrates fell most markedly in the working class: among Class V (unskilled) by some 28 per cent; for Class IV (semi-skilled) 22 per cent; and by 16 per cent for Class IIIB (skilled manual). Among professional and higher-ranking business workers (Class I) fertility fell by 21 per cent but in Class II (lower middle class and service sector) it fell by a massive 38 per cent. Although wider access to cheaper and more reliable contraception was reflected in more widespread knowledge and acceptance of family limitation, there were still substantial differences in family size between different occupational groups that were related to both regional and local birth rates. In 1951 the highest reproduction rates were still, as in 1911, in mining and heavy industrial districts, and the lowest in some rural areas, textile towns, middle-class suburbs and residential (especially retirement) towns and resorts. But the gap between high and low fertility groups had narrowed considerably to a ratio of about 1:1.5 (as compared to 1:2 in 1911) and to less than 1:1.2 for marriages contracted in the 1940s. Less wide-ranging variations in levels of fertility among those married in the 1930s suggest progressive adoption of family limitation despite continuing social differences in marriage practices and attitudes to the family and in methods of birth control.

One of the most potent reasons for this major social revolution was undoubtedly the rapidly changing role of women in society. The enfranchisement of women and increased education of middle-class girls for the professions gave them a bigger voice in public affairs. Changed social attitudes permitted a more independent, though not yet equal, role for women in society. Working women tended to marry later and had fewer pregnancies so that growing employment of women reinforced the tendency to smaller families. A second, and closely related factor, was the much wider range of advice and literature on family care in which the virtues of smaller and better cared-for families were extolled. A third set of factors was economic. As more children survived infancy and the school-leaving age was progressively raised (from 12 before the First World War to 14 between the wars) the cost of raising children and the deferral of their wage-earning prospects persuaded most parents to have fewer children. In middle-class families and then, after the First World War, in all sectors of society rising material expectations and increasing expenditure on the home and leisure activities led to possessions and pleasure being preferred to children. Secure and more widely known contraceptive methods made for surer control, though many working-class people still relied on *coitus interruptus*. Changed moral attitudes to contraception were reflected in much wider adoption of birth control both within marriage and, more generally from the Second World War, outside marriage. Progressively the concept of the family changed.

Mortality

As the expectation grew that healthy children would survive to adult life so the planning of a family of desired size became more certain. The rapid fall in child and, from the turn of the century, infant mortality continued apace as maternity, welfare and medical services improved (Figure 8.4). Better living standards, hygiene and child care produced a 70 per cent fall in child mortality between 1900 and 1935 and a parallel reduction of almost one-third in infant mortality in England and Wales and almost one-half in Scotland. The gap between urban and rural mortality had narrowed from the late nineteenth century and by the first decade of the twentieth century life expectancy in large cities was 51 years at birth as compared with 53 in smaller towns and 55 in rural districts. Yet the health gulf within cities remained wide: in Liverpool, for example, mortality was twice as high in overcrowded slums as in the prosperous suburbs. There remained a clear regional gradient between healthier areas of south-eastern England and higher mortality in urbanized northern and western Britain reflecting environmental differences and the persistent contrast between more and less prosperous areas.

Nevertheless the decline in mortality during the twentieth century owed much to improvements of housing and the urban environment between the wars (Chapter 16). The level of amenity in most homes – sewerage, separate water supply, gas and electricity – improved standards of hygiene and reduced crowding lessened the risk of environmentally-communicated diseases (for example the various fevers) and those associated with overcrowding (such as tuberculosis and infectious diseases of childhood). Enhanced welfare services, even during the two wars, brought more people under regular medical scrutiny in the forces and factories. One major improvement, especially in the towns, was in wider access to ante- and postnatal clinics, part of the better midwifery and child care services which greatly aided infant and child health. School medical services helped through regular eye, dental and hair inspections (head lice were a universal scourge in poorer areas), while provision of free milk for underweight or poorly nourished school children helped improve standards of nutrition. Following the recommendations of the Beveridge Report of 1942, the creation of the Welfare State between 1944 and 1948 embraced not only the National Health service of 1948, but also social security, housing and education (Chapter 16).

However, none of these measures eliminated poverty, and class differentials in mortality and morbidity (experience of disease) remained (Table 13.1). Despite overall reductions in mortality, life expectancy remained substantially higher among the middle classes (I and II). Where class differentials combined with adverse environmental, economic and social conditions, as in the inner areas of large cities, these discrepancies widened. Even after the Second World War the average risk of death in Salford (with one of the highest mortalities) was 33 per cent higher than in Bournemouth (one of the lowest), while that for Bootle was some 50 per cent higher than across the Mersey in Wallasey and more than twice that of the youthful population of Formby, a rapidly-growing middle-class residential area. Class-differentials in infant mortality remained, until after the Second World War, 2 to 2.5 times higher in the families of unskilled workers than in Class I. The reasons included poorer ante- and post-natal care, poorer housing, more pregnancies and greater exposure to and poorer treatment of such ailments as gastro-enteritis and bronchitis. Thus, despite the general fall, class and spatial differences in infant mortality remained substantial and were clearly linked to poverty.

Table 13.1 Standardized male mortality levels in Great Britain, 1921–51

Year	Index of mortality by socio-economic class, ages 20–64 (national rate = 100)				
	I	II	III	IV	V
1921	82	94	95	101	125
1931	90	94	97	102	111
1951	97	86	102	104	118

Class I = professional; II = employers and managers; III = intermediate and junior non-manual; skilled manual; IV = semi-skilled manual; V = unskilled manual
Source: *Social Trends* **6** (1978) Table 7.1

By the mid-twentieth century the final stages in the conquest of epidemic disease, the main 'killers' of the nineteenth century, were being reached. Whereas improvements of the late nineteenth and early twentieth century were due largely to evironmental, economic, nutritional and social improvements – including better health knowledge and greater personal and institutional hygiene and health care (Chapter 8) – those of the early twentieth century also reflect increasing scientific and medical knowledge about the causes, transmission and treatment of disease. In the case of non-infectious conditions, advances in surgery and hospital care reduced the risks of surgery and improved accident treatment after the First World War. However, it was not until the 1940s following the discovery (in 1935) and widening use of sulphonamide drugs and, even more so, of penicillin (used medicinally from 1941) and of antibiotics from the 1950s that many serious diseases, including pneumonia and bronchitis, could be routinely treated.

By then mortality from most infections was controlled by public and personal health care leaving only the extension of vaccination (for example against tuberculosis) and immunization (for example against diphtheria, whooping cough and measles) to further reduce mortality to the present low levels. The principal killers of the late twentieth century are degenerative diseases of old age.

Migration

A number of significant changes in early twentieth-century British society had a major impact on the pattern and process of internal migration in the fifty years after 1890. Some broadened horizons and made migration easier; others constrained migration opportunities for certain people, especially in the depressed areas. Cheap mass transportation increased population mobility. Up to the late nineteenth century most ordinary working people walked to most destinations, with only occasional recourse to the railway. By 1900 not only were most towns and many villages connected into the national rail network but, more importantly, the cost of travel could be met by a larger proportion of the population. Increasing provision of third class coaches with better facilities, after 1883 at a standard rate of only 1d per mile on most trains, led, by the early twentieth century, to nearly 95 per cent of all railway passengers travelling third class.

The accessibility of the railways to ordinary people was further extended by the introduction of concessionary fares for both long-distance and daily travellers. In 1933 monthly return tickets at single fare plus one-third were introduced, and by 1936 some 93 per cent of passengers travelled on reduced fares compared with 68 per cent in 1924. By 1936 the average discounted fare per passenger mile had dropped to 0.6d. Such changes combined to make long-distance movement easier; increased opportunities for daily travel, thus broadening horizons and increasing information about job vacancies; and made it easier to keep in touch with friends and kin elsewhere in the country.

Although railway fares fell in real terms, in the inter-war period the railways suffered competition with road transport which was cheaper and more convenient for short journeys. By the 1930s long-distance buses could carry passengers at a standard fare of less than 1d per mile (compared to 1½d for a standard third class rail fare), offered substantial discounts on return tickets and served a wider range of locations than the railways. Although the private motor car remained a relative luxury, costs of buying and running a car fell rapidly during the inter-war years and by 1938 there were 1,944,394 private cars in use in Britain compared to only 8,465 in 1904 (Figure 16.6). The motor car offered a new dimension in daily mobility and, for those unable to afford a car, motorcycles (637,000 on the road in 1926) provided other opportunities for mobility.

If travel increases migration propensity, then the First World War must also have had a significant impact. Some five million men (approximately 22 per cent of the total male population of Britain) were recruited into the armed forces in 1914–18: most travelled widely in Britain and Europe, for many the first long journey from home. At the end of the war they were more aware of opportunities and many were dissatisfied with conditions at home. During the late 1920s and 1930s such broadened horizons combined with the realities of economic recession, unemployment and poor housing in many areas led many to choose migration or emigration.

However, a number of factors pulled in the opposite direction. For those in the expanding sectors of the economy – especially in south-eastern and midland England – the inter-war years were a period of rising real incomes and increasing material possessions: even the poor were better off than in the nineteenth century. More people became homeowners with mortgages and bought more furniture and consumer durables for their homes. Such trends provided a brake on migration: those who benefited from the restructuring of the inter-war economy found it both more disruptive to move and had fewer reasons to relocate. Negative restraints on migration in depressed areas were the support and familiarity of a neighbourhood and friends in the face of economic insecurity. Local working-class communities provided support systems helping families through unemployment in the hope that traditional industries would eventually recover. Many families preferred to sit out the recession in familiar territory rather than risk unemployment in a strange environment. The fact that some unemployed workers were reluctant to move was recognized in the establishment of a government transference scheme which encouraged some 200,000 workers to migrate from depressed areas in the 1930s, and through various government and charitable agencies which encouraged emigration to the dominions and colonies.

Though Britain lost more population through out-migration than it gained between 1890 and 1940, after 1918 rates of emigration were lower than in the pre-war years – a consequence of world-wide recession – and in the 1930s the net flow was reversed. Internal migration rates were probably much lower than in the nineteenth century when many made frequent short-distance moves; but twentieth-century migration tended to be longer-distance and potentially more disruptive to families and communities.

Friedlander and Roshier's (1966) analysis of inter-county migration flows in England and Wales showed three main trends. First, there was consistent out-migration from northern England to eastern, south-eastern and southern counties; secondly, there were high levels of out-migration from South Wales to the English midlands and south-eastern England; thirdly, there was out-migration from the urban area of London into all adjacent counties through the process of suburbanization (Figure 13.2). Scots continued to move from the Highlands to Glasgow and Edinburgh, and there was a general loss to expanding industrial areas in midland and southern England and, especially, overseas. Suburbanization affected population movements within all large towns, nowhere more dramatically than in London, and during the inter-war years many small towns and rural commuter villages became urban-oriented, attracting city dwellers to countryside living.

The dramatic outflow of migrants from depressed areas to southern and midland England between the wars helped the South East account for 60 per cent of the total population increase in England and Wales; the latter's population actually fell by 200,000 between 1920 and 1939 (Figure 13.3).

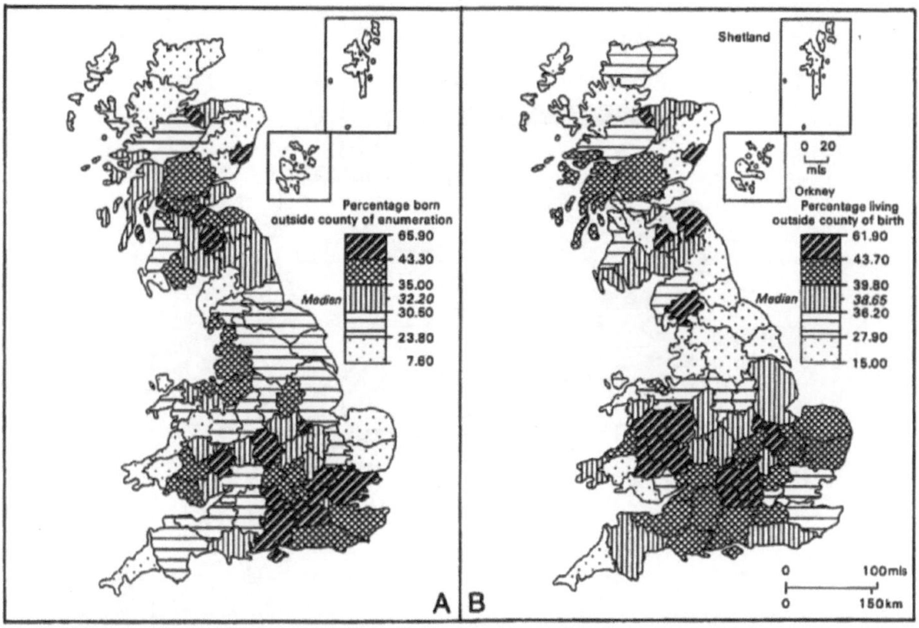

Figure 13.2 In-migration (A) and out-migration (B) by county in Great Britain, 1911.
Source: Census of England and Wales and of Scotland, 1911.

Most people moving at this time were in search of work, both young single men and women and family groups. Much of the movement was long-distance and into relatively unknown territory, but it was not undertaken without careful thought and planning. Oral evidence of Welsh female migrants to England shows that many utilized existing friendship or kinship networks to find lodgings and work; some had work arranged before they left home; and most – although travelling alone – were met by friends or kin when they got off the train in Birmingham or London.

Carefully planned voluntary migration utilizing friendship networks was usually successful, but migrants encouraged to move under government transference schemes had more difficulty in finding work and adjusting to a new environment. An official government scheme, begun in 1929, resulted in substantial movement, for example from South Wales, central Scotland and North East England to the Kent coalfield and to such expanding engineering towns as Oxford, Luton, Slough and Banbury. Others, encouraged by government publicity, made opportunist moves to the South East and midlands. Although some did find work, it was often not well-paid: as the Pilgrim Trust studies showed, many migrants who took part returned disillusioned to their home area. Such bad publicity led to the failure of the scheme and possibly reduced levels of out-migration from depressed areas. As one northern 'trainee' in a Government Training Centre in southern England wrote to a friend back home, 'Don't think of coming down here, it's lousy. The bloke who came down with me has gone home . . . don't advise anyone to come' (Pilgrim Trust, 1938, p. 226).

Suburban migration was most marked around London. The population of

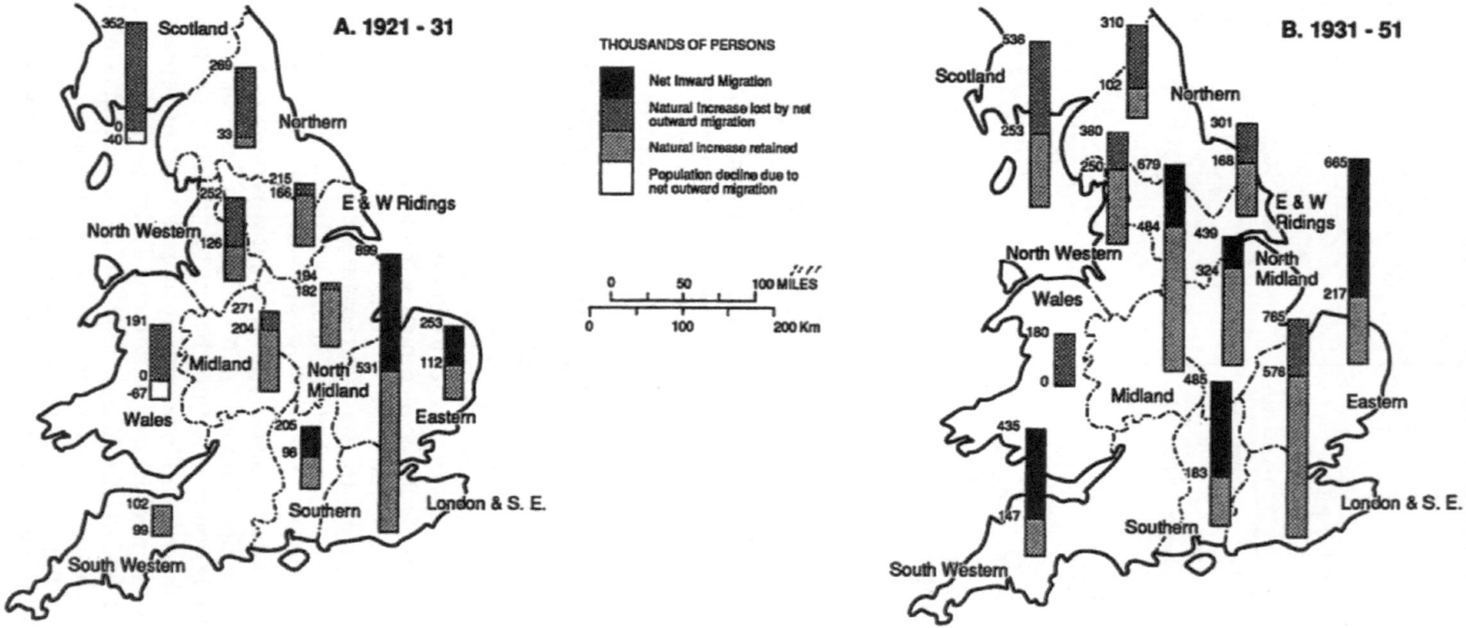

Figure 13.3 Natural and migrational change in standard regions of Great Britain, 1921–51.
Sources: Census of England and Wales and of Scotland; Annual Reports of the Registrars General.

suburbs such as East Ham and Walthamstow increased by over 100 per cent between 1891 and 1901, whilst the central city lost population; middle-class suburbs such as Leyton, Willesden, Croydon and Tottenham expanded rapidly in the early years of the twentieth century; and rings of outer suburban growth spread further along the main communication routes as semi-detached private housing expanded rapidly in the 1920s and 1930s (Chapter 16). Public transport was essential to such growth. From the 1860s, with the opening of the Metropolitan Railway (1863) and the South London Line (1867), railways linked the City with residential suburbs and by 1910 a basic underground network stretched from Clapham to Finsbury Park and Hampstead. Cheap workmen's fares allowed working-class commuters to move to inner suburbs and helped to push the middle classes further into the countryside.

Elsewhere, suburban railways were less important, although large provincial towns such as Manchester, Liverpool, Birmingham, Glasgow and Newcastle had quite well-developed suburban rail services by the First World War. But in most smaller towns suburban dwellers relied on the omnibus and especially the tram to connect home and workplace. The Tramways Act of 1870 marked the beginning of tramway networks in most towns, and from 1891 municipal authorities could take over the running of the tramway system. Even in London the tram and the omnibus competed effectively with suburban railways, and by the twentieth century most large towns had an effective network linking new residential suburbs with the city centre (Figure 13.4). However, Richard Hoggart's 'gondola of the working classes', the tram, was still mainly used by the more affluent working-classes and middle class.

In the 1920s and '30s some of the middle class turned to the motor car for their journey to work, especially in less congested northern towns like Liverpool or Manchester. In the South East the motor car might be used to travel from a semi-rural home to suburban railway stations such as Hounslow, Kingston, Croydon or Eltham, and high rates of car ownership meant that the motor car was used increasingly for peripheral commuting journeys. Journeys to work and patterns of daily mobility were becoming more complex, and this trend increased as suburbanization was partly replaced by the rapid growth of small towns and villages around London and other large cities (Chapter 16).

Families moving from terrace housing to a new semi-detached suburb of the early twentieth century were a distinct sub-set of the total population: they had skilled jobs, with regular hours; they often worked in an expanding sector of the economy; in more distant suburbs most were white-collar workers, sensitive to their perceived higher status although incomes may not have been much higher than blue-collar workers. Many migrants to private suburban housing estates acquired a new way of life and a set of values as well as a pleasant environment. These characteristics distinguished suburbanites from the working classes in the city centre and inner Victorian suburbs.

Not all suburbanites moved into private housing. Suburban local authority housing estates were built around most towns from 1919 (Chapter 16) and, although initially most council houses were expensive and allocated

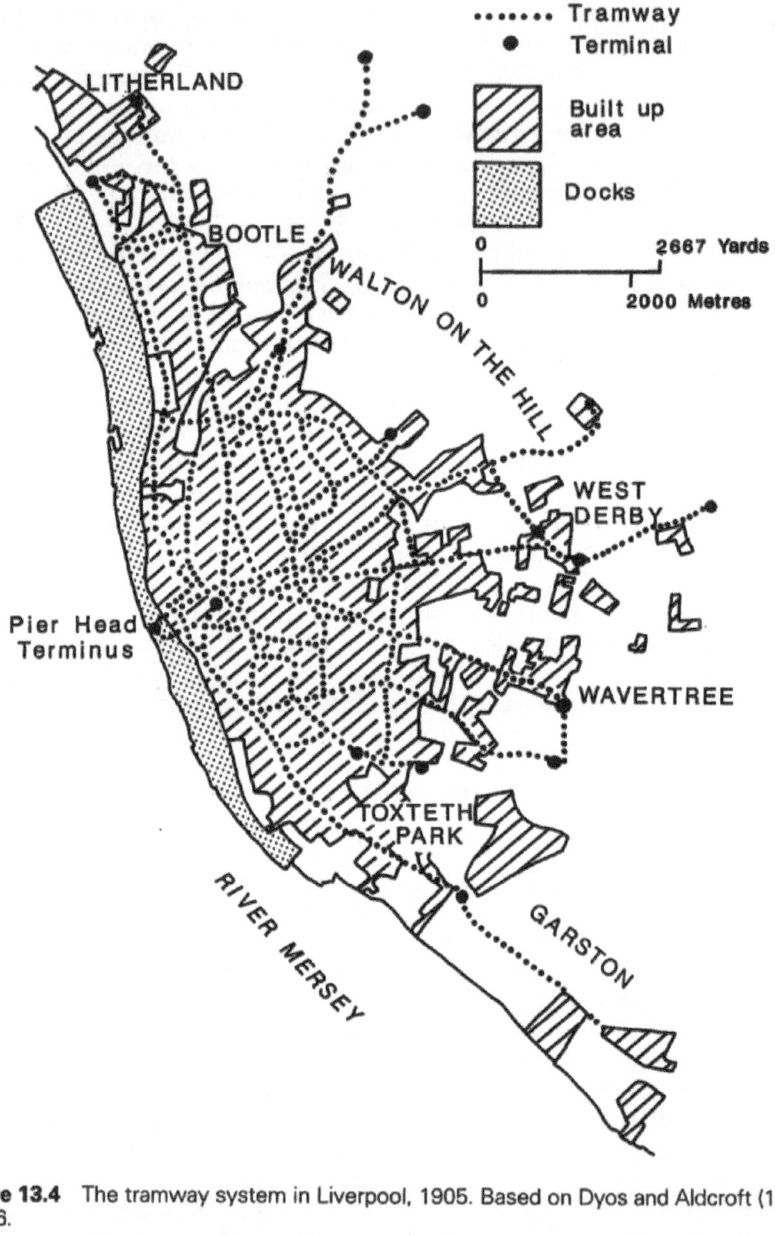

Figure 13.4 The tramway system in Liverpool, 1905. Based on Dyos and Aldcroft (1974) p. 236.

only to the more affluent working class, by the 1930s an increasing number of inner-city slum clearance tenants were rehoused in suburban cottages. Many of these involuntary migrants were unhappy with their new surroundings. There were frequent complaints about the monotony of new vast housing estates; the lack of facilities and community spirit; the high cost of rent, locally-bought food and journeys to work; and the shortage of work nearby not least for women. Stress produced increased mortality rates amongst migrants to some slum-clearance estates (for example in Stockton-on-Tees) and on Merseyside the five miles journey to dockside employment produced high rates of unemployment on the estates and was one factor encouraging families to move back to homes in the central city slums.

Although the total volume of twentieth-century migration was less than that in the nineteenth, more was over long distances and into wholly new environments: even short-distance suburban moves often led to new social values which involved people in increased living expenses which they could ill afford. Such new patterns of migration varied in their impact on families and communities. In depressed industrial regions of high out-migration previously close-knit families could be broken up, communities split and extended family support systems fractured. One woman in Barrow in Furness, which lost 16.2 per cent of its population 1921–31, said of her brothers:

> 'They all just scattered and I wouldn't know where they are. They left Barrow in the twenties looking for work . . . I'm the only one who stayed at home' (Roberts, *A woman's place*, 1984, p. 181).

The extent to which families broke up, and the degree to which this was new in the 1920s should not be exaggerated in comparison with the effect of long-distance migration in the nineteenth century or the impact of the Great War on families and neighbourhoods. Long-distance migrants could and did keep in touch with family and friends through letters and visits. For instance, Welsh women moving to English towns kept in touch with Welsh culture in general and their own friends and communities in particular. Although such long-distance contacts could not provide day-to-day support, cheaper transport meant that families could come together in times of crisis or celebration.

In the suburbs family life became more inward-looking and private. Separation from relatives, longer hours spent in commuting, preoccupation with home and garden, and limited resources for entertainment outside the home conspired to throw many families back on their own resources. The small nuclear family could become more insular, relying on in-house entertainments such as the radio, gramophone and bridge parties. The mutual aid of inner-city working-class communities could be lost, and families began to turn towards state welfare systems rather than to community-based solutions when help was needed. Such changes were gradual, and did not affect all communities especially if there were good local facilities in the form of cinemas, sports and youth clubs. But one result of migration to the suburbs was the acquisition of a lifestyle which made traditional working-class community networks less necessary.

Population structure

The rapid fall in fertility and progressive mortality decline improved life expectancy throughout society during the twentieth century and combined to produce a population with progressively higher proportions of older working-age and elderly people. That tendency was not markedly changed by the post-war baby boom despite hopes of a reversal of the 1930s trends that pointed to future decline in Britain's population.

Falling birth rates before the First World War saw an increase in the proportion of 45–64 year-olds (Table 3.2) and by the inter-war years there was a sharp fall in the under-15s and a growth in those over 45. With rising life expectancy the proportion of over-65s rose sharply after the Second World War and, combined with a falling birth rate, led to a progressively ageing society by the 1950s. These trends led many commentators to express concern over the impact of longer living and fewer children on national economic and social life and over the consequences of a population which the report of the Royal Commission on Population in England and Wales (1949) estimated would be no more than 52 million by the end of the century, possibly as low as 41 million. While some saw advantages in a falling population and reduced juvenile dependency, there were clear implications for different needs in health, social welfare and economic provision in a generally older population.

The population structure in different parts of the country also reflected differential age-selective migration. In large cities the outward movement of young families to the suburbs tended to increase the proportion of children in outer city areas, especially of London. But many older people remained in the inner city and, as their family married and left home and then one partner died, elderly single-person households became a feature of many inner residential areas. Regional differences reflect fertility and economic and social forces affecting migration. Prolonged out-movement from declining industrial areas produced an ageing population with large numbers of retired people even by the 1930s. Most rural areas had experienced this from the mid-nineteenth century. Where birth rates remained high – as in central Scotland, North East and parts of North West England, Yorkshire and Humberside and South Wales – above-average younger age-groups and older age-groups emphasized losses of working-age population. The main beneficiaries were the actively growing economies of the midlands and Greater London. One of the main features of the inter-war map of age structure, and increasingly after 1945, was the growth of retirement areas along the coasts of southern England, North Wales and the North West, and in the south-eastern fringes of the Scottish Highlands. Such resort and retirement areas had to cater for increasing numbers of retired people, many of them widows reflecting higher mortality among men.

The fall in average household size from 4.6 persons in 1891 to 3.2 by 1951 is a clear indication of smaller families. The increase in the total number of households of 30 per cent between 1911 and 1931 and by another quarter in 1931–51 – over three times the rate of population growth – reflects both longer life, often requiring separate accommodation for three rather than two generations, and an increase in marriage. It also reflects better housing provision between the wars and rising real wages with more people able to afford homes of their own (Chapter 16). The wish of both young single persons and of the elderly to be independent (even when widowed) boosted the total and lowered the average size of households and substantially increased the proportion of people living alone from nineteenth-century levels.

Thus, while population mobility and expanding cities were breaking up the close-knit communities of the nineteenth century, a home-centred family life was achieving a major social revolution in most classes of twentieth century society (Chapters 15 and 16).

Bibliography and further reading for Chapter 13

For overviews of demographic change see:

HOBCRAFT, J. and REES, P. (eds) *Regional demographic development* (Part 1, population history) (London: Croom Helm, 1977)

HUBBARD, E. *The population of Britain* (West Drayton: Penguin, 1947).

MITCHISON, R. *British population change since 1860* (London: Macmillan, 1977).

TRANTER, N. *Population and society 1750–1940* (London: Longman, 1985).

WINTER, J. M. 'The decline in mortality in Britain 1870–1950' in Barker T. and Drake, M. (eds) *Population and society in Britain 1850–1980* (London: Batsford, 1982), pp. 100–120.

On migration, mobility and its impacts see:

DYOS, H. J. and ALDCROFT, D. *British transport: an economic survey from the seventeenth century to the twentieth* (Harmondsworth: Penguin, 1974).

FRIEDLANDER, D. and ROSHIER, J. 'A study of internal migration in England and Wales' *Population Studies* **19** (1966), pp. 239–79.

HOLMES, C. 'The impact of immigration on British society 1870–1980' in Barker, T. and Drake, M. (eds) *Population and society in Britain 1850–1980* (London: Batsford, 1982), pp. 172–202.

JACKSON, A. *Semi-detached London: suburban development life and transport 1900–1939* (London: Allen and Unwin, 1973).

LAWTON, R. 'The journey to work in England: forty years of change.' *Tijdschrift voor Economische en Sociale Geographie* **34** (1963), pp. 61–9.

PILGRIM TRUST, *Men without work* (Cambridge: Cambridge University Press, 1938).

POOLEY, C. G. and WHYTE, I. D (eds) *Migrants, emigrants and immigrants: a social history of migration* (London: Routledge, 1991).

SAVILLE, J. *Rural depopulation in England and Wales, 1851–1951* (London: Routledge, 1957).

WILLATTS, E. C. and NEWSON, M.G.C. 'The geographical pattern of population changes in England and Wales, 1921–1951.' *Geographical Journal* **119** (1953), pp. 432–50.

14

The countryside from the 1890s to the 1940s

Introduction

Although agriculture recovered from the depths of its late-Victorian depression (Chapter 9) and farm incomes rose between 1891 and the First World War as overall output recovered its 1871 level, its role in the British economy was much reduced. At the end of Victoria's reign the contribution from the land to the National Product was one-third of that at the beginning, and it declined further from 6.1 to 4.7 per cent between 1901 and 1955. Despite the boost to farm prices and arable production during the First World War, agricultural depression continued into the interwar years and total agricultural output 1936–39 was below its mid-1870s peak (Figure 14.1). However Britain's declining self-sufficiency in food and fodder (except for milk, eggs and some cash and fruit crops) was not arrested until the Second World War and after (Table 14.1). In general terms Britain provided only one-third of its food and half its stock fodder in the 1930s; by the late 1950s these proportions were 47 and 61 per cent respectively, and they continued to rise into the 1980s. Nevertheless, increasing productivity saw the decline experienced between 1885 and 1904 (−0.8 per cent p.a.) arrested and

Table 14.1 Self-sufficiency in selected foodstuffs in the United Kingdom, 1905–64

	Percentage home-produced				
Product	1905–7	1937	1946–47	1953–54	1963–64
Wheat	25	23	30	41	40
Barley	60	46	96	67	94
Oats	74	94	95	97	97
Potatoes	92	N.K.	99	100	100
Butter	13	9	8	9	9
Cheese	24	24	8	28	44
Eggs	32	61	51	86	96
Beef and veal	53	49	58	66	73
Lamb and mutton	52	36	24	35	43

N.K. = not known
Source: Grigg (1989), Table 2.3

DOI: 10.4324/9781003174783-14

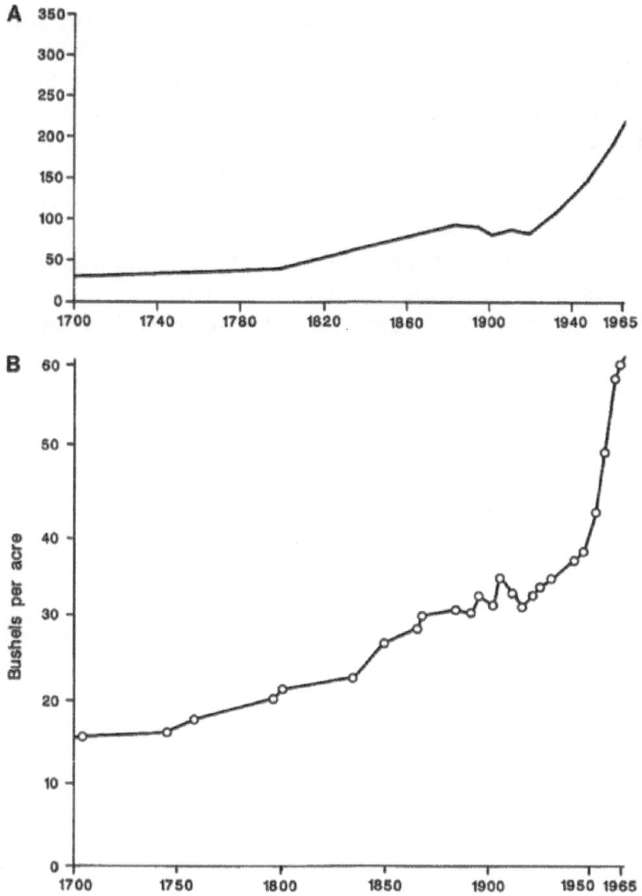

Figure 14.1 (A) Index of agricultural output in Great Britain, 1700–1965; (B) wheat yields in England, 1700–1965. Based on Grigg (1989) p. 7 (A) and p. 65 (B).

reversed by a 1.6 per cent p.a. growth in output by 1922–39 which accelerated during and after the Second World War. The reasons, explored in the following pages, were three-fold: increased intensification in use of labour with machinery; increased yields through greater fertilizer inputs and improved farming methods; diversification and focusing on the products that gave the best return under changing economic circumstances, a process assisted from the 1930s by government help.

The decline in farm labour from about six to under four per cent of the total work force between 1911 and the 1950s was mainly due to continuing decline of agricultural labourers; farmers and working relatives regained their 1871 level by 1911, and increased to one-third of the work force between the wars as compared with one-quarter in the 1890s. Renewed movement from the countryside between the wars halved farming populations in mainly agricultural areas (for example from one-quarter to one-eighth in the Scottish borders, 1921–51) and, as rural industry capitulated to mass-production and city-centred systems reduced rural services, persistent population losses in remote rural areas of upland Britain and parts of eastern England accounted for most of the half-million decline in rural districts of England and Wales, 1921–39 (Chapters 13 and 15).

However, the urbanization of South East England, the corridor from the English midlands to south Lancashire and south Yorkshire, and many parts of central Scotland promoted rural growth as more people commuted to nearby towns from rural-urban fringes (Chapter 16). Urban services, recreation and lifestyles permeated the countryside: so too did the urban 'discovery' of rural Britain – for residence and recreation; as an idyllic retreat for second homes, from holiday shack to sporting estate or hobby farm. A new rural lobby increasingly figured in planning and conservation move-

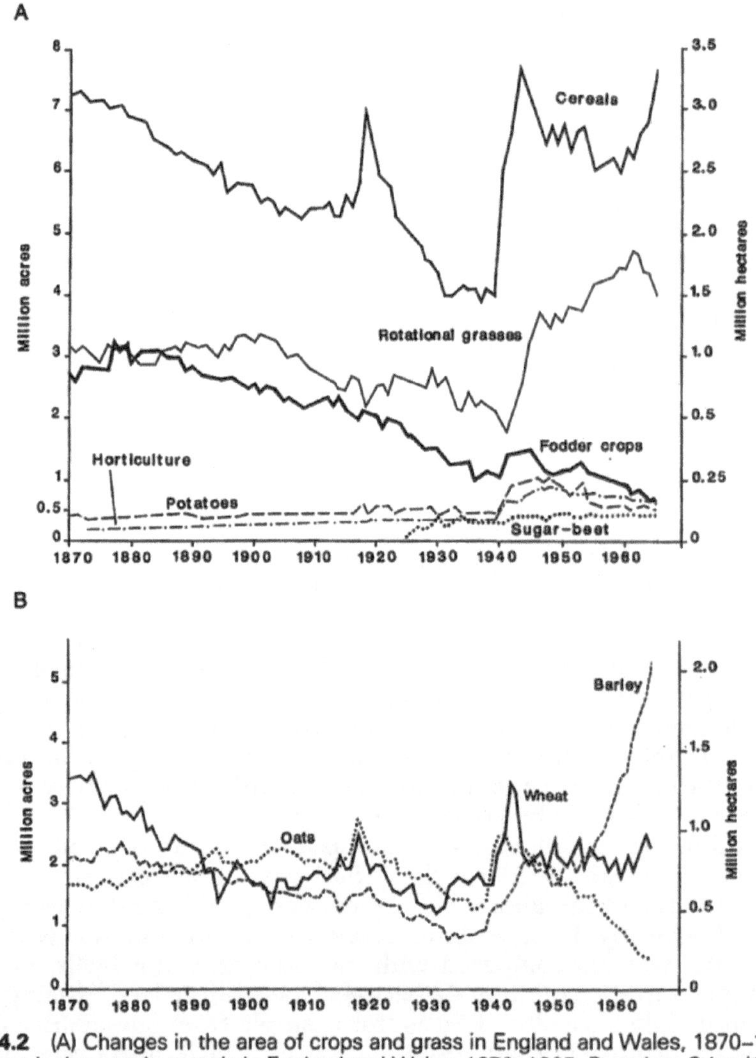

Figure 14.2 (A) Changes in the area of crops and grass in England and Wales, 1870–1965; (B) changes in the area in cereals in England and Wales, 1870–1965. Based on Grigg (1989) p. 39 (A) and p. 55 (B).

ments. Rural isolation and distinctive regional ways of life were eroded and rapidly disappeared in the face of modern communications and the penetration of all aspects of rural life by an urban-dominated economy, society and culture.

Changes in agricultural production and farm policy

The general context of farming change is epitomized in the contraction of the area of tillage and expansion of grass and livestock since the 1870s (Figure 14.2). Arable acreages, in particular of wheat, fell continuously until prices stabilized in the decade before the First World War though barley (for brewing and fodder) suffered less and oats held its own. Despite a brief revival in the face of submarine attacks on merchant shipping in 1917, the government plough-up campaign which aimed to convert the three million acres put down to grass since the 1870s, increased tillage only by 1.9 million acres above the pre-war level. Faced by a depressed international market with massive cheap grain surpluses, this brief upsurge was reversed as tillage – especially wheat, barley and fodder roots – declined after the war. Despite lower rents, falling prices led to many bankruptcies: as land went out of cultivation the proportion of farm output from crops and horticulture fell to a historic low of 30 per cent in the early 1930s.

For many stock farmers and producers of perishable commodities, however, the picture was less gloomy. Rising real wages, a more varied diet, improvements in supplies to retail outlets and health and welfare measures such as free school milk (Chapter 12) helped to increase demand for fruit and salads, vegetables, dairy and poultry products and fresh meat. Improvements in food processing, especially through quick-freezing of meat and in butter- and cheese-making, exposed British farmers to severe competition: half of Britain's cheese and 80 per cent of butter were imported in 1939 and improvements in margarine manufacture leading to mass production by such firms as Van den Berg and Jurgens (later part of the Lever Brothers' empire) took much of the cheaper end of the market and, with it, went further erosion of a hitherto rural industry. Even so, beef producers held on to about half of the British market and egg producers improved to three-fifths; but lamb and mutton producers (challenged by high-quality chilled New Zealand lamb) retained only one-third by the mid 1930s. Only for liquid milk and fresh vegetables (including potatoes) was home production virtually dominant.

Although the brief wartime drive towards self-sufficiency was abandoned, between the wars the beginnings of government intervention in the economy initiated a measure of assistance for agriculture through twin policies of protection and modest price-support systems that covered some five per cent of the total value of agricultural output in 1937–38. These were varied: rate relief on farms in 1924; subsidies for beet-sugar production in 1925; deficiency payments to wheat farmers in 1932, under which the difference between an agreed price and the market price was made good by the government; and subsidies to beef farmers in 1934. The development of Marketing Boards (notably for hops, milk and potatoes) under the Agricultural Marketing Acts 1931 and 1933 significantly assisted farmers in collecting and marketing their produce at a fair price, giving dairy farmers in particular a steady monthly income. The imposition of tariffs (1931) and import quotas on wheat marked a partial retreat from free trade and the 1932 Ottawa Agreement gave preference to Empire producers of such commodities as sugar, wheat, dairy products and meat.

Such measures provided a basis for more interventionist agricultural policies during the Second World War which were maintained under the 1947 Agricultural Act and its successors. Faced by the threat of a massive U-boat blockade, wartime governments took immediate steps to increase home food production and switch the emphasis to essential foods, to ration basic foods (such as butter, cheese and meat) in short supply, and to control marketing, prices, labour and supplies of fertilizer, fuel and machinery. Thus, after a lengthy period of stagnation since the 1880s tillage increased by 55 per cent between the 1935–39 average and 1945, especially on heavier soils in midland and western counties where conversion to grass had been most marked (Chapter 9). Wartime yields maintained the upward movement of the late 1930s: wheat increased by 19 per cent between 1936–39 and 1946–47; oats by nearly one-half; potatoes and sugar beet production almost doubled and barley more than doubled. In contrast, other than milk, fodder and stock products – especially meat and eggs which had been substantially raised on imported fodder – were reduced.

Farming methods

As in the High Farming years, increased output and intensification of methods of cultivation owed much to mechanization and increased inputs of fertilizers. Before the Second World War most farmers still relied mainly on farmyard manure. There was a surge in fertilizer consumption (especially of nitrates) after 1945: whereas, pre-war, potatoes and sugar beet tended to receive most artificial fertilizer, from the 1940s grain and grasses also benefited. Whatever their long-term environmental impacts, use of chemicals contributed substantially to greatly increased yields of all crops from the 1930s (Table 14.2).

Improved seed and control of pests and plant diseases through pesticides and herbicides also contributed much to increased yields. In the early twentieth century treatment by natural sprays (such as pyrethrum, derris or nicotine) was mostly reserved for fruit and vegetables (e.g. the protection of potatoes against blight). From the 1920s and especially after 1945 chemical sprays on cereals, then other crops and, finally, grassland increased yields and changed many farming techniques.

Although efficient horse-drawn machinery was reflected in the number of farm-horses which peaked in 1901 at around 1.1 million out of a maximum horse population of over 3.4 million, petrol then diesel driven tractors developed from 1902 were more powerful and efficient. It was only with the introduction of the Ferguson coupler, that linked tractor power to farm implements, that they began to displace horses in the late 1930s. Many farmers did not use tractors until after 1945, but such were their advantages

Table 14.2 Yields of selected crops in England and Wales, 1885–1974

Crop	Yield (hundredweight per acre)					Index of yield (1885–90 = 100)			
	1885–9	1921–5	1933–7	1950–4	1970–4	1921–51	1933–7	1950–4	1970–4
Wheat	16.2	18.6	17.8	22.7	35.6	115	110	140	220
Barley	13.0	14.6	16.2	21.1	30.0	112	125	162	231
Oats	13.8	13.8	16.2	19.4	30.8	100	117	141	223
Potatoes	119.9	126.4	136.1	162.8	254.3	105	114	136	212
Sugar beet	NA	148.2	188.7	231.7	277.0	–[1]	127	156	187
Turnips and swedes	238.1	238.1	213.0	321.6	386.4	100	89	135	162
Hay	27.5	22.7	21.1	20.3	34.8	83	77	74	127

[1]Sugar beet yields indexed to 1921–5 = 100

Source: Grigg, (1989), Table 6.1

in speed and power in ploughing, cultivating and harvesting that the internal combustion engine became a crucial part of farming. The tractor could tackle heavy soils and slopes that were difficult for horse-teams, helping to bring land back into cultivation in the wartime plough-up campaign. Most of the machinery pulled by tractor had been available from the late nineteenth century, but refinements such as the drilling of seed and fertilizer together lessened costs of cultivation and raised yields. Harvesting was another matter. Self-binding reapers had accelerated the taking and storage of grain from the 1890s, but only from the 1930s and especially after 1945 when combine harvesters, widely used in North America in the 1920s, were adopted in Britain. Together with new cultivation techniques, they revolutionized cereal farming though economies in growing root crops were less striking before 1945. The ten-fold increase in power on British farms between 1908 and 1939, with a further three-fold increase by 1948, led to the 3.3 hours needed to produce 1 acre of grain in 1930 being only one-seventeenth of the 56 hours needed a century before: moreover, labour requirements fell by a further three-quarters by 1960.

Changing patterns of farming

Although the worst of the depression for arable farmers was over by the mid-1890s, by which time wheat prices had fallen by 50 per cent over twenty years (Figure 9.3), their position was, at best, stabilized up to 1914. Contraction in grain production was most marked in low-output areas and where land could be successfully converted to grass as in northern and western Britain and many parts of midland England. Wheat and barley remained significant in eastern England (Figure 14.3) where good farmers on the better soils made it pay. On heavy land, which was costly and difficult to work, many arable farmers continued to go out of business: in Essex, for example, Scottish incomers switched to stock to make low-rental farms pay. By 1913 permanent grassland occupied over one-third of the county's cultivated area and over one-sixth was agriculturally unproductive.

The switch to grassland was most marked on midland clayland and in western districts (Figure 14.4): nearly 70 per cent of Wiltshire's farmland was under permanent pasture by the First World War and one-fifth of its arable in grass ley. Upland counties depended even more on grass, much of it long ley in extended arable rotations as in Cornwall and mid and South Wales. One of the virtues of such flexible cropping was the ability to switch the emphasis from food to stock-feeding through grass or fodder crops, and to introduce more profitable cash crops.

Increases in fruit, market garden and field vegetables were reflected in intensive cultivation of specialist crops such as the fruit growing of the Vale of Evesham, the Wisbech area of the Fens or the raspberry crop of the Carse of Gowrie, and in arable rotations producing potatoes, brassicas, peas (for canning as well as fresh) of south west Lancashire, the Fens and the Lothians. Where climatic conditions favoured production of early vegetables (as in Cornwall) or potatoes (e.g. south Pembrokeshire and the

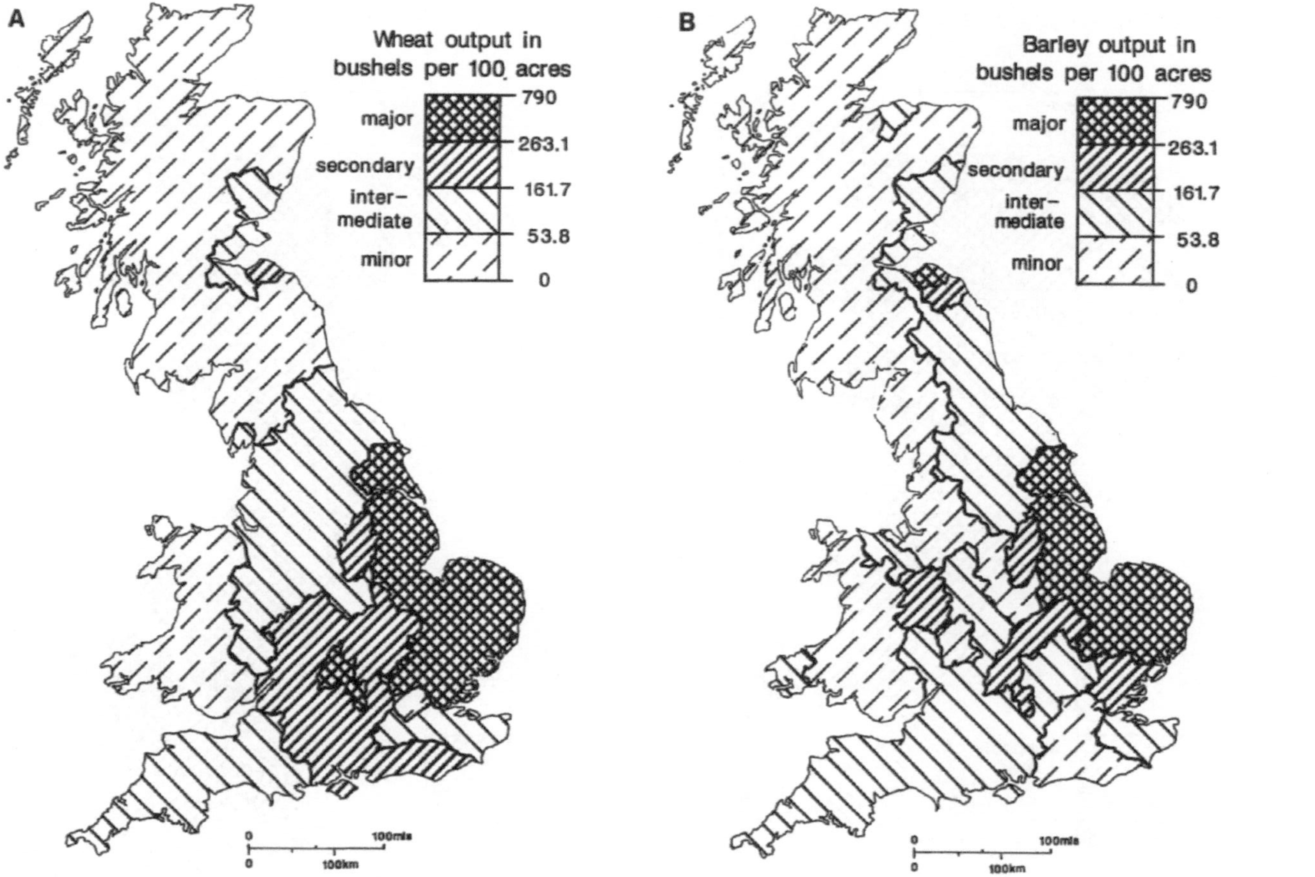

Figure 14.3 Principal areas of grain production [Wheat (A) and Barley (B)] in Great Britain, 1911. Based on Langton and Morris (1986) pp. 37 and 39.

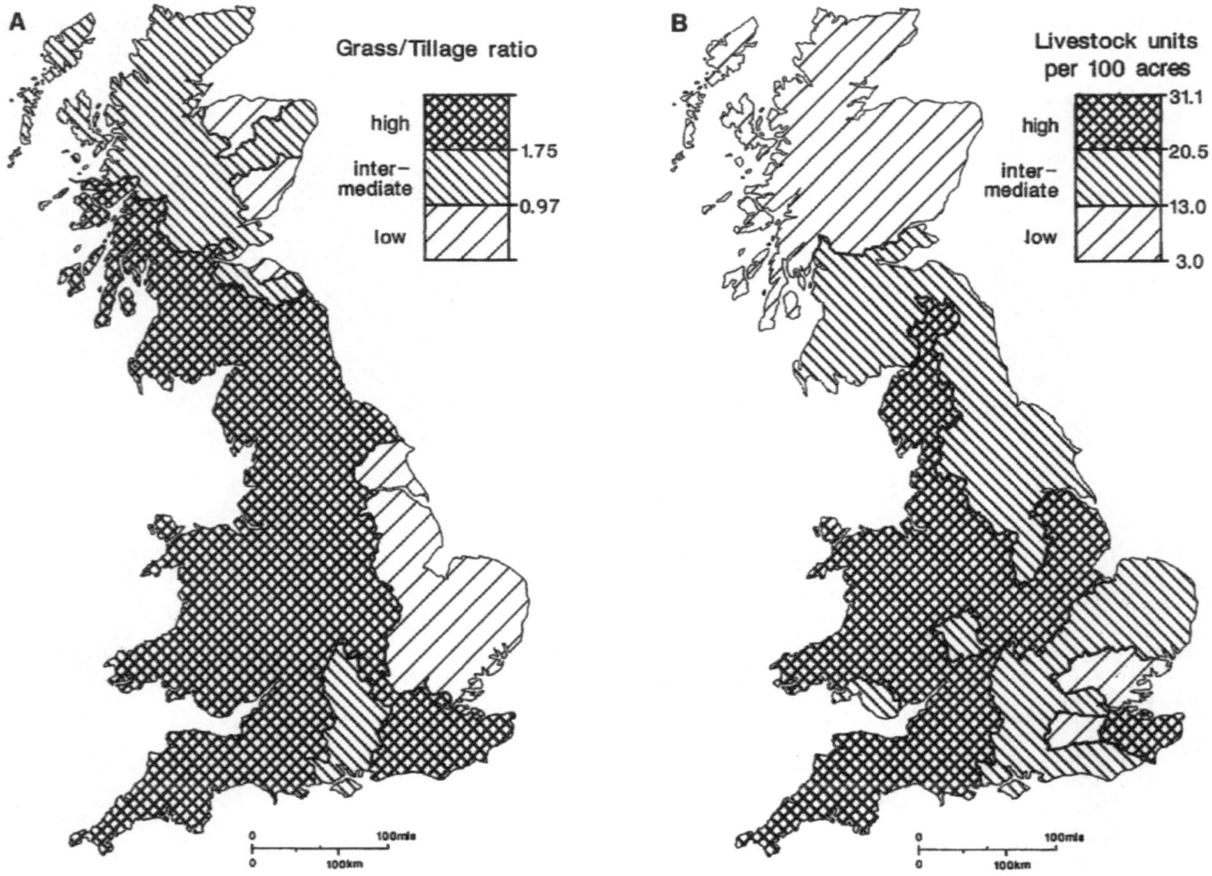

Figure 14.4 Principal areas of grassland (A) and livestock (B) in Great Britain, 1911. Based on Langton and Morris (1986) p. 47.

Ayrshire coast) there were pockets of intensive cultivation in the midst of a mainly pastoral economy.

Although in general stock products fared better under foreign competition than arable, increasing imports of wool, frozen then chilled meat, and butter and cheese increasingly made liquid milk production the most profitable form of stock farming. Numbers of dairy cattle rose steadily from the 1870s to the 1930s, beef cattle fell slightly, and sheep flocks were considerably reduced. By the First World War London's demands extended grass and arable dairying from southern and midland England into Cheshire and the South West for winter supplies. All major cities had a similar influence on surrounding farmland. By 1914, 70 per cent of milk production went direct to the consumer, increasing to three-quarters between the wars: by then one-third was produced in the main pastoral regions and one-sixth to one-fifth in western areas, chiefly from small farms with herds of less than 20. Improved feeding from home-produced fodder and imported grain and oil-cake helped to extend lactations and increase yields, though development of specialist, let alone pedigree, dairy herds was limited. Shorthorns were the main dairy cow. Friesians, introduced in the 1930s, formed only 7 per cent of dairy cattle in 1938: they dominated after the Second World War with 41 per cent of herds by 1955 (as against 25 and 19 per cent, respectively, Dairy Shorthorns and Ayrshires); in the 1990s they provide nearly nine-tenths of British dairy cattle. With mechanized milking and feeding systems bigger herds with large-scale bulk production of milk have become a major feature of British farming since the 1950s.

Although beef cattle were less significant in British farming from the turn of the century, they continued to be important in such arable-fattening systems as those of eastern England and Scotland and on the fattening pastures of the East Midlands. Better feeding, management and breeding (for example of dual-purpose breeds such as the Shorthorn, or the specialist Hereford and upland black beef cattle) raised production levels in both mixed and specialist areas. Sheep fared less well, though the sheepfold persisted in traditional rotations on downland, wold and lowland limestone until recently: on upland grazings increased lamb production was a response to both changing market tastes and the capabilities of and support given to hill farmers.

While most crops and stock were present in farming systems of the late nineteenth century, one major innovation was the introduction of sugar beet into lowland arable rotations in the 1920s. Eastern England, where the sunshine levels and cool autumns promote the best sugar content, was the main beneficiary. A labour-intensive crop, sugar beet could not compete with Empire cane sugar until government subsidies (in 1925) and more intensive cultivation led to higher yields and a sevenfold increase in acreage by 1934. It provided a valuable cash crop with an assured market and cheap and useful fodder from the tops and processed pulp. At a time when industry was leaving the countryside, the employment offered at the 17 English and one Scottish sugar-beet factories was a welcome diversification of rural employment.

The retreat of cultivation from the mid 1870s stabilized from the mid 1890s but, apart from the sharp upturn of tillage in 1917–18, farming was dominated by grassland up to 1939. Progressive government intervention and the adaptation of farmers to economic and agrarian conditions led to substantial and lasting changes during the Second World War. Despite labour and fuel problems, the drafting of women into the workforce and increased use of tractors saw a sharp increase in tillage of some 5.5 million acres (45 per cent) between the late 1930s and 1945, mostly at the expense of permanent grassland which declined by 6.5 million acres (38 per cent) particularly on

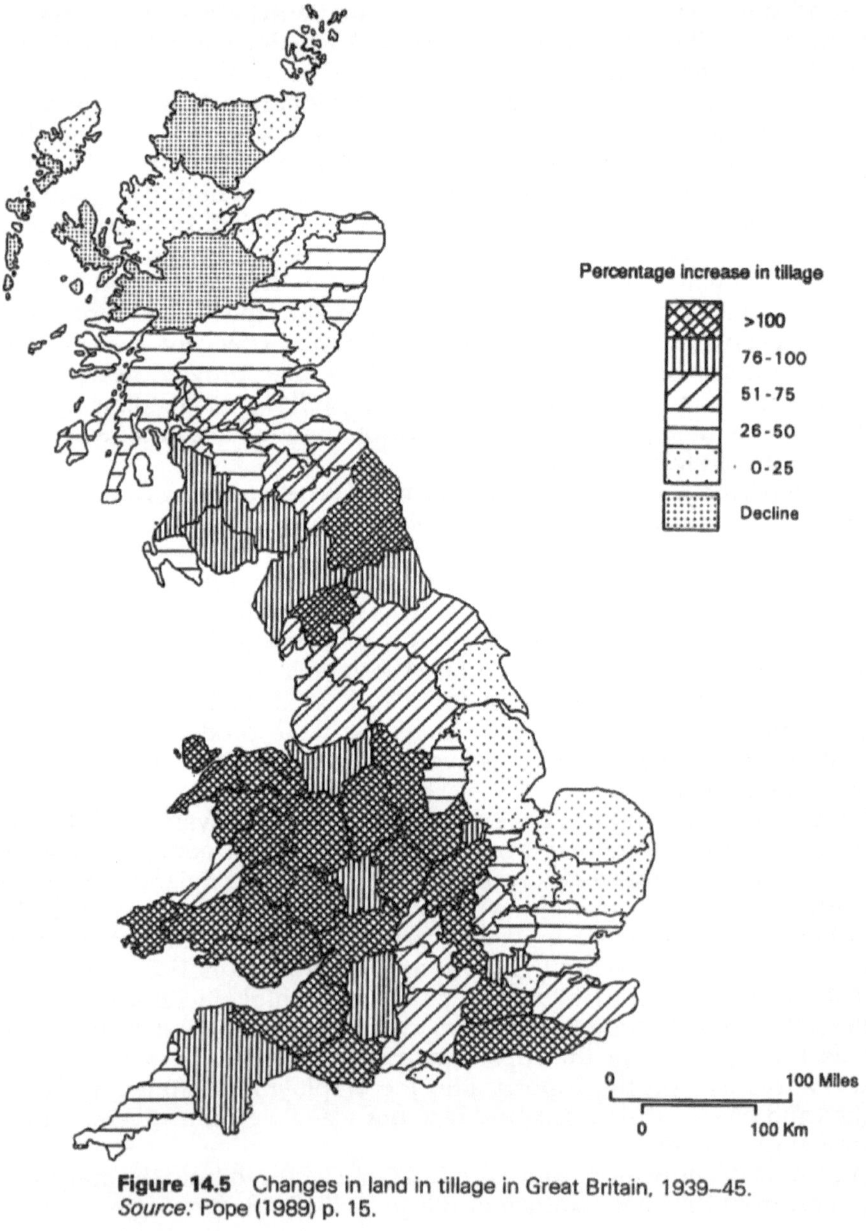

Figure 14.5 Changes in land in tillage in Great Britain, 1939–45.
Source: Pope (1989) p. 15.

heavy land in the midlands and in the uplands of Wales and northern England (Figure 14.5). The focus on essential crops increased self-sufficiency to 40 per cent: after 1945 price support and subsidies such as those to hill farmers and for land improvement helped maintain those trends. Renewed emphasis on tillage as the basis of mixed farming initially emphasized rotation grass but from the 1950s increasing use of fertilizers, herbicides and pesticides ushered in a new 'high farming' that has substantially altered British farming, rural landscapes and environments.

Land use, land quality and types of farming

At the beginning of the century three-quarters of the land in England and Wales, but less than one-quarter in Scotland, was cultivated, with one-tenth and over half, respectively, in rough grazing. Woodland (5 per cent), occupied little more than urban land. Wartime apart, the amount of cultivation in Britain declined and rough grazing increased to around 52 to 29 per cent respectively by the 1950s when both woodland and the urban areas occupied 7.4 per cent (Table 14.3 and Figure 14.6). These proportions varied widely between regions: in Norfolk 70 per cent of land was tillage in the 1940s; in Carmarthen it was one-tenth between the wars, 30 per cent in 1944–45 at the peak of the plough-up campaign before falling back almost to pre-war levels in the mid-1950s; rough grazing and commons dominated upland Wales, and, especially, the Scottish Highlands and much of the Southern Uplands and northern Pennines.

Although reclamation of waste, notably by wartime and post-war drainage and by extension of ploughland in downland areas, resumed from 1940 most increases in cultivation came from conversion of grassland and through improvement in hill grazings, substantially assisted by subsidies and grants for underdrainage (revived in 1939), ploughing and reseeding, and for lime and fertilizer.

The centuries-old decline of woodland was reversed in the 1920s. Though improving landlords in the eighteenth and nineteenth centuries planted for landscaping, sport and shelter, only with the extension of planting on lowland heaths and some upland areas from the late 1870s (Chapter 9) did systematic afforestation and commercial management of woodland develop. In addition some remaining ancient woodlands, such as Epping Forest and Burnham Beeches, were acquired for public recreation and conservation. From 1870 to 1904 the acreage of woodland increased by 28 per cent though this was more than offset by half a million acres felled in the First World War. The 1924 Census of Woodland showed that half the new planting was coniferous – Scots pine and such exotics as Corsican pine, Douglas fir, spruce and larch. However, while lowland areas often presented a wooded appearance from the ground, with many copses and hedgerow trees, the amount of true woodland was still limited and most of upland Britain was bare.

Growing awareness of the value of timber as a crop on poor land and the strategic necessity to replenish stocks led to the establishment of the Forestry Commission in 1919. Since then substantial planting and better

Table 14.3 Major land uses in Great Britain *c.*1960

Type of use	Total (million acres)[1]				Percentage of area			
Area	England	Wales	Scotland	Great Britain	England	Wales	Scotland	Great Britain
Cultivated land (tillage and grass)	21.79	2.61	4.31	28.72	68.0	51.1	22.6	51.1
Rough grazing and commons	3.29	1.67	11.0	15.96	10.2	32.7	57.6	28.4
Woodland	2.10	0.46	1.61	4.17	6.5	9.0	8.5	7.4
Urban	3.80	0.20	0.51	4.51	11.8	3.9	2.7	8.0
Total area	32.04	5.10	19.07	56.21	100	100	100	100

[1]Excluding some residual land not categorized: rows 1–4 do not total to row 5

Source: Watson and Sissons (1960) Table 11

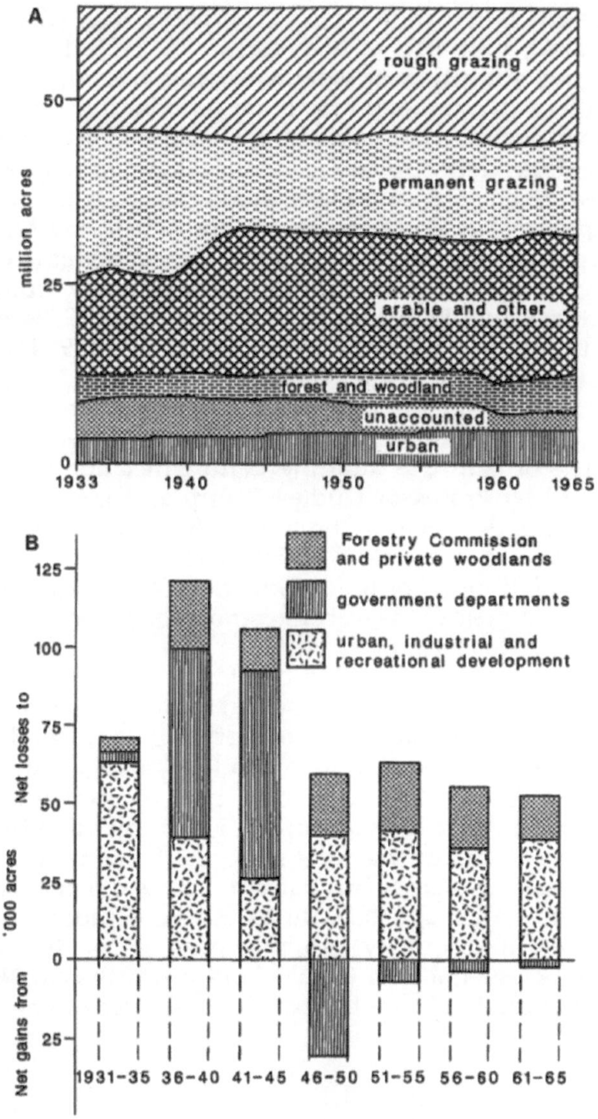

Figure 14.6 (A) Land-use changes in the United Kingdom, 1933–65. In 1932 and 1959 rough grazing was redefined with an increase of 889,560 acres and 1,482,600 acres respectively. Annual data for agriculture and woodland; decadal estimates for urban areas. (B) Transfers of agricultural land in England and Wales to urban and other non-agricultural uses, 1933–65. Based on M. Parry in R. Johnston and J. Doornkamp (eds) *The changing geography of the United Kingdom.* (London: 1982) p. 17.

management of both public and private woodlands have increased both the area under woodland and the timber crop, though Britain remains largely dependent on imports for construction, furniture-making and pulp for paper and chemicals industries (Chapter 15). The original objective of the Forestry Commission was to grow enough timber for three years' demand by 2000 AD – some 1.75 million acres as compared with 1.9 million acres of existing woodland, only half in high forest. By 1938 it had acquired 1.1 million acres, including Crown woodland, and had afforested one-third (over nine-tenths with conifers) mostly on moorland grazings in Scotland and Wales, or on such lowland wastes as Thetford Chase, Norfolk and the Culbin Sands, Morayshire. Upland reservoir catchments were also substantially afforested. Such activity brought some new jobs to the countryside, often at the expense of upland farming, and the debate on alternative uses of land for forestry, water supply, farming and recreation remains an issue for rural planning of upland areas. Nevertheless, steady Commission and private planting reached 40,000 acres per annum in the 1950s which, despite depletion of Britain's woodlands by one-fifth in 1939–45, brought the proportion of woodland to 7.4 per cent by 1960.

The debate on land quality and land use in a nation under increasing pressure from urbanization (Chapter 16) was hampered by the lack of a complete picture of land use until the Land Utilisation Survey of Britain (1930), organized by Professor Dudley Stamp and largely carried out by volunteer geographers, surveyed the country field by field between 1931–39 using a six-inch to one mile scale. The land-use maps and county reports of the Survey provided much of the information for the assessment of land quality and farming basic to the wartime reports on which postwar land-use and rural planning was based.

The 17 types of farming identified by Stamp, simplified in Figure 14.7, epitomize regional contrasts in land use and farming systems from the turn of the century to 1940 (Chapter 9). Mixed farming dominated. While arable characterized most of eastern Britain, and pastoralism the upland and lowland areas of the west, a substantial part of midland England was 'intermediate' and likely to shift to crops or grass according to prices. High-quality soils close to urban markets supported intensive cash cropping within all three major systems. This pattern reflects soil quality, climate and land management. Although difficult to compare productivity from different enterprises, high quality fattening pastures of the East Midlands equate with first-class arable in the Fenland or north-east Norfolk, while Cheshire's dairy grasslands are the equal of south-west Lancashire's cash crop districts.

Tenure

After the Second World War, improved farming and inputs of capital, equipment and skill determined farming success and increasingly transformed a depressed industry. Although much investment in marginal areas might have been more profitably spent on good land, all farmers benefited from price-support and national marketing systems. As the number and proportion of farm workers continued to fall, wages and working conditions improved. There have also been major changes in landownership and farm tenure. After the First World War the proportion of owner-occupiers in England and Wales increased threefold to 26.6 per cent by 1938 and to more than half by the late 1950s. Institutional landowners partly replaced large

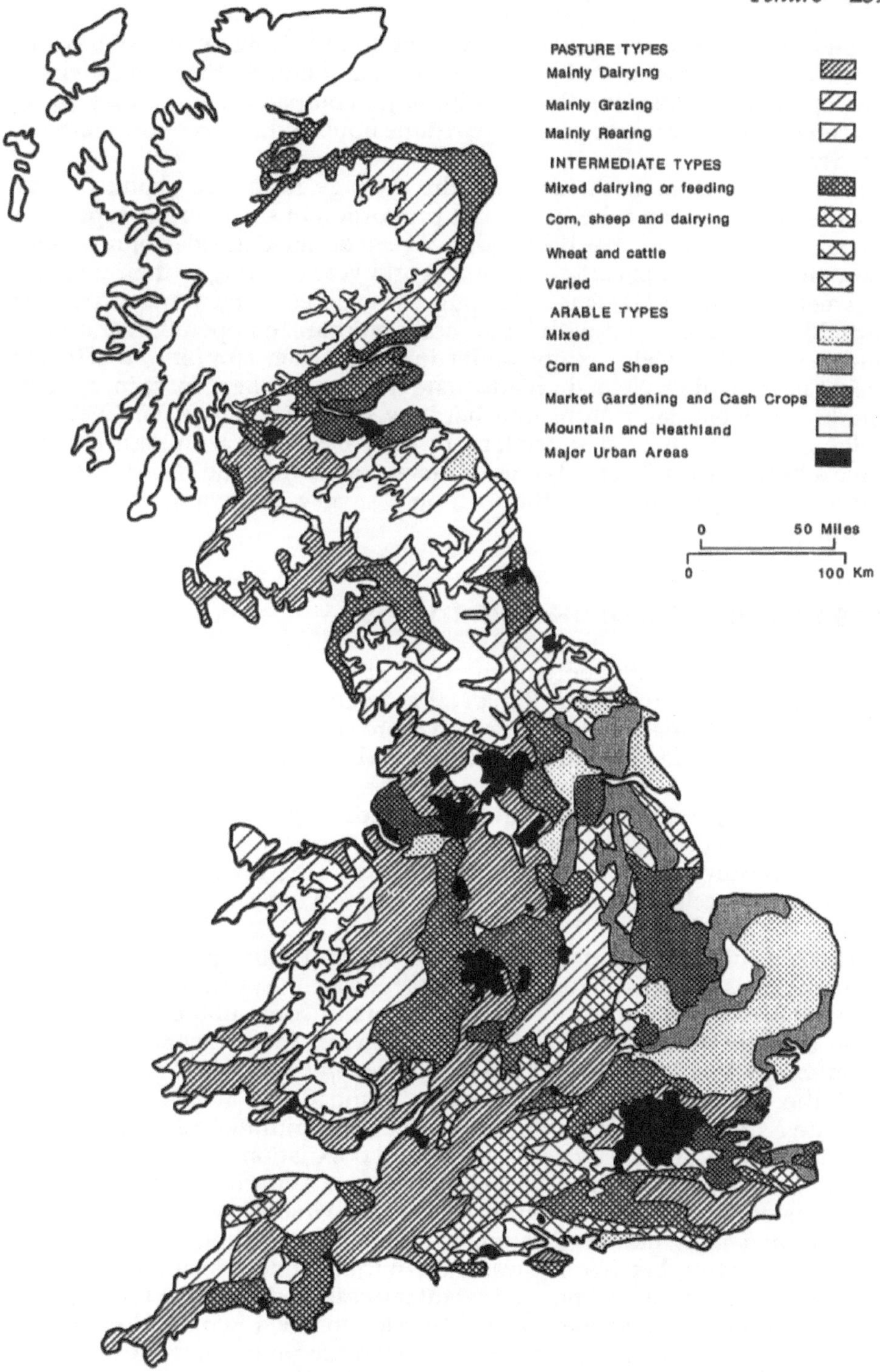

Figure 14.7 Types of farming in Great Britain in the 1930s. Based on Stamp (1962) pp. 301 and 318.

landed estates through government-owned land, from small holdings to defence establishments and practice ranges, and latterly through purchase and operation of big arable farms by large companies. In contrast one-quarter of post-war holdings were part-time hobby farms or country retreats for city people.

Despite growing mechanization, the average farm size changed little between the 1890s and the 1950s. The proportion of small farms, especially those under 200 acres, has declined, not least because the idea of resettling men on small-holdings after the First World War (leading to their increase between 1913 and 1931) was not a long-term success. The number of farms over 300 acres increased with larger-scale mechanized operation and farm mergers. By the 1950s, except under intensive cash cropping, a 100-acre arable farm neither yielded an adequate income nor kept machinery fully employed. Most dairy farms and herds were very small in the 1930s but milking machines have changed that. By the 1960s a typical unit was likely to have a herd of over 20 on 50 acres or more, rather than the two-thirds with less than 20 cows (half less than 10) on a small acreage after the Second World War.

The urbanization of the countryside

Even in the nineteenth century few rural areas were wholly remote from urban influences (Chapter 9). In the twentieth century their inter-connections became more obvious and more complex. Many rural areas of Britain were affected by suburbanization and urban sprawl, made possible by a combination of improved transport and greater affluence. In the commuter belt around London and other cities, large and small, the distinction between countryside and town became increasingly blurred as more people lived in the countryside and worked in the town. Although such trends have accelerated since 1945, their effects were clearly evident by the 1920s.

The urbanization of the countryside was increasingly recognized in planning legislation as planners sought to recreate the perceived advantages of rural living within new and extended urban communities. One of the most obvious interactions is seen in the Garden City movement. In 1898 Ebenezer Howard published *Tomorrow: a peaceful path to real reform* and in 1900 the Garden City Association was founded. Its aims were both to provide a new urban environment in small communities of around 6,000 people within a garden city with a maximum population of 30,000 and, at the same time, to revitalize the countryside. Howard's vision sought to combine urban amenities with rural beauty. In 1903 the First Garden City Company was formed and, mainly through the initiative of Thomas Adams and Raymond Unwin, Letchworth was built in the first decade of the twentieth century. Although it was not an instant success, it did lay the foundations for the garden city movement which later led to new town developments – urban incursions into the countryside – which sought to retain the planner's vision of what was good about the countryside.

The harsh and continuing reality of rural poverty was conveniently forgotten in such schemes. Despite increasing mechanization agricultural work was still hard and poorly regulated, while rural labourers worked longer hours for less pay than most other workers. A cowman in East Anglia who started full-time work at 14 in the 1920s remembers living in a two-bedroom cottage with his parents and eight other children. He began work at 5.30 each morning but, with a 3½ mile walk to the farm where he worked, he left home at 4.45 a.m. and got back after 6.00 each evening, permanently tired from long hours and lack of nutritious food. In the 1920s young labourers earned only 5s–10s for a seven-day week and adults 25s per week. Although the National Union of Agricultural Workers was founded in 1872, union membership was low and much of the welfare legislation did not apply to or was ignored by the agricultural sector. Not surprisingly half a million farm workers left the land during the 1920s and labour shortages were a factor in gradual mechanization from the 1930s. By the late 1930s, when tractors and combine harvesters were more common on farms, labourers' wages were around £1. 15s per week.

Most of the legislation affecting rural Britain in the first half of the twentieth century reflected urban attitudes to the countryside and was rarely specifically related to rural needs. The Town Planning Act of 1909 and especially the Town and Country Planning Act of 1932, meant to stimulate land-use planning by local authorities, had relatively little effect in rural areas: indeed by the 1930s there was growing concern about the loss of agricultural land and the effects of suburban sprawl in the countryside. Although the Green Belt Act of 1938 established a 25,000 acre ring of protected countryside around London, it was not until the 1944 and 1947 Town and Country Planning Acts that effective planning control over development in rural areas was established.

As in the nineteenth century, legislation concerned with housing and sanitation was mainly urban-orientated, although the countryside eventually benefited. For instance, rural authorities built council houses after 1919, which gradually helped relieve pressures on rural housing, but their impact was much less significant than in towns. Indeed, living conditions in the countryside suffered because of the urban bias of most legislation which failed to take account of rural needs. Private housing in commuter villages was well beyond the pocket of most rural, especially agricultural workers. Even in the 1950s many rural areas had proportionately more sub-standard houses than many large industrial towns, although the gradual spread of electricity and other services to the countryside cumulatively improved rural life.

The city exploited the countryside in many other ways. Not only did rural areas contribute vitally to the national food supply, as reaffirmed by the 1942 Scott Report, but the countryside was increasingly important for timber and water resources. Forestry Commission planting added 240,000 hectares of woods 1919–39 and further Acts in the 1940s increased its powers of land purchase and planting. The use of rural areas for water catchment was initiated in the nineteenth century when many local authorities purchased land and built reservoirs (for example Manchester Corporaton constructed the Thirlmere reservoir in the Lake District in the early 1890s). These activities continued largely uncontrolled into the twentieth century and a second Lake District reservoir was built by Manchester Corporation at Haweswater in the 1930s. Yet there was no significant legislation to improve water supplies or control pollution until the 1940s. The 1945 Water Act established central government control over water resources, with the Minister of Health responsible for water conservation and supply, and in 1948 a system of 32 river boards was established with powers over water pollution and abstraction. The Forestry Commission and the river authorities both became major rural landowners. This trend has accentuated since 1945 as large financial and commercial institutions have invested in agricultural land. Hence an increasing proportion of rural land is not controlled by the people who live and work there, a trend that accelerated from the 1920s.

Even remote rural areas such as the Highlands and Islands of Scotland were affected by urban-based government intervention from the 1890s. Rapid out-migration in the mid-nineteenth century was followed by strong moves to protect the interests of crofters who became better organized and more visible to government. The report of the 1884 Royal Commission on Crofters promoted greater security of tenure, fairer rents and compensation on enclosure for crofters and, in some areas, created new holdings and enlarged existing ones. Support for crofting came at a time when sheep farming by large landowners – which threatened the crofters' existence in the nineteenth century – was becoming less profitable and many large landowners were expanding their deer forests. Indeed the 1895 Royal Commission examined the extent to which land taken for sport could be returned to smallholders. However, lack of land for crofters remained a major problem, although the 1897 Congested Districts Board made some progress in the worst affected districts and initiated improvements in communications and local industry. The Forestry Commission also had a major influence on the Highlands, creating many new jobs. In the 1930s the Scottish Economic Committee reviewed the problems of the Highlands and an Advisory Panel was established in the 1940s to assist in the formulation of government policy. In 1943 the North of Scotland Hydro Electric Board was set up and in 1949 a Highland Development Area was established to aid the development of the Highland economy. The most remote rural area of Britain was thus both brought within the framework of national regional planning, and its resources (especially water and timber) were increasingly exploited.

Finally, the countryside has been increasingly linked to urban-dominated values and activities through tourism, leisure activities and the growth of conservation movements. Even in the nineteenth century the most scenic rural areas (such as the Lake District) increasingly became playgrounds for the rich and leisured, and the pursuit of rural recreation imposed middle-class, urban-based values throughout the population. In the twentieth century such values also became institutionalized in government legislation and the conservation movement. The National Trust, now a major landowner in the Lake District, was established in 1895 and its Scottish counterpart in 1931; the Council for the Preservation of Rural England was set up in 1926; and in 1939 an Access to the Mountains Act was passed. Although ineffective, this latter recognized the need to deal with the problem of public access fuelled by increasingly mobile urban populations with the time and money to visit the countryside. By the 1920s, the working classes were also gaining access to the countryside through cycling and hiking. New groups such as the Ramblers Association and the Youth Hostels Association represented broadly-based leisure interests and put pressure on government to provide access to the countryside (which most large landowners resisted) and to protect rural areas from urban and industrial development. The mass trespass of hikers (mainly from Manchester and Sheffield) on Kinder Scout in the Peak District in 1932 was symbolic of this movement and influenced not only the Scott Report (1942) on land use, but more specifically the work of Dower and Hobhouse both of whom recommended the establishment of National Parks. The 1949 National Parks and Access to the Countryside Act, which only related to England and Wales, eventually led to the establishment of ten National Parks and also created protected Areas of Outstanding Natural Beauty.

By the late 1940s the values espoused by nineteenth-century writers such as Wordsworth and Ruskin had thus become enshrined in government legislation and, effectively, had become the national view of landscape conservation. The fact that many citizens could still not afford the time or money to participate to any significant degree in recreational activities, let alone visit protected upland landscapes, and that problems of poor housing and high unemployment severely affected much of the resident population of those areas visited by urban weekenders, went unnoticed by legislators and most of those who used the embryonic national parks for recreation. Nevertheless, the impact of Michael Dower's 'fourth wave' breaking over many parts of rural Britain after the First World War was felt socially as well as economically and environmentally in both country and town.

Bibliography and further reading for Chapter 14

Many of the references in Chapter 9 are still relevant.

In addition on agricultural and economic change see:

BECKETT, J. V. *A history of Laxton: England's last openfield village* (Oxford: Blackwell, 1984).

BEST, R. *The changing use of land in Britain* (London: Faber, 1962).

COPPOCK, J. T. *The agricultural geography of Great Britain* (London: Bell, 1971).

GRAHAM, P. A. *The revival of British agriculture* (London: Jarrold, 1899).

GRIGG, D. *English Agriculture. An historical perspective* (Oxford: Blackwell, 1989).

MAXTON, J. P. (ed.) *Regional types of British agriculture* (London: Allen and Unwin, 1936).

PERRY, P. J. (ed.) *British agriculture, 1875–1914* (London: Methuen, 1973).

PERRY, P. J. *British farming in the great depression, 1870–1914. An historical geography* (Newton Abbot: David and Charles, 1974).

SIMPSON, E. S. 'Milk production in England and Wales: a study in the influence of collective marketing' *Geographical Review* **49** (1959), pp. 95–111.

STAMP, L. D. *The land of Britain: its use and misuse* (London: Longman, 3rd edition, 1962).

STAPLEDON, R. G. *The hill lands of Britain: development or decay* (London: Faber, 1937).

WHETHAM, E. H. *The London milk trade, 1900–1930* (Reading Institute of Agricultural History, Research Paper 3, 1970).

WHETHAM, E. H. *The agrarian history of England and Wales*, vol viii, *1914–1939* (Cambridge: Cambridge University Press, 1973).

On rural life and the urbanization of the countryside:

AMBROSE, P. *The quiet revolution: social change in a Sussex village, 1871–1971* (London: Chatto and Windus, 1974).
ARMSTRONG, A. *Farmworkers: a social and economic history 1770–1980* (London: Batsford, 1988).
BEST, R. and ROGERS, A. *The urban countryside* (London: Faber, 1973).
GILG, A. *Countryside planning* (London: Methuen, 1979).
HAGGARD, SIR H. RIDER, *Rural England* (London: Longman, 1902).
HORN, P. *The changing countryside in Victorian and Edwardian England and Wales* (London: Athlone Press, 1984).
PATMORE, J. A. *Land and leisure* (Newton Abbot: David and Charles, 1970).
PHILIPS, D. and WILLIAMS, A. *Rural Britain: a social geography* (Oxford: Blackwell, 1984).

15

Industry and industrialization from the 1890s to the 1940s

Structure and growth of the national economy

A decline in British manufacturing and trade competitiveness from the Great Depression of 1873–96 saw a transition to a broader-based economy the impacts of which varied from industry to industry and, not least, regionally. Older staple industries – cotton, iron and steel, shipbuilding and railway engineering and heavy chemicals – declined during this period, but later, in the world depression of the early 1930s, they collapsed alarmingly. Yet substantial growth occurred in the electrical and light engineering, motor vehicle and consumer industries. Even in textiles, new types of yarn brought hope. But in the third Kondratieff cycle all industries faced a growing challenge from other manufacturing countries.

The changing international position underlined Britain's structural weaknesses. Even a new wave of post-1945 prosperity (the upswing of the fourth Kondratieff) was less successful than past experiences of industrialization. Growth in the working-age population (which had fallen by one million between 1911 and 1926) to 22.1 million by 1936 and 23.1 million by 1939 did not help the unemployment queues, which were never less than one million between 1920 and 1939 and nearly three million in the deepest trough of the depression in the winter of 1932–33. A changing workforce with more women, especially in the service sector, and many more 'semi-skilled' workers reflected changes in the nature and location of economic activity. New industries in large production units focused on markets and abundant supplies of labour rather than on the raw materials, old infrastructure and unionized craft labour of the great Victorian industrial districts. In the 1880s the top one hundred firms provided under one-tenth of British output: by 1909 after a number of big industrial mergers they produced 15 per cent and, from the 1930s to the 1950s, one-quarter of output. Such giants as Unilever, Imperial Tobacco and ICI dominated their respective industrial groups in the twentieth century. In London and the major cities ancillary services – industrial, transport, financial and professional – were advantageous in setting up new industrial complexes. Such factors underlay the growing disparity in regional employment (Table 15.1).

 DOI: 10.4324/9781003174783-15

Table 15.1 Regional employment changes in Great Britain, 1921–61

Standard Region	Employment 1921 Total (000's)	1921 Percentage of GB	1931 Total (000's)	1931 Percentage of GB	1951 Total (000's)	1951 Percentage of GB	1961 Total (000's)	1961 Percentage of GB	Changes in employment (per cent per annum) 1921–31	1931–51	1951–61
South East	5436	27.9	5958	32.3	7005	31.7	7628	32.8	1.0	0.9	0.9
East Anglia	483	2.5	471	2.5	559	2.5	586	2.5	−0.2	0.9	0.5
South West	1133	5.8	1160	6.3	1399	6.3	1470	6.4	0.2	1.0	0.5
West Midlands	1584	8.1	1585	8.6	2125	9.6	2293	9.9	0.0	1.7	0.8
East Midlands	1272	6.5	1259	6.8	1558	7.0	1684	7.2	−0.1	1.2	0.8
Yorkshire and Humberside	1730	8.9	1649	8.9	1889	8.5	1942	8.4	−0.5	0.7	0.3
North West	2897	14.9	2607	14.1	3027	13.7	3015	13.0	−1.0	0.8	−0.04
North	1203	6.2	1029	5.6	1320	6.0	1349	5.8	−1.5	1.4	0.2
Wales	1057	5.4	897	4.9	1040	4.7	1066	4.6	−1.5	0.8	0.3
Scotland	2149	11.0	1852	10.0	2193	9.9	2202	9.5	−1.4	0.9	0.04
Great Britain	18944	100	18457	100	22115	100	23235	100	−0.2	1.0	0.5

Source: based on Law (1980), Tables 8 and 9 and Censuses of Population 1931 and 1961

Older industrial regions concentrated principally on staple heavy industries, textiles, port processing industries, coal and steam power, which were vulnerable to both cyclical fluctuations in international trade and to local or regional structural problems arising from over-dependence on a single or narrow range of industries. They were victims, in part, of intense specialization creating frictional immobility of labour (in skills and residentially) that bottled up workers in dying industries (Table 15.2).

In contrast, new 'growth' industries – electrical, motor vehicles, light engineering, consumer goods (e.g. furniture, clothing), food processing and branches of the chemicals and paper industries – were healthy and, together with the war industries, provided most of the economic growth of 3.0 per cent per annum between 1935–55, as against a modest 1.7 per cent between 1907–35. Whilst engineering areas in the North West and West Yorkshire claimed their share of electrical and other engineering trades, the bulk of such new jobs went to the South East and West Midlands. Even where, as in the latter region, there was a strong engineering tradition new sites were sought leaving old centres vulnerable: for example sewing machine, bicycle, then motor manufacture were sited at Coventry and Birmingham rather than in the Black Country. South-eastern Britain claimed 56 per cent of new jobs after the First World War, as against their 51 per cent share of total employment by 1951.

The enormous scale of national and regional economic and employment problems has been interpreted by economists in different ways. Many contemporaries saw it as a time of decline and collapse in manufacturing. Others, with the benefit of hindsight, argue that with growth rates of 2–3 per cent per annum – up to 3–4 per cent in the late 1930s – the problem was one of regional structural adjustment rather than national collapse.

Political pressures for corrective action led, ultimately, to a turning away from nineteenth-century free trade. Policy was focused on three areas. Fiscal measures sought to keep inflation in check and improve the balance of payments by strict controls on public expenditure, despite the growing force of Keynesian arguments for investment in the economy; some degree of protection was sought through bilateral, mainly imperial, trade agreements (e.g. the 1932 Ottawa agreement). The structural problem was recognized in various attempts to rationalize key industries such as coal, and the railway amalgamation into four regional companies; there were measures to stimulate industrial training and labour mobility. The regional problem, expressed so poignantly in the levels of unemployment, was seen in the 1920s chiefly as a maladjustment to changing labour markets best solved by labour migration and market forces including the lure of better wages in the new industries.

However, underinvestment in both labour and plant in declining areas was gradually and reluctantly acknowledged by government recognition that the real and substantial differences in regional prosperity were unlikely to be remedied by private investment. The Board of Trade surveys of 1932 and the designation of four Special Areas (West Cumberland, North East England, South Wales and Central Scotland) in 1934 (Figure 15.1) were a beginning (Table 15.3). Government-assisted investment in industry also helped: the continuous production steel strip mill at Ebbw Vale and the

Table 15.2 Occupational structure for selected industries by region, Great Britain 1921 and 1951

Region	1921											1951										
	Mining	Metals	Engineering	Electrical engineering	Metal mfg.	Ship building	Vehicles	Chemicals	Textiles	Clothing	ALL selected	Mining	Metals	Engineering	Electrical engineering	Metal mfg.	Ship building	Vehicles	Chemicals	Textiles	Clothing	ALL selected
ENGLAND AND WALES South East (T)	6.6	16.0	154.1	69.4	63.1	65.1	103.0	65.9	30.2	280.7	4838.6	14.2	44.0	318.8	233.5	107.6	61.8	218.5	131.8	41.3	226.6	6998.8
(percentage of GB)	0.0	3.3	21.6	41.3	16.0	16.4	28.1	28.4	2.3	33.3	27.7	1.7	7.7	30.5	41.9	21.4	22.3	19.7	30.3	4.2	32.9	31.6
East Anglia	0.7	0.5	12.9	1.5	3.9	2.8	5.7	2.0	5.0	25.1	436.6	0.8	2.7	18.1	10.3	2.2	2.0	4.7	5.0	4.2	17.5	563.9
	0.0	0.1	1.8	0.9	1.0	0.7	1.6	0.9	0.4	3.0	2.5	0.1	0.5	1.7	1.8	0.4	0.7	0.6	1.1	0.4	2.6	2.5
South West	35.7	4.1	28.0	2.4	8.8	22.7	21.3	8.5	17.4	58.8	1022.2	19.5	6.2	46.0	14.5	10.0	21.7	59.1	12.7	13.4	28.7	1400.5
	2.8	0.8	3.9	1.4	2.2	5.7	5.8	3.7	1.3	7.0	5.9	2.3	1.1	4.8	2.6	2.0	7.8	8.0	2.9	1.5	4.2	6.3
West Midlands	97.2	96.3	66.5	27.6	171.2	0.6	103.6	10.4	33.2	46.9	1474.8	66.2	128.8	121.2	102.3	182.7	0.7	205.0	19.7	40.3	27.9	2216.9
	7.7	19.6	9.2	16.4	43.3	0.2	28.3	4.5	2.6	5.6	8.4	7.9	22.6	11.6	18.4	36.3	0.3	27.9	4.5	4.1	4.1	10.0
East Midlands	139.6	29.3	56.8	5.2	10.0	3.0	31.6	7.4	126.3	116.9	1174.2	122.4	53.2	86.7	17.9	18.1	3.1	54.8	21.3	139.1	98.2	1568.9
	11.0	6.0	7.9	3.1	2.5	0.8	8.6	3.2	9.8	13.8	6.7	14.6	9.3	8.3	3.2	3.6	1.1	7.4	4.9	14.1	14.5	7.1
North West	120.4	34.9	160.3	39.0	34.7	46.7	42.5	67.2	622.8	121.9	2673.6	61.4	42.8	153.4	98.0	50.2	36.7	76.6	106.3	382.0	121.1	3029.2
	9.5	7.1	22.2	23.2	8.8	11.8	11.5	28.9	48.1	14.4	15.3	7.6	7.5	14.7	17.6	10.0	13.3	10.4	24.4	38.8	17.9	13.7
Yorkshire and Humberside	159.8	88.2	81.3	7.1	64.2	10.2	23.4	28.9	292.7	77.0	1604.2	145.8	93.0	107.1	21.3	71.0	7.2	40.2	36.6	217.3	71.6	1890.4
	12.6	17.9	11.2	4.2	16.2	2.6	6.5	12.4	22.6	9.1	9.2	17.3	16.3	10.3	3.8	14.1	2.6	5.5	8.4	22.0	10.6	8.2
North	246.4	65.8	43.1	4.7	9.6	105.3	9.1	13.9	8.0	23.8	1092.5	179.0	57.2	55.5	27.3	13.2	60.3	18.3	43.3	14.7	32.2	1320.5
	19.4	13.4	6.0	2.8	2.4	26.6	2.5	6.0	0.6	2.8	6.3	21.3	10.0	5.3	4.9	2.6	21.8	2.5	9.9	1.5	4.8	6.0

South Wales	226.7 17.8	51.3 10.4	5.2 0.7	1.4 0.8	3.8 1.0	10.2 2.6	5.3 1.4	7.8 3.4	0.7 0.5	14.2 1.7	607.0 3.5	108.1 12.9	63.2 11.1	16.2 1.6	12.6 2.3	15.0 3.0	5.2 1.9	12.8 1.7	15.1 3.5	6.6 0.7	13.4 2.0	660.3 3.0
Mid and North Wales	56.2 4.4	16.3 3.3	2.6 0.4	0.3 0.2	3.1 0.8	4.6 1.2	2.3 0.6	1.1 0.5	3.4 0.3	9.9 1.2	344.4 2.0	24.2 2.9	16.8 2.9	6.0 0.6	1.0 0.2	4.1 0.8	1.3 0.5	8.1 1.1	5.6 1.3	9.0 0.9	2.4 0.4	380.6 1.7
SCOTLAND Strathclyde	92.4 7.3	70.0 14.2	88.8 12.3	5.7 3.4	12.7 3.2	98.2 24.8	10.7 2.9	10.5 4.5	61.8 4.8	39.4 4.7	1082.3 6.2	37.5 4.5	44.2 7.7	88.3 8.5	11.4 2.0	19.3 3.8	58.2 21.0	9.4 4.0	22.2 5.1	55.7 5.7	30.9 4.8	1093.3 4.9
East Central	85.2 6.7	17.7 3.6	17.7 2.4	3.3 2.0	6.0 1.5	21.4 5.4	3.9 1.1	6.3 2.7	68.2 5.3	17.7 2.1	707.1 4.0	58.6 7.0	17.3 3.0	20.0 1.9	7.1 1.3	8.2 1.6	14.7 5.3	5.8 0.8	13.3 3.1	42.0 4.3	7.1 1.1	722.6 3.3
Highlands	1.0 0.1	1.1 0.2	4.4 0.6	0.4 0.2	3.0 0.8	5.1 1.3	1.9 0.5	1.3 0.6	8.4 0.6	8.6 1.0	301.5 1.7	1.1 0.1	1.1 0.2	4.6 0.4	0.2 0.0	1.4 0.3	3.9 1.4	1.1 0.1	1.6 0.4	7.1 0.7	1.5 0.2	268.3 1.2
Southern	2.5 0.2	0.4 0.1	1.5 0.2	0.2 0.1	1.0 0.3	0.3 0.1	1.9 0.5	0.8 0.3	16.1 1.2	3.3 0.4	117.1 0.6	2.3 0.3	0.0 0.0	1.7 0.2	0.1 0.0	0.5 0.1	0.0 0.1	0.3 0.0	1.1 0.3	13.8 1.4	0.7 0.1	110.5 0.5
Not specified	142.5	6.4	10.2	7.2	9.5	12.2	9.2	8.8	14.3	30.5	1530.5											
GREAT BRITAIN	1412.9	498.3	739.4	175.4	404.6	408.5	375.4	241.0	1308.3	874.7	19006.6	841.0	570.6	1043.7	557.3	503.6	276.8	734.6	435.5	985.6	675.9	22134.7
percentage of GB	7.4	2.6	3.9	0.9	2.3	2.4	2.0	1.3	6.9	4.6	(34.3)	3.8	2.6	4.7	2.5	2.3	1.3	3.3	2.0	4.5	3.1	(30.1)

Totals (T) in 1000s; regional percentages in each column are for each occupation as a proportion of those in Great Britain as a whole. The last column gives selected occupations as a percentage of all occupied in GB. The regions are as in Table 10.3.

Occupational categories are as in the Registrar General's (census) 1911 series B and are closely but not precisely comparable with those in Table 10.3. 11 – Vehicles includes also motor vehicles. The following are added: 9 – electrical engineering; 4 and 5 – coal and petroleum by-products and chemicals

Source: C. H. Lee, *Regional statistics of Great Britain 1841–1971* (1979)

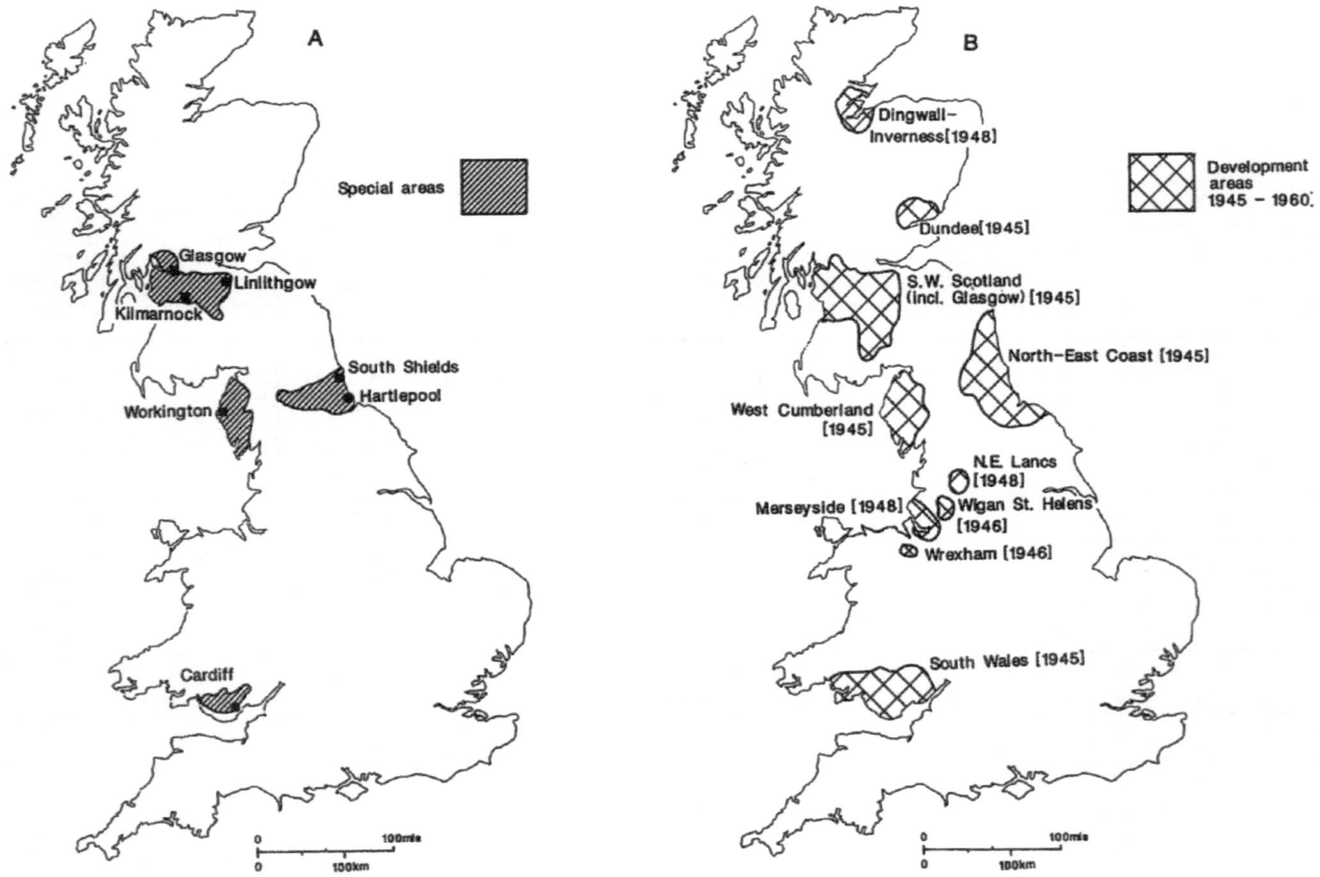

Figure 15.1 (A) Special areas in Great Britain, 1934; (B) development areas in Great Britain, 1945–60. Based on Law (1960) p. 46.

Table 15.3 Summary of British regional policy, 1928–1963

Date	Legislation	Purpose and/or effects
1928	Industrial Transference Board	Assisted migration of workers
1932	Board of Trade surveys	Surveys of areas of high unemployment,
1934	Special Areas (Development and Improvement) Act	which led to the scheduling of four Special Areas (West Cumberland, North East England, South Wales and Central Scotland) eligible for government aid for industrial development
1936–37	Special Areas financial aid	For industrial expansion and building of Trading Estates
1937–40	Royal Commission on the Distribution of the Industrial Population ('The Barlow Report', Cmd 6153 HMSO, 1940)	Reviewed the nature and implications of the regional inequality of industrial activity and made recommendations for 'a proper distribution of industry'
1945	The Distribution of Industry Act	Designated Development Areas (DAs) (this combined the pre-war Special Areas together with, from 1949, Merseyside)
1947	Board of Trade controls on the location of new industries	Via industrial Development Certificates (IDCs) and building licences (to 1950)
1958	The Distribution of Industry (Industrial Finance) Act	To provide financial incentives for firms moving to DAs
1960 and 1963	Local Employment Acts	To provide financial incentives focused on Development Districts (of very high unemployment) and increased incentives for larger districts

For the location of Special Areas and Development Areas (Districts) see figure 4
Source: chiefly C. M. Law (1980) Chapter 2

Corby steelworks and tube mill; prestige projects, such as the Cunard liners Queen Mary and Queen Elizabeth built at John Brown's yard, Clydebank; and, partially influenced by evidence to the Barlow Commission (1937–40), defence contracts to northern and western industrial areas. Local authority legislation (for example the Corporation of Liverpool Act, 1934) and the Special Areas Act and, after the war, the Distribution of Industry Act (1945) also promoted employment through public works, labour mobility schemes and industrial development on new trading estates.

However, the bulk of private investment went into the growth industries of the South East and Midlands (Figure 15.2). As a result the inter-war years furthered changes begun in the late nineteenth century in both the structure and location of industry, though the legacies of the past continued to manifest themselves into the depression accompanying the downswing of the fourth Kondratieff and the impact on jobs of the electronic revolution of the 1970s.

Energy

Despite earlier misgivings over coal's future the *Royal Commission on Coal Supplies* reported in 1905 that proven coal reserves to 4,000 feet in Britain

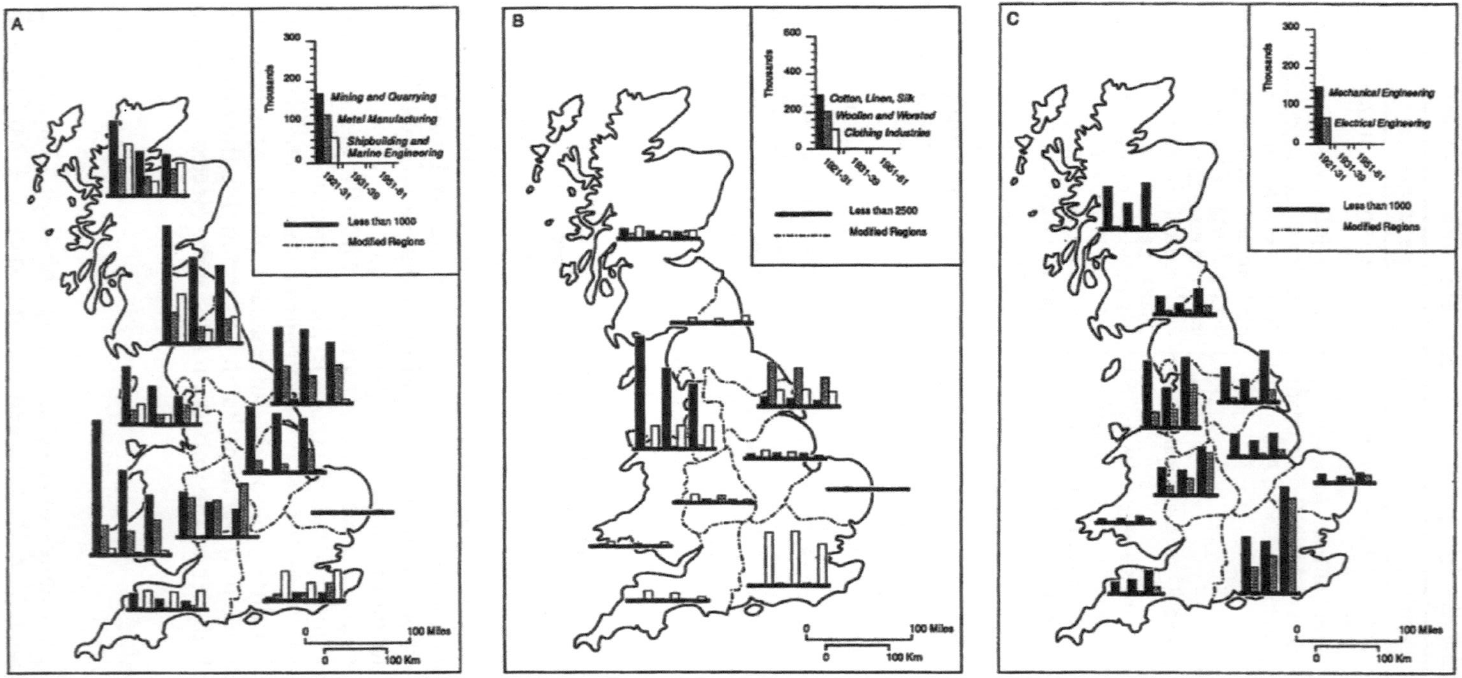

Figure 15.2 The regional distribution of employment in Great Britain, 1921–31 to 1951–61. (A) Heavy industries (mining and quarrying; metal manufacture; shipbuilding and marine engineering); (B) textile industries (cotton, linen and silk; woollen and worsted; clothing); (C) engineering industries (mechanical engineering; electrical engineering). Based on Law (1980) pp. 104, 114–117, 120, 126 and 129.

exceeded 100,000 million tons – over 400 years' supply at the current rates of consumption. That forecast was not over-optimistic. Rapidly improving efficiency of steam engines, including the newly developed Parsons steam turbine of the 1880s; the advent of the electric generator in the 1890s; the development of the internal combustion engine by Benz (1885) and Daimler (1886): all pointed to a future for coal alongside alternative forms of energy. Improved manufacturing and distribution systems for gas and electricity (notably the national grid constructed between 1926 and 1933) put coal under increasing competition, especially for domestic use, some forms of transport and in the location of many light industries. Ease of transportation and cheapness of oil, despite the need to import supplies before the 1970s, gave it growing advantage in transport, some industrial uses and electricity generation.

By the early twentieth century there was a close relationship between reserves and the level of production of coalfields (Table 5.4). Changing demand affected fields differently: steam coals for ships and industry favoured the North East and South Wales coalfields: the latter had important external and internal markets for anthracite for central heating; coking coal from Durham, south Yorkshire, and Nottingham supplied three-quarters of British coke requirements. Falling output per miner and increased costs of hand cutting led to increased mechanization. British mines cut only 8 per cent of output by machine in 1913 as compared with over 90 per cent in Belgium: by 1938 the proportion had risen to 59 per cent and to three-quarters by the late 1940s. Rising production costs and falling demand in international and home markets between the wars were met by substantial wage cuts leading to major strikes in 1921 and 1926 that soured labour relations in the industry, saw a loss of young recruits, left an ageing labour force and posed problems of recovery from the horrendous unemployment of the early 1930s.

Problems of management in a fragmented industry aggravated the difficulties. The coal question entered the political arena in several ways. First, international competitiveness and declining exports were aggravated by cheap, often subsidized, exports from Europe, North America and, occasionally, India and China. Secondly, the problem of small, inefficient operations called for positive intervention to eliminate wasteful competition and over-production. Such amalgamations as Lord Rhondda's Cambrian Group and the Lambton, Hetton and Joicey collieries in east Durham were a step in that direction but failed to resolve the problem. Thirdly, increasing trade union militancy forced government to regulate hours of work (the Eight Hours Act of 1908) and safety conditions (the Coal Mines Act of 1911). The Sankey proposals of 1900, to rationalize the mining industry, anticipated the 1914–18 controls and the post-war debate on a regulated industry that led to the Royal Commission (the Samuel Commission) of 1925 and the Coal Mines Acts of 1930, 1932 and 1938 (The Coal Act) which, though stopping far short of government control, sought to regulate competition through regional marketing associations.

The strategic and economic need for coal during the Second World War and in the bitter aftermath of a post-war fuel crisis in Europe's ruined economies brought what was seen as the final solution: the nationalization and reconstruction of the coal industry by a Labour Government wedded to public ownership. But despite the brief post-war revival, by the 1950s the coal industry was once more in decline as imported oil, nuclear energy and, from the 1960s, North Sea gas and oil began to challenge an uncompetitive, low-productivity, high-wage and high-cost industry (Table 15.4).

These trends are reflected in substantial changes in coal production (Table 5.4). Accessibility and ease of working focused production on larger, newer and better-equipped mines in the south Yorks-Derby-Notts coalfield which also had one-third of British reserves, the best output per man-shift, and the lowest production costs. Cheap rail-freight rates enabled it to capture nearly two-fifths of the London market by the 1930s, to produce coal for substantial sections of the steel industry, and to supply major power stations along the Yorkshire Ouse and Trent valleys.

In contrast increasing difficulties of mining, falling reserves, low output and high production costs were reflected in the rapid fall in output and share of the national market in Scotland, Lancashire, and, especially, South Wales. The latter, along with North East England, suffered badly from progressive decline in steam coal markets as the world's shipping converted to oil and as electricity provided more of industry's energy. While maintaining its pre-1914 share of national coal production the low-cost North East coalfield suffered from the region's industrial decline and the erosion and, after 1939, collapse of its European export market.

Coal production increasingly focused on large, deep pits as epitomized in the changes in colliery employment in the Yorks-Derby-Notts fields (Figure 15.3) which produced a traumatic change in the economy, community life and landscapes of coalmining areas in the 1930s.

In west Durham, for example, many small mines closed leaving a scarred landscape of pitheaps and silent pithead gear. Young men moved elsewhere: to new pits in south Yorkshire, Nottingham or east Kent, or to jobs in the new factories of southern England. The girls went to factories, domestic service, shops and offices in London and the big regional cities. The 1921 and 1926 strikes and cruel unemployment of the early 1930s put over half the menfolk of many mining villages on the dole: many of the middle-aged never worked again. As the economic basis of their existence was removed, whole communities slowly died. In South Wales massive migration from the valleys took young men to new industries in midland and southern England – for example the Morris car plant at Cowley – but such migration only absorbed part of the redundant workforce.

Although coal still supplied 90 per cent of Britain's energy in 1950, patterns of domestic coal production changed substantially: in the late 1880s 7 per cent of production was converted to gas; by 1913 11.8 per cent was used as secondary energy, one-quarter for electricity (Table 10.4). Sharp increases in electricity demand, especially following the completion of the national grid (1926–33), saw coal consumption in power stations increase to 6.7 per cent of production; gas and coke works used 8.7 per cent each by 1938. Gas supply doubled and electricity increased almost six-fold between the wars, while imports of petroleum (essentially for road transport), increased 3.5 times. The full implications of this revolution in energy use for coal were realised by 1960 when one-quarter of production went to power stations (82

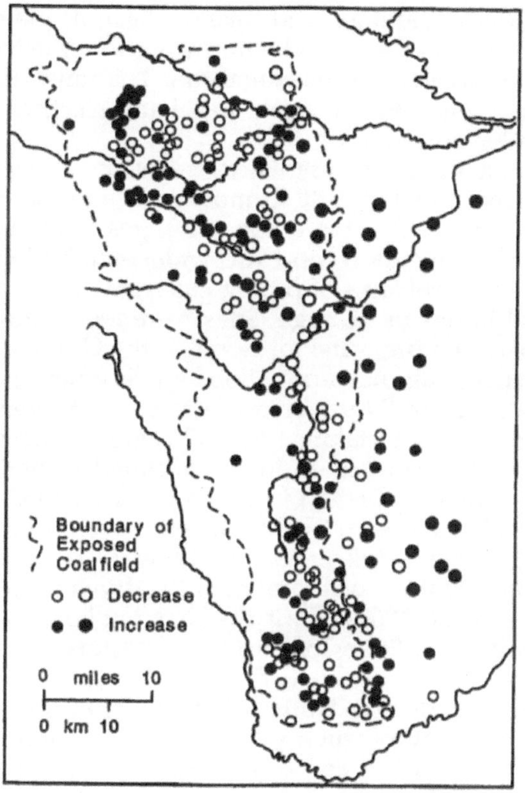

Figure 15.3 Changing employment in the Yorks, Derby and Notts coalfield, 1913–38. Based on Smith (1949) p. 297.

Table 15.4 Energy supply in Great Britain 1890–1960

Year	Coal (m.tons)	Gas (m.therms[1])	Electricity (m.kW hrs.)		Petroleum (m.galls)	
			Thermal	Hydro	Imports	Re-exports
1890	181.6	440	NA	NA	105	1
1910	264.4	851	1432[2]	NA	346	6
1920	229.5	1123	4263	12	3399[3]	253
1930	243.9	1383	10597	320	8932	657
1938	227.0	1500	23384	988	11662	592
1950	216.3	2337	53486	1035	19132	1193
1960	193.6	2612	116309	2539	58417	9134

NA = not available (amounts were negligible)
[1]1 Therm = 2.09569 cb feet of gas
[2]Public supply according to 1907 Census of Production
[3]Refined and Crude and Process oils, 1920–: refined products accounted for 0.5 per cent of imports in 1920, 19.5 by 1938, but 48.3 and 73.7 per cent, respectively in 1950 and 1960

Sources: Mitchell and Deane (1962)
Mitchell and Jones (1971)

per cent of their capacity) and gas and coke works took one-quarter; half the electricity and one-third of gas production went to industry and one-third and one-half, respectively, to domestic users. The railways lost ground to road transport and began to use more diesel and electricity, especially on suburban and underground lines.

Gaslight symbolizes late-Victorian city streets and homes, though paraffin and, later, kerosene lamps still lit most rural and many urban homes. The two catalysts which led to the wider use of gas from the 1880s were the mantle, an Austrian invention, and the widespread adoption of gas for cooking and heating. Problems of piping gas from small gasworks to local markets near coal mines or in large cities were eased through permitted amalgamations between the wars: for example the Gas, Light and Coal Co. supplied 12 per cent of national production by 1945 when 65 undertakings in major cities accounted for 70 per cent of gas sales. However, despite the reorganization of 1050 undertakings into 12 regional Gas Boards after the 1948 Gas Industry Act, the gas industry declined in both domestic and industrial supply until the spread of natural gas after 1967.

Electricity symbolized a new, twentieth-century world of flexible energy supplies. The scientific bases of electricity were established by earlier discoveries – especially the development of the dynamo, principally by the Siemens brothers in the 1860s – but the first public electricity supply in Britain came only in 1881. From 1882 turbo-generators provided for a rising demand from cheap and effective filament lamps – developed by the amalgamated Edison and Swan United Electric Light Company (1883) – and from reliable electric motors (such as the 1888 Westinghouse motor) that powered electric trams (1890) and trains. After 1900, electric motors were used in many industrial and household appliances.

Large-scale electricity generation required efficient distribution. Early power stations supplied limited local markets close to coal supplies. High voltage generation from the late 1880s and high-conduction insulated copper cables led to rapid diffusion of both private and public generating companies for town lighting and transport: by 1900 most towns had electricity and over half the 250 suppliers were municipal. Standardizing of transmission at 132 kV (kilovolts) in a national grid developed between 1926 and 1933 enabled electricity to penetrate rural areas. The massive electrification programme to the mid 1950s (when grid transmission was increased to 275 kV) was reflected in electricity pylons striding across the countryside – to the despair of many conservationists – and near-universal availability of electricity. In 1920 only 12 per cent of Britain's homes had electricity: by 1938 two-thirds were wired: and by 1950, 86 per cent. Furthermore, electricity sales increased by 300 per cent between 1938 and the nationalization of the industry in 1948.

Although electricity replaced coal and gas in many homes and factories, it continued to be largely coal-generated. Initially widely scattered, as transmission became more efficient and coal transportation costs reduced by special rail freight agreements, the size of power stations greatly increased: Ferranti's 1889 Deptford power station generated only 50,000 volts; the 1900 Carville station, Newcastle operated at 6000 kilowatts; but the 1933 Battersea power station – that architectural and technological tribute to modernism –

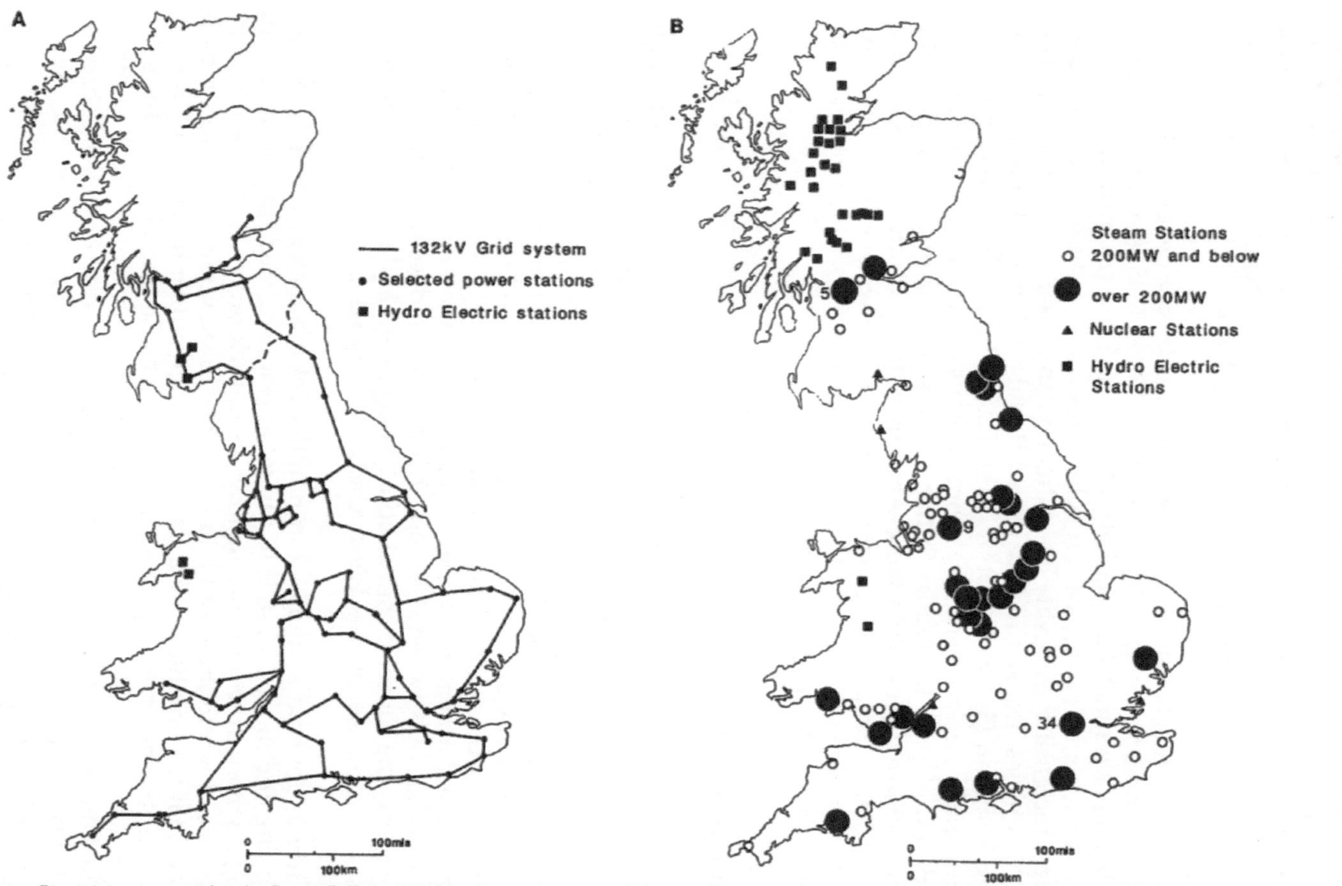

Figure 15.4 Electricity generation in Great Britain. (A) The national grid in 1934 (132kV only, with selected primary and some secondary stations); (B) power stations in 1962. Based on Pope (1989) p. 90 and Rawstron (1964) p. 305.

had a capacity of 105 MW (megawatts). Production was increasingly in large power stations at major collieries or in clusters along such major waterways as the Thames, Trent and Ouse with cheap transport, abundant cooling water and gravel pits or off-shore dumping for ash disposal (Figure 15.4): their massive cooling towers are one of the legacies of mid-twentieth century technology.

Despite the revolution in transport heralded by the development of the internal combustion engine by Otto (1876), Benz (1885) and Diesel (1886), petroleum is essentially a twentieth-century fuel and raw material. Motor-buses, lorries and private cars revolutionized life between the wars and assisted increasing flexibility of industrial and residential location. The 420,000 vehicles licensed in Britain in 1914 (about 7 per 1,000 population): rose to 3 million (59 per 1,000) by 1939 (Figure 16.6). The revolution in oil refining and petro-chemicals has come into its own only since the Second World War. Up to 1950 most British oil was refined prior to import (for example at Anglo-Iranian's large Abadan refinery on the Persian Gulf), although a number of refineries were developed between the wars, notably at such estuarine sites as Grangemouth, Ellesmere Port, Thameshaven and Shellhaven, Fawley and Llandarcy.

Heavy industry

Like coal, British pig iron production peaked in 1913 (Table 5.5). Although there had already been a substantial shift to steelmaking (in a ratio of 3:4, pig: steel by 1910–14) Britain's share of world steel production was then one-third as compared with half of world production in the mid-Victorian heyday of British pig iron. Britain's steel-workers were outstripped by the USA and Germany and technological innovation and investment in high-quality production (for example of acid open-hearth steel) in large-scale integrated and automated plants lagged, though the average blast furnace capacity doubled between 1882 and 1915. Moreover, despite the efficiency of many firms, the conservatism of both management and workforce in small firms, for example of the West Midlands, placed them increasingly in jeopardy.

Wartime demand stimulated iron and steel production but delayed re-equipping and limited post-1918 capital investment, and the sensitivity to unemployment prevented rationalization and relocation. As the proportion of Jurassic ores and imported and home-produced scrap in furnaces grew, the fate of inland coalfield sites was increasingly under threat. Yet forces of industrial inertia, existing plant, capital, expertise and cheap rail-freight charges kept increasingly anachronistic production going. Tariff support clinched the choice of Ebbw Vale for the new steel works of Richard Thomas and Baldwin (RTB) in 1938 and the 1935 Corby Jurassic orefield site of Stewart and Lloyd, to provide basic Bessemer steel for tubes hitherto principally imported from Belgium. But despite such rationalization, 19 of the 22 integrated steelworks of the 1960s were on pre-1900 sites (some going back to the eighteenth century) and only two were post 1930.

The changing emphasis in production saw pig and, especially, Bessemer

Table 15.5 Pig iron and steel production in Great Britain, 1890–94 to 1960–64

Type of product	Output (in million tons)							
	1890–94	1890–04	1910–14	1920–24	1930–34	1940–44	1950–54	1960–64
Pig iron	7.3	8.6	9.5	6.1	4.7	7.5	10.3	13.2
Steel, ingots and castings	3.1	5.0	7.0	7.1	6.7	12.7	16.9	23.1
Bessemer Acid	1.3	1.2	1.0	0.4	0.2	0.2	0.3	0.2
Bessemer Basic	0.3	0.6	0.6	0.2	–	0.7	0.9	2.6
Open Hearth Acid	1.4	2.7	3.4	2.2	1.7	1.6	1.2	0.5
Open Hearth Basic	0.1	0.5	2.1	4.2	4.9	9.3	13.5	17.7
Electric, etc.	–	–	–	0.1	0.1	0.9	1.1	2.0

Sources: Mitchell and Deane (1962)
Mitchell and Jones (1971)

steel production fall dramatically between the wars (Table 15.5) but steel held its own in the industrial revival and rearmament from the mid 1930s and dramatically doubled its output to over 13 million tons by 1939, a trend which continued into the post-war years. Demand for plate, for example from the motor industry, saw firms such as John Summers of Hawarden Bridge, Deeside (originally a producer of galvanized plate) develop steel rolling mills and then a fully integrated open-hearth plant with continuous operation from blast furnace to finished product. Improved technology and cheap ore imports, as British production declined (Table 10.6), increasingly led to the economies of large-scale production at coastal sites, a rationalization not to be fully realized until the reshaping of British Steel in the 1970s and '80s.

Regional production of iron and steel was reshaped first by the movement of blast furnaces then steelworks to the Jurassic ore-fields, notably at Scunthorpe then Corby. By 1913 one-tenth of British steel came from this area, increasing to around one-eighth by the late 1930s. Despite the declining locational significance of Coal Measure ores and coal, strong regional markets and the use of scrap kept many small West Midlands steelworks going. High quality specialist Sheffield steel maintained its relative position. The biggest producers were those with large integrated plants, principally at coastal locations notably in Teeside and South Wales. Inland sites in the North East, South Wales and Scotland (for example, Consett, Ebbw Vale and Clydebridge) dependent on cross-subsidization and government grants, became increasingly uneconomic. The continuing widespread distribution had more to do with specialist skills in making steel for and cross-links with engineering and with government regional policies than with maximising profitability.

Engineering

Changes in manufacturing techniques, an increasing range of raw materials and new sources of energy, especially electricity, transformed engineering from the 1890s. Heavy engineering was still closely linked with steel production: light engineering was much less tied by raw materials and energy and attracted by major markets or entrepreneurial initiative. After 1918 mass-production methods increasingly used semi-skilled labour to assemble parts that were often produced in existing areas of skilled engineering. The differing fortunes of the engineering industries reflected their changing position in international markets. Heavy engineering, shipbuilding, steam locomotives and boilers, and textile machinery were badly hit by the slump. The Midlands and South East, with a bigger share of engineering 'growth industries' fared much better. As a result the balance of employment in this industrial sector shifted (Figure 15.5).

At its peak in 1913, when Britain launched 60 per cent of world tonnage, shipbuilding was concentrated in areas with cost advantage in steel plate, boiler and engine manufacture. The Clyde's 36 per cent of production had attracted John Brown, the Sheffield steel masters, to the production of large vessels at their Clydebank yard. The old-established industry of North East England – which had pioneered iron ships, boilers, engines and steam turbines – had 51 per cent of ship construction in 1913, roughly shared between Tyne (19 per cent), Wear and Tees (16 per cent each). Barrow and Birkenhead, specialist producers of naval vessels, were dominated by the firms of Vickers and Cammell Laird, respectively.

Such dangerous dependency on a single industry became abundantly clear as British shipbuilding slumped from half to one-third of world production between the wars: output in the North East and on the Clyde declined by between one-third and two-thirds from the 1913 level; Birkenhead's shipyards which employed 11,807 in 1921 had only 2,768 workers in 1931, a major cause of the town's 40 per cent male unemployment. Despite wartime and post-war demand for British ships (half the world's new tonnage, two-fifths for export), failure to modernize for the production of new types of vessels such as supertankers and bulk cargo carriers cost its leadership and, from the 1960s, virtually the whole industry to European and, especially, Far Eastern competitors.

Locomotives and railway rolling stock, the most valuable sector in British engineering at the time of the 1907 Census of Production, was spread across railway workshops at major company headquarters such as the GWR in Swindon and the Midland Railway in Derby and in private companies in major engineering districts of Glasgow, North East England, south Lancashire, Leeds, south Yorkshire, and Birmingham. But increasing competition that led former customers in Latin America, India and China to seek other suppliers or to build their own railway stock produced a sharp fall in exports.

In contrast motor vehicles and aircraft were major growth industries from the First World War. Motor manufacture, the twentieth-century equivalent of the early nineteenth-century cotton industry, originated mainly in France and Germany, but grew fastest in the USA. 280 of Britain's 393 early car

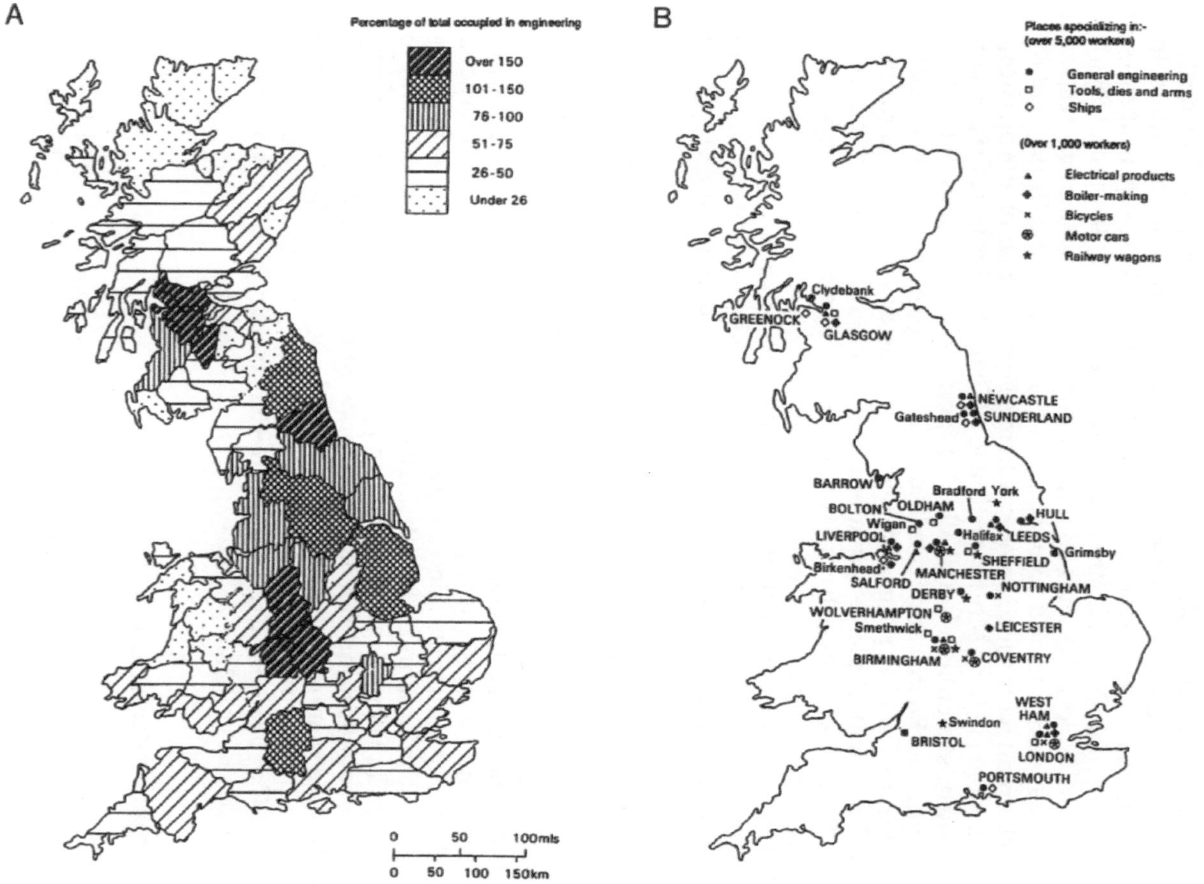

Figure 15.5 Employment in the engineering industries in Great Britain, 1911. (A) percentage of the total workforce in all engineering; (B) major centres of employment in specific engineering industries.
Source: Langton and Morris (1986) p. 139.

manufacturers went bankrupt: but by 1914 it had a workforce of 34,000 with several firms producing over 1,000 vehicles per year, far behind the USA's 145,000 workers producing 373,000 vehicles. The biggest firm, Ford, produced over 6,000 cars per year at its assembly plant on the Trafford Park Estate, Manchester. Other pioneers – Rolls Royce of Derby and Manchester and Singer of Coventry – capitalized on engineering skills nurtured in industries such as cycles and sewing machines. In the 1920s car manufacture became an assembly-line industry focused on the West Midlands and South East England (Table 15.2) with 56 and 38 per cent of production, respectively, in 1929, nearly 45 and 55 per cent in 1938, and 39 and 59 per cent by 1961. Small urban workshops gave way to mass-production plants on cheap green-field sites such as Morris at Cowley (Oxford), Fords at Dagenham with its own steelmills, Vauxhall at Luton, Wolseley and Austin in Birmingham, and Rootes and Standard at Coventry. As production shifted from workshops making parts and engines for custom-built cars to heavily capitalized automated plants, standardized components often came from specialist firms: rolled steel from Summers or RTB; electrical components from Lucas of Birmingham, originally oil lamp makers; car instruments from Smith's the former London clockmakers. With its assembly methods and electrical power, the car industry is the classic footloose industry drawn to more prosperous regions by labour, plant site advantage and major internal markets.

Similar locational attributes attracted early aircraft manufacture to London's airfields, though Manchester (A. V. Roe) and Bristol were pioneers. Engineers such as Rolls Royce and Napier were also drawn into air engine manufacture. Strategic considerations also saw government encourage works such as those of English Electric (near Preston), Armstrong Whitworth who moved their aircraft company from Newcastle to Coventry in the 1920s, and Napier close to the Kirkby Industrial Estate near Liverpool.

The electrical industries, one of the most significant and locationally interesting of the 'new' industries, increased from 175,000 workers in 1921 to 557,000 in 1951. They spanned a range of products including heavy cables, turbines, switch-gear and industrial motors – often manufactured in such old-established engineering areas as Greater London, the West Midlands (e.g. BTH, Rugby and Siemens (English Electric, Stafford), south Lancashire (BICC at Prescot), Greater Manchester (Westinghouse and Ferranti) and Tyneside (Reyrolle).

Consumer industries

Electrical domestic products, from cookers and cleaners to light bulbs and radio valves, were mostly manufactured in the South East or Midlands, often in suburban factories or 'new' industrial towns such as Slough, Watford or Luton. These industries epitomize twentieth-century technology and its market- and labour-orientation. Their many products for the home reflect greater suburban spending power, higher living standards and a changing pattern of expenditure (Table 15.6).

Consumer industries (furnishing, clothing, food, drink and tobacco) were

Table 15.6 Consumer expenditure in Great Britain 1900–60 (at 1958–63 prices)

	Percentage of GNP expended by			Percentage of consumer expenditure						
	Consumers	Public authorities	Fixed capital formation	Food	Drink	Tobacco	Furniture and durables	Clothing	Cars	Other
1900	83.6	9.3	10.0	27.3	20.8	3.7	2.5	10.0	0.0	35.7
1910	84.7	7.6	5.8	28.7	16.0	4.0	2.5	10.1	0.2	38.5
1921	85.9	9.0	8.9	29.9	12.4	6.0	3.4	9.4	0.2	37.7
1930	87.9	9.7	9.5	30.2	8.4	5.7	4.7	9.8	0.8	39.4
1938	84.7	14.4	11.1	29.0	7.2	6.3	4.7	9.3	1.2	42.3
1950	80.5	17.5	14.4	31.1	6.2	6.6	4.0	9.8	0.7	41.6
1960	74.1	18.6	18.0	24.9	5.9	7.0	4.8	9.6	2.7	45.1

The rows of columns 1–3 do not total 100 per cent because of overlap between categories of expenditure

Source: Halsey (1972) Tables 3.4 and 3.6

included among the growth industries of the twentieth century. Their emphasis on the home market, new power sources – especially electricity – and mass-production techniques in large factories led to their location close to the major markets.

Many clothing and footwear firms remained in their existing locations. Factory production dominated the boot and shoe industries from the 1880s and employed 83 per cent of their workforce by the early twentieth century. By the Second World War it was largely a factory trade, the West End's handmade production excepted. Nearly three-fifths of the industry was concentrated on Northampton, Leicester and small industrial towns of the Nene Valley such as Kettering and Rushden where half to three-quarters of the workforce was employed in shoe factories. Other important small-town producers also dominated employment – Clarks at Street, Somerset; K Shoes at Kendal; and in Norwich, Bristol and some of the Rossendale towns where women's footwear manufacture took over from declining textile trades.

In contrast, clothing manufacture was more widespread and ranged from the individual tailor to big factories. Traditionally a handicraft domestic or workshop industry, the advent of the sewing machine in the 1850s initially reinforced the tendency for small-scale production by pieceworkers in 'sweatshops' in London and the big textile towns with much of the hand-finishing put out to poorly-paid outworkers. Even the adoption of electric power left much of the industry, especially women's clothing, in small firms.

Tailoring switched more easily to mass-production by the adoption of the bandsaw to cut several layers of cloth to a standard pattern. Faster, more efficient electric blades permitted large-scale mechanization and, by dispensing with tailors' and dressmakers' skills, prepared the way for cheap factory-made clothing. By the interwar years more sophisticated sewing machines had replaced much of the hand finishing. Cutting out the middleman, firms such as Burton ('the tailor of taste'), Weaver to Wearer, and Alexanders sold good quality ready-made clothing in a variety of designs, material and sizes at a cost that all but the poorest could afford.

Such developments substantially changed clothing manufacture. First, through mechanization and a mainly young female labour force: for example, in Victorian times nearly 90 per cent of tailors were men; by the 1930s nearly 60 per cent were women. Secondly, although bespoke and handmade garments were still made in small establishments for example in London's quality West End shops and in cheap East End workshops, two-thirds of the clothing firms in the 1935 Census of Production were wholesale factories which accounted for an increasing proportion of output. Finally, there was growing concentration: Greater London was the principal market and fashion centre for all types of clothing; the cotton textile area, especially Greater Manchester specialized in dress, shirt and overall manufacture; the Leeds area dominated wholesale tailoring. The industry remained widespread, but specialized in particular garments – the East Midlands mass-produced hosiery, underwear and knitwear for example. It was highly competitive in materials, methods of manufacture, and scale and organization of production and marketing. A growth industry in investment and output, if not in workforce, it was, however, increasingly challenged in the 1930s and especially after the Second World War by cheap imports from East Asia and the Third World.

Textiles

In contrast textiles, especially cotton and linen, were challenged from the late nineteenth century. Exports declined relatively in value, then absolutely. Between 1925 and 1938–9 cotton exports fell by 75 per cent, woollens by 63 per cent and linen 57 per cent. Employment in these industries peaked before the First World War and, while wool held up relatively well until after 1960, cotton had a catastrophic collapse of jobs in the 1930s, with devastating impact on the regional economy of east Lancashire.

In 1911 over half a million cotton workers – 85 per cent of the British total 60 per cent of them women – were in Lancashire and north east Cheshire. Despite falling exports and profits, a surge of investment in new mills and machinery in late-Victorian times increased output, especially of fine high-quality yarn and cloths. Intensified specialization sharpened the division between the yarn spinning districts of south east Lancashire and the weaving of Rossendale. Although there was a good deal of combined production in some firms by the mid-1930s, 83 per cent of spinning and 72 per cent of weaving were carried on in separate works and firms. Despite diversification – with engineering, papermaking, clothing and footwear prominent – employment in most textile towns remained dangerously dependent on the mills. In smaller towns half or more of males and over four-fifths of females were cotton workers: even in the bigger boroughs, such as Blackburn, Bolton, Bury, Oldham and Rochdale, up to one-quarter of men and half to three-quarters of women were textile workers.

As competition from Asia grew, exports of unfinished cloths (one-third of production in 1913) declined to one-tenth of their pre-war level by 1935: finer printed and dyed cloths fell by more than half from their peak. Mills spinning coarse yarn (mostly in east Lancashire and north east Cheshire) declined faster than the fine spinners of Bolton and Leigh; fine woven cloths of Burnley, Nelson and Colne held up better than the cheaper ones of Blackburn and Accrington. There was no major switch to the more productive ring spindles widely used in Japan and the USA, though their proportion rose as 22 million spindles (mostly in old mills) were scrapped: despite the loss of quarter of a million looms, there was little move to automatic looms. Enforced rationalization and closure of up to half the mills in many towns limited reinvestment. The failure to create bigger vertically integrated firms spanning all branches of production (as in Japan) was to prove fatal.

Wartime boom and the postwar export drives gave only temporary respite. The early-1950's recession precipitated a further collapse which halved the workforce between 1951 and 1961, though, ironically, a declining and ageing workforce caused the import of Indian and Pakistani labour. As the yardage of cheap cotton imports topped British exports, only fine quality yarns and light-weight cloths, which often included synthetic fibres, survived. Massive dismantling of obsolete machinery removed half the remaining spindles and two-fifths of the looms in the North West's mills in the radical reorganization under the Cotton Industry Act, 1959 which virtually ended the long reign of King Cotton. The clack of the loom and clatter of clogs was silenced in many mill towns as the remnant industry focused on those with the most valuable and specialized production – for example Bolton's and Leigh's fine yarns and Preston's fine cottons.

Increasing amounts of man-made fibres were used from the 1930s by Lancashire and Yorkshire weavers and in the knitwear and stocking trades of the East Midlands. Viscose rayon, invented in 1892, was manufactured by Alfred Courtauld at Coventry (an old silk ribbon-weaving centre) from 1907, though production was limited before the interwar years. The major British plants for staple fibre rayon were built in the 1930s in the former chemicals area of lower Deeside and at Preston. Elsewhere local authorities and industrial associations were crucial in attracting the great Viscose Artificial Silk Company (SNIA) to Aintree near Liverpool in 1926 while the American du Pont nylon patent was brought to Britain by British Nylon Spinners at Pontypool in the South Wales Development Area in 1946. Thus while artificial fibres were increasingly used by textile manufacturers, especially in knitwear and stockings, their technology often drew them to new locations with linkages to either chemicals or textile manufacture.

One of the major problems following the catastrophic demise of cotton was the inability to easily adapt multi-storey spinning mills to other uses though weaving sheds were more easily divided into small industrial units or converted to storage and distribution, for example for mail order firms. Moreover, the level of job replacement lagged far behind some 600,000 mill redundancies between 1921 and 1971.

Woollen manufacture was less badly hit by the interwar slump: the 11 per cent job losses, 1912–37, were only one-quarter of those in cotton. The real collapse came in the 1970s. Nevertheless, decline in the market for heavier materials and a fashion shift away from worsted cloths affected some places badly. Bradford and Keighley lost some 20,000 textile workers (30 per cent of the 1939 workforce) by the end of the 1950s. Even so, 80 per cent of all woollen workers were still found in the West Riding though the Scottish industry and specialist West Country firms remained important for knitwear and certain fine cloths.

Services

Women were more affected than men by the post-1918 decline in textiles and, to a lesser extent, clothing, but the range of jobs undertaken by women widened substantially and their proportion in the workforce remained relatively steady (Tables 10.2 and 15.8). Between the 1880s and the First World War service employment grew by 42 per cent, but over 50 per cent for women. Much of that increase was in domestic and personal services, banking and commerce, the professions, government service and, above all, distribution. By 1911 London and the South East was dominant in the provision of all service sectors, particularly professional, banking, commercial and miscellaneous services (Table 15.7), though many big ports and

Table 15.7 Regional employment in services in Great Britain, 1921–51

Standard Region	1921 Total	1921 %	1951 Total	1951 %	1921–51 % increase
South East	3645	38.8	4588	37.8	25.9
East Anglia	239	2.6	328	2.8	31.2
South West	626	6.4	884	7.4	41.2
West Midlands	593	6.3	890	7.4	50.1
East Midlands	486	5.2	695	5.7	43.0
North West	1242	13.2	1497	12.4	20.5
Yorkshire and Humberside	668	7.2	851	7.1	27.4
North	489	5.2	653	5.4	33.5
Wales	443	4.7	542	4.5	22.4
Scotland	956	10.2	1156	9.6	20.9
Great Britain	9398	100	12083	100	29.6

Source: C. M. Law (1980), Table 25

commercial cities provided substantial commercial and government functions as well as administrative services.

After 1914 the number of women in domestic service declined sharply, but those in commercial, professional and administrative services and distribution increased. From one-tenth of the labour force in the late nineteenth century employment in services for men rose to nearly one-quarter by the 1930s and from half to over three-fifths for women: whereas the index of employment for all such workers doubled between 1911 and 1951 it was 2.8 times for clerical workers. Britain, especially southern Britain, witnessed a white-collar revolution.

The factors in this remarkable growth were three-fold: first, progressive government involvement in administrative, economic and social affairs; secondly, still labour-intensive transactional work (organizing, buying and selling, providing services); thirdly, the increasing proportion of women employed in offices and shops. Two world wars played a major role in the change in female employment, though it was only from the 1950s that the growing tendency of married women to remain in the workforce led to a substantial increase in their overall participation rate.

The office sector

Whereas the proportion of women in industry fell from the 1890s, except in unskilled and some semi-skilled work, their employment in higher professional, shop and clerical work increased. The telephone and typewriter revolution from the 1880s saw an army of male clerks progressively replaced by female office workers. The growth of the office sector, a quintessentially twentieth-century phenomenon, reflects the scale of governmental, commercial and industrial activity and substantially influenced the structure of the workforce and patterns of employment. Business became concentrated into fewer, bigger firms in the large CBDs of London and the major cities.

A 'whole new psychology of business', as Sydney Checkland (1964) describes it, led to the corporate economy and state and the professionalized society of the twentieth century. Big company headquarters' offices were concentrated in specialized buildings served by specialist professional services. Similarly the role and number of professional administrators and office staffs in the Civil Service and town halls increased with progressive government involvement in all aspects of life (Chapter 12).

In the early nineteenth century banking was widespread. Country banks discounted their bills through banks in large cities, especially London where specialist financial businesses focused on the Bank of England and the Stock Exchange. By the late-Victorian period local banks were incorporated into bigger regional joint stock banks and in the 1920s amalgamations of clearing banks concentrated business on a few major companies. The more specialized merchant banks, mainly London-based, fell from 168 to 66 between 1891 and 1914. After the First World War, the City increasingly dominated Britain's financial life and attracted wealth and a massive number of jobs.

Elsewhere commerce became more specialized, focusing on particular commodities or on particular parts of the world: for example by late-Victorian times, Liverpool dominated trade with West Africa and Latin America, and had substantial links with India and China: Manchester controlled the cotton trade; Birmingham many of the metal trades. While there was locational advantage in such regional specialization, it increased vulnerability to the trade cycle. By the 1890s Liverpool's important commercial sector, which had nearly 3,000 firms in mid-Victorian times, focused on fewer, larger firms in an impressive business district with many specialist office buildings (some dating back to the 1840s). Though many remained small, the bigger firms such as marine insurance had 40 to 50 staff and some of the biggest offices (e.g. the Mersey Docks and Harbour Board) employed over 300 clerks even in the late 1870s. Many were young office lads (almost half of the clerical workers according to one mid-Victorian estimate) who were poorly paid and easily dispensed with. In slack trade or industrial depression the commercial arm of the commodity, financial and shipping interests suffered also. Against the national trends office employment in central Liverpool suffered a 30 per cent decline in commercial jobs in 1924–30 and a collapse in 1930–34 from which many activities did not recover, leading to a relocation of some headquarters, such as Cunard, after the Second World War.

In a society in which the 'meritocracy' was displacing both old, 'landed' influence and individual industrial wealth, there was opportunity for bright young men (and some women) to rise through the ranks of the professions, the Civil Service and industrial corporations to the very top. But there were many casualties in the lower ranks of commercial organizations.

Retailing and wholesale distribution

One of the major beneficiaries of the growth in real wages in the nineteenth and twentieth centuries was retailing. From late-Victorian times mass production and new technologies lowered the real cost of most goods and made more available to the mass market a wider range of both home-produced and imported foodstuffs, consumer durables, household fittings, furnishing, and clothing. Advertising, improved transport and distribution – including regular store and mail order deliveries by road – gave people better and cheaper access. As the working week shortened between the wars shopping became part of the leisure activities of everyone, not just the middle classes.

Mid-Victorian food distribution was focused largely on rural and urban covered markets and specialist retailers (e.g. tea merchants, cheese factors, poulterers and butchers) who provided mainly for the better-off. From the late nineteenth century shopping provision was more diversified with a wide range of specialist shops and large, well-stocked department and chain stores, especially in the large towns. Though rural provision declined with falling populations (Chapter 13) there was easier and cheaper access by train, tram and then bus to local and regional shopping centres. By the turn of the century all suburbs had shopping areas supplying day-to-day and many occasional needs, often replacing city-centre food markets.

Liberal service – at the cost of low wages, insecure employment and long hours for shop assistants, one-third of whom were women by 1911–created a golden age for the shopper. Cut-throat competition and a constant search for cheap supplies led suppliers to by-pass the producer/wholesaler/retailer route to the customer. Bulk, brand-named products marketed through countrywide outlets and delivery vans characterized food chains such as the River Plate Meat Co., Thomas Liptons (which had 450 shops by 1900) and the United Dairies; clothiers such as Burtons (tailors), Dunns (hatters); Freeman, Hardy and Willis, Stead and Simpson (footwear manufacturers); W. H. Smith (newsagents and booksellers); and Boots (cash chemists). From under 1,000 outlets in the mid 1870s, such multiples grew to nearly 23,000 by 1914.

Department stores developed from mid century. Their market was initially mainly middle-class, though one of the earliest, Lewis's of Liverpool – clothiers from 1856 and with a burgeoning group of general stores from 1877 – dubbed themselves 'Friends of the People'. Many of the great London stores were mid-Victorian in origin – Harrods (1849) were grocers until the late 1870s; William Whiteleys became the so-called 'Universal Provider' in 1872 – but their most rapid expansion in individual size, number and geographical distribution came from the 1890s. Selfridges opened its Oxford Street store in 1909 with 1,800 staff; in 1914 Whiteleys and Harrods employed four and six thousand respectively. The Cooperative movement established its Wholesale Society (CWS) in 1863, producing brand products for sale in the shops of 1,385 societies, mostly in the industrial north and working-class areas of the large towns where it remained one of the major 'store' outlets until after the Second World War.

In the interwar years stores and specialist shops provided a widening range of furnishings, electrical goods and domestic equipment, and sales were stimulated by a shopping boom. Small shops – newsagents, butchers and greengrocers, tobacco, sweets and drinks, hardware, retail chemists and hairdressers – also responded to growing and widening purchasing power in most classes of society (Table 15.6). Independent shops including local chain grocers increased and were incorporated both into the frontages of main roads of late-Victorian terrace areas and the small shopping 'parades' of inter- and post-war housing estates; they still dominated small-town and village shopping.

Just before the First World War 20,000 out of approximately 600,000 shops were branches of multiple firms, principally in the grocery, meat and footwear trades in which they supplied one-tenth, one-eighth and one-third respectively of the whole market. Departmental and Cooperative stores together supplied only one-tenth of the national retail market, mainly in women's wear, grocery and household goods. Moreover, the 1950 Census of Distribution recorded 583,000 retailers (mostly single owners), who had 45 per cent of the trade: 12 and 5 per cent, respectively, were Co-ops and department stores. The collapse of the independent shopkeeper, and a sharp decline in the number of retail outlets, came mainly from the 1960s.

Changes in the labour market and working conditions

The major restructuring of the British economy brought significant changes in working conditions and the operation of the labour market. Women played an increasingly important role in the workforce, new technology and machinery created different jobs demanding new and often less individually-crafted skills. But for those unwilling or unable to acquire new skills long-term unemployment became a reality. Older workers, particularly in heavy industry, often found it difficult to adjust to new jobs and work practices. These changes affected all sectors of the economy from agriculture to ship-building, and all parts of the country from rural Devon to industrial Clydeside, but their overall effect was to restructure working skills and work-place relations especially in the changing economy of the inter-war period. The years 1890–1914 were in effect a transitional period which retained many characteristics of the nineteenth-century economy whilst signs of the new work patterns of the inter-war years began to develop. These changes are highlighted by focusing on selected areas of the economy and different parts of the country.

The role of women in the workforce was affected both by a general tendency to compete for jobs in a wider range of occupations and by regional restructuring of the economy. The experience of women workers in depressed areas of Lancashire was quite different from that in new types of employment in South East England. Although the number of women in commerce and many industries increased between 1891 and 1951, the proportion of women in paid employment hardly changed and remained at around 35 per cent (Table 10.2). But the characteristics of female employment changed substantially. Before 1914 domestic service was still the overwhelming source of employment for women and girls, although the clothing and textile trades employed more women than men. Women however were also beginning to infiltrate the lower grade clerical and service occupations. For instance in 1901 13 per cent of clerks were women, but by 1911 this had risen to 21 per cent, although higher clerical grades were mostly men. Nevertheless the employment status of women remained

Table 15.8 Occupational status of male and female workers* in Great britain, 1911 and 1951

Occupational category	Males		Females	
	1911	1951	1911	1951
Self-employed and higher-salaried professionals	1.5	2.8	1.0	1.0
Employers and proprietors	7.7	5.7	4.3	3.2
Administrators and managers	3.9	6.8	2.3	2.7
Lower-grade professionals and technicians	1.4	3.0	5.8	7.9
Inspectors, supervisors and foremen	1.8	3.3	0.2	1.1
Clerical workers	5.1	6.0	3.3	20.3
Sales personnel and shop assistants	5.0	4.0	6.4	9.6
Skilled manual workers and self-employed artisans	33.0	30.3	24.6	12.7
Semi-skilled manual workers	29.1	24.3	47.0	33.6
Unskilled manual workers	11.5	13.8	5.1	7.9
Total per cent	100	100	100	100

Expressed as a percentage of all workers
Source: Stevenson, J. (1984) p.184

inferior to that of men: in 1911 52.1 per cent of women occupied semi-skilled or unskilled manual jobs compared with 40.6 per cent of men (Table 15.8).

By 1951 women had entered a wider range of occupations but the numbers employed in domestic service and textiles, their traditional areas of employment, had declined dramatically and, were among the hardest hit by economic restructuring. Although 40.5 per cent of women workers were still in unskilled or semi-skilled manual occupations (compared with 38.1 per cent of men), 29.9 per cent were now in white-collar clerical or shop work (10 per cent for men) and 10.6 per cent were administrators or lower grade managers and professionals (9.8 per cent men). The proportion in skilled manual work had declined much more quickly for women than it had for men (mainly due to the decline in the textile trades), and women had moved freely into expanding clerical, service and administrative jobs. By and large men still retained the senior and best paid positions, but nationally the restructuring of the economy towards the service sector had had a much bigger effect on women than it had on men.

The decline of the cotton industry had a major impact on both men and women in Lancashire's workforce. Until 1914 the cotton textile industry had operated in much the same way that it had since the mid-nineteenth century. Conditions of work in spinning mills and weaving sheds were unpleasant, hot, noisy and dusty – and minor injuries from the machines were common. There was a clear hierarchy of male labour with boys and younger men doing the more menial jobs (cleaning and oiling machines, repairing broken threads) and new spinners chosen from the older and more reliable piecers. However, opportunities for promotion were limited as a spinner might work into his eighties. Noise was the worst problem in weaving sheds where up to 500 looms were crammed into a single building. Female weavers were usually responsible for four looms, younger girls would be placed with a more experienced weaver to learn the trade, and both spinners and weavers were on piece rates. Supervision of the weaving sheds was almost always in the hands of male 'overlookers'.

Although work in cotton mills was hard and dangerous it was well paid and regular before 1920. In the 1920s recession, small firms who could not afford to modernize and re-equip shed labour and cut wages. Lancashire spinners suffered constant wage reductions during the 1930s as those in work found themselves slipping rapidly down the national league table of wages (Table 15.9). Working conditions also deteriorated and in 1931 Lancashire weavers struck over plans to increase the number of looms worked by each weaver. Gradually during the 1930s the formation of the Lancashire Cotton Corporation and other government measures provided funds for a partial restructuring of the industry. This created problems as the industry had great difficulty in responding to the demand placed upon it by the Second World War. Skilled labour had also disappeared as women moved from weaving into white-collar jobs and men from the dole queue into the armed forces in 1939.

Although working conditions in the textile industries (and other old staple industries) were bad and exacerbated by the effects of recession, they were by no means the worst in Britain. During the 1920s legislation had been passed to control working hours so that in most factories an 8 hour working day had been negotiated with unions. Until 1937, however, it was still legal for women and young people to work up to 60 hours a week in non-textile factories, and in many small firms and casual trades very long hours were common. It was easy for employers to avoid legislation, and the economic pressures of recession made employers reluctant to improve working conditions.

Long working hours and low pay continued to be the norm in retailing. In the sector dominated by the small independent trader butchers slaughtered animals on the premises, bakers made their own bread and grocers dealt with a vast range of goods which came loose or packed in sacks. Although many of these family-run businesses survived until the 1950s, from the 1920s multiple chains were establishing themselves in the high streets of most large towns.

These structural changes affected working conditions and the status of the shop workers. Independent traders normally worked at least a 12 hour day, six days a week in the 1920s and some shop assistants could work from 7.00 a.m. until midnight with only short breaks. Although the Shop Act of 1912 imposed half-day closing, and Acts of 1923 and 1928 in theory brought a general closing time of 8.00 p.m. the legislation was difficult to enforce. Working conditions for many shop assistants were little different from those of the nineteenth century.

Most multiple and department stores had shorter working hours than small independent traders and, although most shop assistants were poorly paid, floor managers and chief assistants in department stores could acquire status as well as a decent income. Their growth thus provided increased opportunities for retail employment in mostly better conditions than small independent shops but, even in the chain stores, there was a clear hierarchy of occupations from the errand boy to the floor manager and discipline could be harsh. Shop assistants could be fined for repeatedly failing to make a sale

Table 15.9 Indices of weekly wage rates in selected occupations (United Kingdom), 1920–1945

Year	Agricultural workers (England and Wales)	Agricultural workers (Scotland)	Coal miners	Iron and steel workers	Engineering workers	Cotton factory workers	Building workers	Railway workers
1920	161	143	187	203	152	162	146	128
1925	109	99	99	99	101	100	103	100
1930	113	94	87	93	104	94	100	98
1935	113	88	84	92	103	86	93	95
1940	156	127	113	131	125	110	111	113
1945	247	209	167	164	166	154	138	158

Source: B. R. Mitchell, *British Historical Statistics* (1988), p.160
(1924 = 100)

and floor managers could intervene to claim sales commission from the shop assistant. There was little union activity among shop assistants and the union itself had only 30,000 members in the 1920s. Even if wages were low and hours were long the expanding retail trade at least offered the opportunity for work; many men and women moved out of declining manufacturing industry into retailing in the 1920s and '30s.

The motor car and electrical industries were among the most rapidly-expanding sectors of the economy. Located mainly in the West Midlands and southern England – although electrical factories also developed in some northern towns where they could recruit female labour from textile mills – they offered better working conditions in well-organized and modern factories. Wages were quite high (up to £5 per week for a shop floor worker in the car industry in the 1930s), but there were considerable variations caused by irregular trade. Ford's new continuous assembly line plant at Dagenham opened in 1931, and this innovation (rapidly followed by other manufacturers) had a major impact on factory working conditions. The need to keep pace with the assembly line, the strict discipline involved in timekeeping and production rates, the threat of seasonal lay-offs and the noise on the factory floor combined to create a harsh and authoritarian environment. Although work was much more regular and well-paid than in declining heavy industries, working conditions could still be unpleasant. For the large numbers of women on the electrical industries' production lines work was monotonous, piece rates were demanding and work could be stopped without pay if there was a drop in demand or a shortage of components. These new industries of the twentieth century were relatively slow to abandon some of the worst aspects of nineteenth-century working conditions. The introduction of the production line was also partially responsible for the gradual de-skilling of the workforce which has occurred in the twentieth century, epitomized by the reduction in apprenticeships.

Any job in retailing or on a car assembly line was, however, preferable to unemployment: throughout the inter-war years the spectre of unemployment enabled many employers to keep wages low and working hours long. The structural impact of unemployment was the most significant factor in regional differences in prosperity. In 1932 unemployment in coal mining was 35 per cent, in cotton 31 per cent, in ship-building 62 per cent and in iron and steel 48 per cent. Despite periodic unemployment in newer industries such as car manufacture, chemicals and electronics, the position was never so severe as it was in the staple industries. The regional effect also worked alongside structural factors (Table 15.10). Regions such as Merseyside, Tyneside, South Wales, south Lancashire and Clydeside suffered particularly badly because of heavy reliance on traditional industries such as coal, shipbuilding, iron and steel and textiles (Table 15.11). High unemployment – for instance in shipbuilding – meant that there was less money circulating in the economy. Retailing suffered knock-on effects, as did office and other employment partially dependent on heavy manufacturing industry. Thus, on Merseyside, unemployment was most above the national average in the service, distribution and retail trades. Consumer trades which were expanding in the south were hard hit by the effects of recession in northern manufacturing industry. Overall in 1932 unemployment in

Table 15.10 Percentage unemployment in Great Britain by region, 1923–1939

Year	English Regions								
	London	South East	South West	Midlands	North East	North West	Wales	Scotland	Great Britain
1923	9.9	9.2	10.4	9.9	11.5	14.2	6.3	13.8	11.2
1924	8.6	7.1	8.7	8.3	10.4	12.3	9.0	13.3	9.9
1925	7.1	5.5	8.0	8.5	14.6	10.9	16.9	14.7	10.7
1926	6.3	5.0	7.8	10.4	16.8	14.0	18.2	15.8	11.9
1927	5.3	4.7	7.1	8.1	13.5	10.3	20.0	10.1	9.4
1928	5.2	5.2	8.1	9.8	15.5	12.5	23.2	11.6	10.8
1929	5.3	5.6	8.3	9.2	14.1	12.3	19.9	12.3	10.6
1930	7.9	8.0	10.5	14.9	20.8	23.9	26.7	18.7	16.3
1931	11.6	11.8	14.4	20.7	27.4	28.2	32.4	26.5	21.4
1932	12.9	14.3	17.2	19.6	28.8	25.6	36.8	28.2	22.1
1933	11.4	11.8	16.0	17.0	26.7	23.6	35.3	27.1	20.2
1934	8.9	9.0	13.8	12.7	23.0	21.3	33.5	24.5	17.2
1935	8.2	8.5	12.3	11.1	21.8	20.4	33.3	23.2	16.1
1936	7.7	7.5	9.8	8.9	17.7		31.8	20.2	13.6
1937	6.0	7.2	8.3	7.1	11.4	14.4	24.2	17.4	11.2
1938	7.8	8.9	8.8	10.1	14.1	18.4	26.6	17.9	13.3
1939	7.4	8.6	7.2	8.1	11.8	15.4	22.0	16.1	11.1

Source: B. R. Mitchell, *British Historical Statistics* (1988), p.125

Table 15.11 Percentage of insured workers unemployed by selected industries in Great Britain, 1924–1938

	Year							
Industry	1924	1926	1928	1930	1932*	1934	1936	1938
Coal mining	5.8	9.5	23.6	20.6	34.5	29.7	22.8	16.7
Chemicals	9.9	10.9	6.1	10.0	17.3	11.3	9.2	7.4
Iron and steel	22.0	40.4	22.4	28.2	47.9	27.3	17.4	19.5
General engineering	16.9	15.1	9.8	14.2	29.1	18.4	9.6	7.0
Electrical engineering	5.5	7.5	4.8	6.6	16.8	9.6	4.8	4.7
Motor vehicle construction, etc.	8.9	8.2	8.1	12.1	22.4	10.8	6.9	7.2
Shipbuilding/repairing	30.3	39.5	24.5	27.6	62.0	51.2	33.3	21.4
Cotton textiles	15.9	18.3	12.5	32.4	30.6	23.7	16.7	23.9
Wool textiles	8.4	17.4	12.0	23.3	22.4	17.8	10.3	21.3
Bread/cake manufacture	9.8	8.4	6.6	9.1	12.2	10.7	10.3	8.8
Printing/publishing	5.6	5.4	4.6	6.0	11.0	9.2	8.1	6.9
Building	12.5	12.1	13.9	16.2	30.2	21.3	19.8	16.7
Gas/water/electric supply	6.3	6.0	5.8	7.0	10.9	10.1	9.7	8.3
Railway service	6.4	12.3	6.5	6.9	15.7	12.3	8.5	7.3
Tram/omnibus service	3.1	4.3	3.1	3.8	5.9	5.6	4.0	3.2
Dock service	25.8	29.7	29.1	33.4	33.3	31.1	29.3	25.0
Local government service	6.9	9.0	8.6	10.7	18.2	20.3	19.7	17.2

*For most industries unemployment peaked in 1932. Industries with a 1931 peak include chemicals (17.6%); cotton textiles (43.2%); wool textiles (33.8%); bread/cake manufacture (12.4%); dock service (39.8%)

Source: B. R. Mitchell, *British Historical Statistics* (1988), p.129

South East England was only half that in the north. Men were more likely to be unemployed than women (although married women had higher unemployment rates than single women); and the elderly suffered particularly from long-term unemployment; youths under 18 could often find work, but unemployment was high in the 18–24 age range as employers dismissed workers when they were forced to pay an adult wage. Irrespective of location, unemployment was most common for males over 40, with skills which were no longer in demand.

The effects of unemployment on individual families could be devastating. The Pilgrim Trust (1938) demonstrated the extent to which reliance on unemployment benefits led to poverty. They suggested that incomes dropped to 45–66 per cent of previous wages during periods of unemployment and that some 41 per cent of unemployed families were living in poverty. In comparison with the nineteenth century the impact of unemployment was mitigated by the effects of inter-war welfare measures (Chapter 16) but poverty continued to be worst in those areas and industries which were traditionally affected by low wages and underemployment. It is difficult to assess the precise extent to which unemployment in the 1930s led to worse housing and health for those affected, or whether poor living conditions were mainly a continuation of nineteenth-century circumstances exacerbated by depression. However, many local studies did show a deterioration in both mental and physical health among the unemployed.

Arguably, those most severely affected were workers who had previously enjoyed relatively high and stable incomes. A skilled worker on Merseyside in the 1920s who moved to a suburban council house with a high rent found that unemployment both severely damaged his pride and status and imposed an intolerable economic burden. In contrast, a dock worker probably had fewer financial commitments, would be more used to irregular earnings and might manage the impact of unemployment more effectively in the short term. In the long term, however, the skilled worker was more likely to regain employment quickly. Nevertheless, one serious effect of the inter-war depression was to create an under-class of long-term unemployed which was only removed when the Second World War required the mobilization of all labour into the war effort and the war-time boom economy.

Bibliography and further reading for Chapter 15

Many of the references for Chapters 5 and 10 are still relevant. Additional references include:

On general industrial change:

CHANDLER, A. D. *Scale and scope: dynamics of industrial capitalism* (Cambridge, Mass.: Harvard Univ. Press, 1990).

HEIM, C. 'Industrial organization and regional development in inter-war Britain. *Journal of Economic History* **42** (1983), pp. 931–52.

TAYLOR, E. G. R. et al. 'The geographical distribution of industry' *Geographical Journal* **92** (1938), pp. 22–40 and pp. 409–26.

See also references on industrial change in Chapter 12.

Coal and energy:
ASHWORTH, W. *The history of the British coal industry*, vol. 5, *the nationalized industry* (Oxford: Oxford University Press, 1986).
BYATT, I. C. R. *The British electrical industry, 1875–1914* (Oxford: Clarendon Press, 1929).
CHURCH, R. A. *The history of the British coal trade*, vol. 3, *1830–1913, Victorian pre-eminence* (Oxford: Oxford University Press, 1986).
HANNAH, L. *Electricity before nationalization: the electric supply industry in Britain to 1948* (London: Macmillan, 1985).
KIRBY, M. W. *The British coalmining industry, 1870–1946* (London: Macmillan, 1977).
MORGAN, R. H. 'The development of the electricity supply industry in Wales to 1919' *Welsh History Review* **11** (1983), pp. 317–37.
SUPPLE, B. *The history of the British coal trade*, vol. 4, *the political economy of decline* (Oxford: Oxford University Press, 1987).

Engineering and the motor industry:
BLOOMFIELD, G. *The world automotive industry* (Newton Abbot: David and Charles, 1978).
CAMPBELL, R. H. 'Scottish shipbuilding: its rise and progress' *Scottish Geographical Magazine* **80** (1964), pp. 107–113.
CHURCH, R. *Herbert Austin: the British motor industry to 1941* (London: Europa, 1979).
FOREMAN-PECK, J. 'Diversification and growth of the firm: the Rover company to 1914.' *Business History* **25** (1983), pp. 179–92.
POLLARD, S. and ROBERTSON, P. *The British shipbuilding industry, 1870–1914* (Cambridge, Mass: Harvard University Press, 1979).
RICHARDSON, K. *The British motor industry*, 1896–1939 (London: Macmillan, 1977).
SAUL, S. B. 'The motor industry in Britain to 1914' *Business History*, **5** (1962), pp. 22–44.
SAUL, S. B. 'The market and development of the mechanical engineering industries in Britain, 1860–1914' *Economic History Review*, **20** (1967), pp. 111–130.
SAUL, S. B. 'The machine tool industry in Britain to 1914' *Business History* **10** (1968), pp. 21–43.
THOMAS, D. W. and DONNELLY, T. *A history of the motor vehicle industry in Coventry* (London: Croom Helm, 1985).

Textiles:
COLEMAN, D. C. *Courtaulds: an economic and social history*, vol. 2, *Rayon* (Oxford: Clarendon Press, 1969).
KENNY, S. 'Sub-regional specialization in the Lancashire cotton industry, 1884–1919. A study in organizational and locational change.' *Journal of Historical Geography* **8** (1982), pp. 41–67.
ROBSON, R. *The cotton industry in Britain* (London: Macmillan, 1957).
RODGERS, H. B. 'The changing geography of the Lancashire cotton industry' *Economic Geography* **38** (1962), pp. 299–314.
SINGLETON, J. *Lancashire on the Scrapheap. The Cotton Industry 1945–1970* (Oxford: Oxford University Press, 1991).

Service and consumer industries:
BRIGGS, A. *Friends of the people: The centenary history of Lewis's* (London: Batsford, 1956).
HARTWELL, R. M. 'The service revolution' in Cipolla, C. M. (ed.) *The Fontana economic history of Europe*, vol 3, *the industrial revolution* (London: Fontana, 1973) pp. 358–396.

JEFFREYS, J. B. *Retail trading in Britain, 1850–1950* (Cambridge: Cambridge University Press, 1954).

MARTIN, J. E. *Greater London. An industrial geography* (London: Bell, 1966).

SHAW, G. *Processes and patterns in the geography of retail change. Kingston upon Hull, 1880–1950* (Hull: University of Hull, 1978).

On working conditions and the labour market:

BUXTON, N. and ALDCROFT, D (eds) *British industry between the wars: instability and industrial development 1919–1939* (London: Scolar Press, 1979).

BRAYBON, G. and SUMMERFIELD, P. *Out of the cage: women's experience in two World Wars* (London: Pandora press, 1987).

CONSTANTINE, S. *Unemployment in Britain between the wars* (London: Longman, 1980).

GLYNN, S. and BOOTH, A. 'Unemployment in inter-war Britain: a case for relearning the lessons of the 1930s.' *Economic History Review* **36** (1983), pp. 329–48.

JONES, D. *The social survey of Merseyside* (Liverpool: Liverpool University Press, 1934).

JONES, M. 'The economic history of the regional problem in Britain, 1920–38' *Journal of Historical Geography* **10** (1984), pp. 385–95.

MITCHELL, M. 'The effects of unemployment on the social condition of women and children in the 1930s' *History Workshop Journal* **19** (1985), pp. 105–27.

PAGNAMENTA, P. and OVERY, R. *All our working lives* (London: BBC, 1984).

PILGRIM TRUST, *Men without work* (Cambridge: Cambridge University Press, 1938).

ROBERTS, E. *Women's work 1840–1940* (London: Macmillan, 1988).

SUMMERFIELD, P. *Women workers in the Second World War: production and patriarchy in conflict* (London: Croom Helm, 1984).

WINSTANLEY, M. *The shopkeepers' world 1830–1914* (Manchester: Manchester University Press, 1983).

16

Urbanization and urban life from the 1890s to the 1940s

The urban hierarchy and urban sprawl

Twentieth-century Britain has an overwhelmingly urban culture. Nine-tenths of its population is economically urban-dependent. By 1951 two-thirds of Britain's population lived in towns of over 50,000 and 37 and 35 per cent, respectively, in the conurbations of England and Scotland. Although the 1890s and 1900s overspill of population from declining inner residential areas of London and the larger cities produced below-average population increases, many urban and some rural districts adjacent to larger towns had sharp population increases (Figure 16.1). Moreover, the inter-war decline in many specialized manufacturing towns produced population migration mainly to Greater London and the towns of south-eastern Britain. Seven conurbations – Greater London, the West Midlands, Merseyside, Greater Manchester, West Yorkshire, Tyneside and Clydeside – were defined for the first time in the 1951 Census. They headed an urban hierarchy increasingly dominated from the late nineteenth century by the national capitals and major provincial capitals – Glasgow, Newcastle, Leeds, Manchester (and possibly Liverpool), Birmingham, Nottingham, Bristol and Cardiff. Some 700 other towns in England and Wales and a further 235 in Scotland also offered a wide range of urban services to their surrounding areas. Many drew a growing part of their workforce from adjoining rural areas as reflected in the level of commuting recorded in the 1921 and 1951 Censuses. Some daily migrants were recruited to the growing urban workforce from more accessible rural hinterlands. Others were town-dwellers who moved to outer residential areas of semi-detached suburbia and massive local authority overspill housing estates (Chapter 13).

The growing interplay of town and country, and the progressive dominance of economic, social and cultural life by the towns, especially the large regional cities, is implicit in the greatly increased scale and range of daily mobility. Analyses of journeys to work, the frequency and intensity of bus and train services, and the range and nature of goods and services supplied by urban-based firms have all been used by geographers to define the increasing influence of towns and their evolving patterns of urban regional-

DOI: 10.4324/9781003174783-16

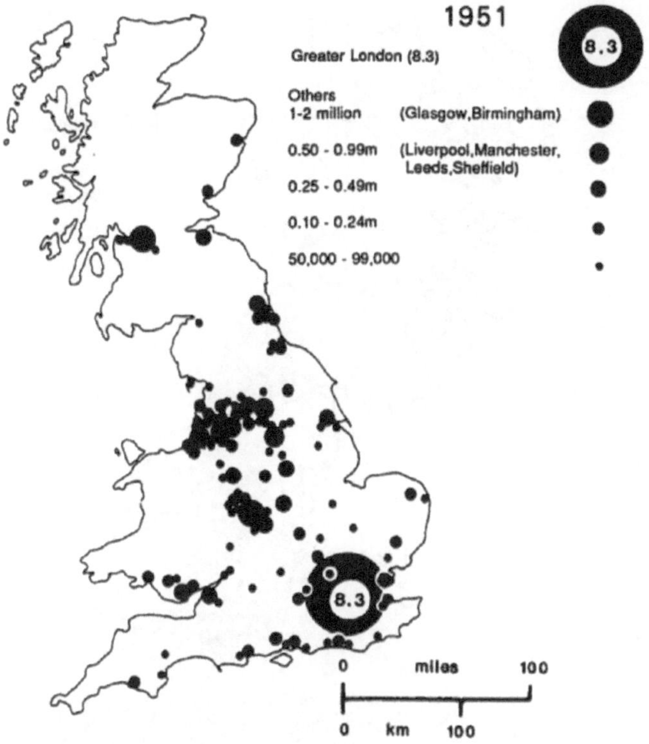

Figure 16.1 Distribution of the urban population of Great Britain, 1951.
Source: Census of England and Wales and of Scotland, 1951.

ism (Figure 16.2). In 1921 one-seventh of the two and three-quarter million workers living in rural districts of England and Wales travelled to work in urban areas. By 1951 over 20 per cent of the resident population in all areas around the large towns were involved in commuting and over 40 per cent in and around Greater London and the major conurbations. Except for those urbanized districts on the periphery of large towns, rural Britain was increasingly characterized by outward travel to work, often over long distances. Much of this reflected the increasing presence in the countryside of what has been termed 'adventitious population', people previously unconnected with rural activities who had moved residence (some on retirement) but who often retained an essentially urban focus (Chapter 14).

The almost total urbanization of British society after the First World War was accompanied by massive urban sprawl. A two-fold job concentration – in the commercial centres and around the periphery of large cities – greatly increased the scale and complexity of intra-urban movement. Daily in-movement to central London increased from around 600,000 in 1921 to nearly 1 million by 1951. In most provincial cities of England the resident occupied population changed little between these dates but there was marked increase in commuting to commercial centres: Manchester, Glasgow, Birmingham, Liverpool, Newcastle and Nottingham gained, in descending order, between 137,000 and 43,000 commuters each day.

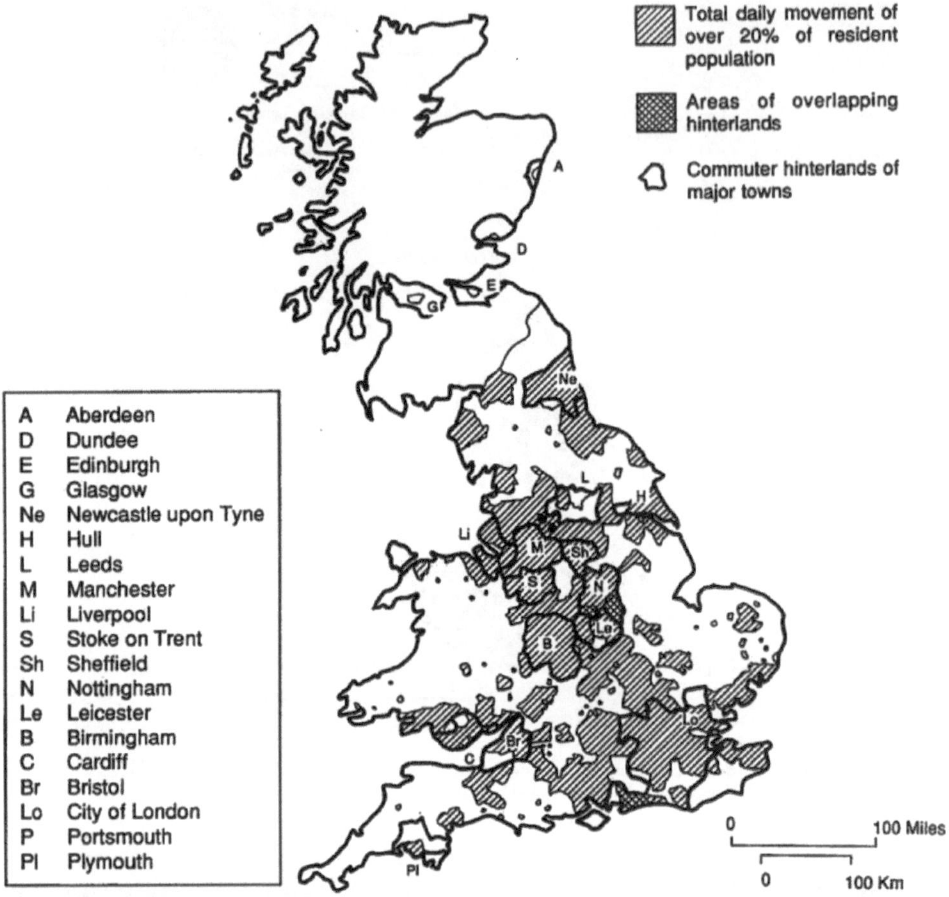

Figure 16.2 Daily journey to work in Great Britain, 1951. Local authority areas to and from which movement to work exceeds 20 per cent of the resident population are shaded. Commuter hinterlands of major towns include all local authorities from which over 100 workpeople are drawn.
Source: Census of England and Wales and of Scotland, 1951.

In contrast, around most big cities new factories, warehouses and distribution industries of a mass consumer assembly line age were drawn to cheap green field sites near good rail and road communications and a suburban, not least female, workforce. From such pre-First World War sites as Park Royal, Wembley and others in north London industry spread along the Great West Road to Slough and the rail corridors to Watford and South Hertfordshire. In the North West there were similar developments along the Liverpool and Manchester ends of the East Lancs. Road. At Salford docks on the Manchester Ship Canal (completed in 1894), the Trafford Park development produced a massive increase in daily movement of workers involving some long journeys from residential areas around the suburban periphery. In the West Midlands, the car industry not only attracted large-scale commuting to plants in Coventry and east Birmingham, but also created complex movements within the car manufacturing zone of the region.

Urban and regional planning

Decentralization of housing reflected land values, social forces and cheaper transport. From the 1870s a growing 'civic gospel' began to create progressive municipal involvement in provision and regulation of housing and such amenities as markets, baths, libraries, art galleries and museums, parks and recreation spaces, as well as gas, electricity and, by the late nineteenth century, transport services. This larger social role, epitomized in Joseph Chamberlain's Birmingham of the 1870s and 1880s, was a prelude to more interventionist planning principles and policies. By the early 1900s most large towns were involved in such 'municipal socialism'.

Other more individual influences also lay behind the growth of the concept of town and then regional planning in Britain from the late nineteenth century. Some, such as the provision of adequate housing for the working classes, were inspired as much by philanthropic ideals as government legislation. Others owed their inspiration to enlightened employers who created model villages: Strutt at Cromford (1771) Owen at New Lanark (1800), Titus Salt at Saltaire (1851–76), William Lever's remarkable garden suburb at Port Sunlight (1887) and George and Richard Cadbury's Bournville Village Estate (1895) (Chapters 5 and 10). Yet others had a vision of a new type of community which married function and form in a greener and less crowded environment such as Raymond Unwin's garden suburb and Ebenezer Howard's garden city (1902), (Chapter 11). Others, such as Patrick Geddes in his *Cities in Evolution* (1915), saw the necessity of planning within the framework of a city and its region. Those who perhaps exerted the greatest influence in the emergence of Town and Country Planning after the Second World War stressed conservation of the countryside in terms of both farmland and scenic quality as well as providing land for housing and industry, as in Patrick Abercrombie's Greater London Plan, 1944.

The first direct state intervention in town planning *per se* – the Housing, Town Planning, etc., Act of 1909 – was limited in scope to building and land-use plans for developing peripheral areas of towns and was permissive rather than mandatory. Where enlightened municipal officials, such as Liverpool's City Engineer James Brodie, and a philosophy of planning (as in the University of Liverpool's Department and Lever Chair of Civic Design, established in 1910) came together the results may be still seen in the quality of layout of suburbs and roads. But little was achieved before 1918.

But there was more to planning than urban design. It also involved land-use planning, provision and co-ordination of transport and communications (new road systems and suburban railways), industrial location, especially for new industry in the depressed areas and around the large cities and, crucially, the relations between town and country. This was implied in the requirement under the Planning Act of 1919 that all urban authorities should produce plans and in the move towards Joint Town Planning Committees in some of the most congested areas. Concepts of low-density housing, recreational amenity (both in private gardens and public open space) and a more spacious layout of civic services and amenities (from schools to swimming baths) placed pressure on land around the relatively tightly constrained urban areas created under the 1888 and 1894 Acts. Urban expansion could be achieved by extending the urban administrative areas, a relatively easy procedure under the Local Government Board, but the encroachment of cities on adjacent county areas contributed to rapid urban sprawl, loss of territory, population and rate revenue by the county authorities, and frequent conflict between urban and rural areas. This, as much as anything, led to the Royal Commission on Local Government of 1923 (the Onslow Commission), following which the 1923 Town and Country Planning Act was extended to all built-up areas and brought all land under planning control, whilst the 1929 Local Government Act obliged County Councils to produce planning schemes.

Although such measures pointed to a growing awareness of the necessity for regional planning, little was achieved before the Second World War either in terms of integrated town and country planning or, still less, the solution of the problem of administrative divorce between urban and rural, and borough and county authorities implicit in the 1888 and 1894 Acts. While the 1926 Local Government (County Boroughs and Adjustments) Act tightened up the procedures for creating and enlarging Boroughs, it left many big cities desperate for land for housing and industry and without proper means to make and implement balanced regional plans. Most large cities expanded their administrative areas in the years after 1890 to include newly expanding suburbs, but the growth of London posed special problems. In 1889 a new London County Council was formed and from 1900 28 new borough councils became the main administrative units within the L.C.C. area. There followed a long period of conflict between the city and its suburbs, with various proposals for political reform and the integration of administrative units, partially resolved only by the formation in 1965 of the now defunct Greater London Council.

London's planning needs were also seen in a broader regional context. The massive nature of its unplanned urban sprawl; the extent of its commuter hinterland (50 miles across by the 1920s and up to 80 by the early 1950s); the need to disperse more of its inner area population and to accommodate the flood of interwar migration to the South East: all were recognized in the setting up of the Greater London Planning Committee under the directorship of Raymond Unwin in 1929. The national context of this manifestation of the regional problem – over-concentration in London and the South East – was recognized by the Royal Commission on the Distribution of the Industrial Population which reported in 1940 (The Barlow Report) and was one of the most powerful arguments for a new approach to regional planning in the postwar years. The Greater London Plan, 1944 was the model for other areas. Recognizing the need to rehouse inner area families at lower (though not over-low) densities, to prevent further indiscriminate suburban expansion, to protect open country and to selectively develop new, satellite towns beyond a 'ring-fence' of restricted development, Abercrombie planned the dispersal of over one million people (with a further quarter of a million to allow for growth and mobility) (Figure 16.3). Thus were born the concept of overspill from large central cities to satellite towns, the designation of green belts to protect the countryside around them, the provision of New Towns, and the development of a framework for integrated regional planning.

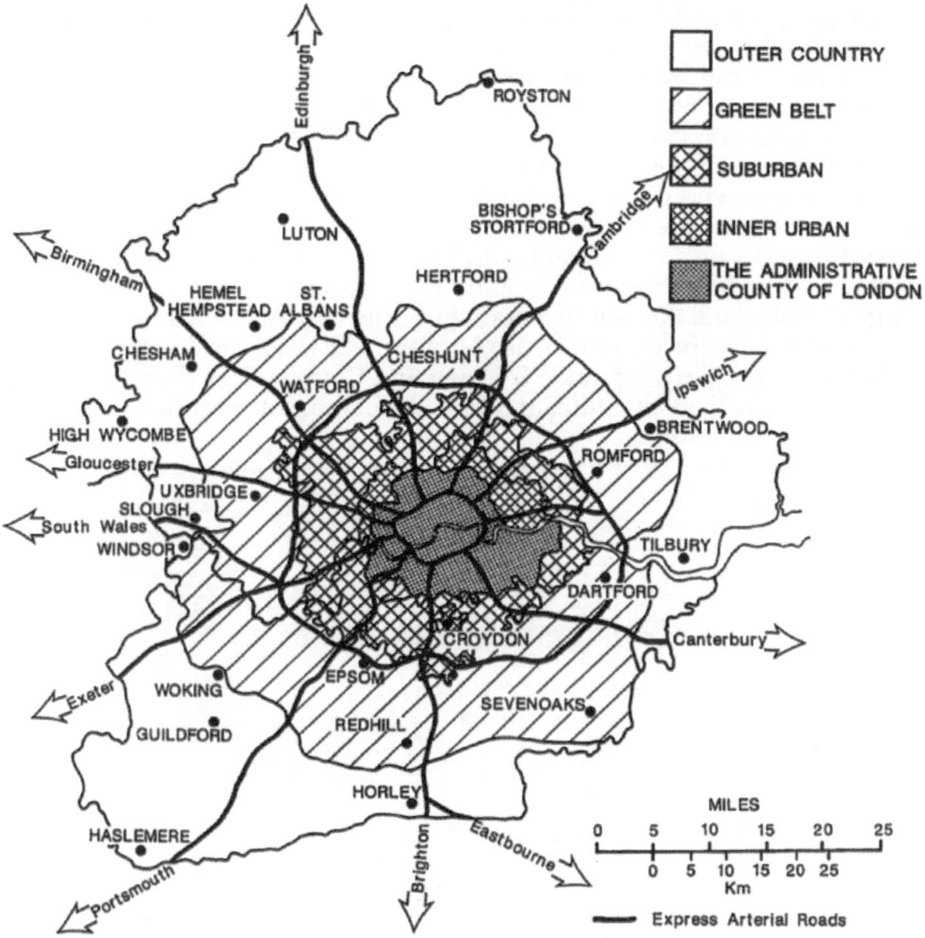

Figure 16.3 Main features of the Greater London Plan, 1944 showing the four planning rings and the major road proposals. Based on Cherry (1986) pp. 125–6.

London's problems were unique in scale but not in essence. South Lancashire's urban belt, including the Merseyside and Greater Manchester conurbations, struggled with equal problems of slum clearance, local authority and private overspill, and land for new industry in areas severely hit by the depression. Their local government boundaries – divided between a host of local authorities, both urban and rural, and split between two Administrative Counties – were complex. Lancashire, on the whole, favoured the case for urban expansion; Cheshire resisted incursions into its territory. This epitomized the difficulties of effective planning and local government for the large cities.

By the end of the nineteenth century Manchester's commercial and service provision extended over the arc of cotton spinning towns from Bolton to Stockport and the new industrial belt from Trafford Park along the Ship Canal, a conurbation of over two million. Its ageing infra-structure and housing, with 80,000 out of its stock of 180,000 dwellings in slum areas, presented major problems. Under the 1919 Planning Act the city expanded into north Cheshire, creating a municipal satellite for 100,000 people at Wythenshawe and new arterial roads with bus and tramway access to the city centre and to industrial areas (Figure 16.4). The need for controlled access to open country in Cheshire and the Derbyshire Peak District that recognized the interests of both town and country-dwellers, led to the District Joint Town Planning Advisory Plan of 1926. Its schemes for a new East Lancs. Road, the site for a civic airport (now Ringway International Airport), provision for new industrial developments and transport were bold and far-sighted. Indeed this and its successors – the City and District Plans of 1945 and the South Lancashire and North Cheshire Advisory Plan of 1947 – were far ahead of planning legislation and beyond what could be sustained under the 1947 Town and Country Planning Act which emphasized land-use planning and local development controls rather than broader regional issues.

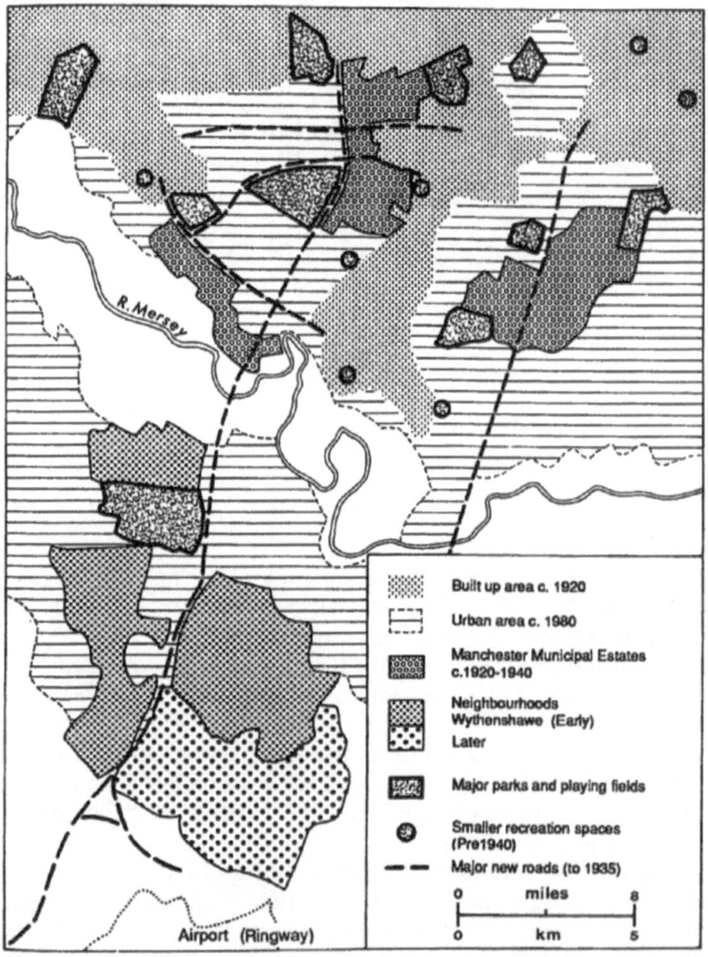

Figure 16.4 Major elements in the planning strategy for south Manchester, 1920–40.
Source: Gordon (1986) p. 47.

Similar problems bedevilled Glasgow's relationship with Clydeside. The Clyde Valley Regional Plan of 1949 saw the need to plan for a distinctive economic and cultural region around Glasgow. The area's problems of housing, environment and employment, reflecting post-1890s decline and catastrophic interwar depression, could only be met in a broad regional framework. Yet the means of doing so, the Strathclyde Regional Council with its extensive territory and two-and-a-half million population, was not achieved until the reorganization of local government in 1975. However when it came, the Wheatley Report (of the Royal Commission on Local Government in Scotland, 1969) was more successful in implementing its radical proposal for this region than was its equivalent (the Redcliffe-Maud Report of the Royal Commission on Local Government in England, 1969) for the English conurbations.

Housing and urban structure

Although the main elements of the physical and social structure of most British towns had been established in Victorian times, a number of factors caused major changes in the twentieth century. None was completely new, but ideas and processes begun in the nineteenth century developed in new ways and at a greater pace after 1890 and especially after the First World War.

The process of residential decentralisation through voluntary out-migration (Chapter 13) and the construction of suburban housing estates by private enterprise gathered new momentum after 1890. This was most clearly seen in London, but similar processes were operating in all large towns. Mid nineteenth-century Ilford was a quiet village on the main line from London to Ipswich, seven miles from Liverpool Street Station. In 1891 there were some 11,000 people in the parish, but by 1901 the new urban district of Ilford had expanded to accommodate 41,240 inhabitants and its population almost doubled again by 1911. Encouraged by good railway communications two London builders, W. P. Griggs and A. C. Corbett, acquired large areas of land and began to develop massive private housing estates. In 1906 on the Griggs estate a four-room house started at £260; a four-bedroom double-fronted house at £375 and a five-bedroom house at £450, good value for money. Both Ilford Council and the builders provided further incentives to move to the suburbs. Corbett gave loans to purchasers to cover some of the cash deposit whilst Ilford Council used the Small Dwellings Acquisition Act of 1899 to give cheap mortgages. Most houses were, however, bought with readily available building society mortgages. Ilford is a classic example of the way in which improved transport, availability of land, willingness of entrepreneurs to invest and demand for suburban living linked to increased affluence combined to restructure the city in the early twentieth century (Figure 16.5). The same process continued in other locations around the metropolis during the 1920s and '30s (for example on the Stoneleigh estate around Ewell in south London) leading to rapid expansion of the built-up area of London along lines of communication.

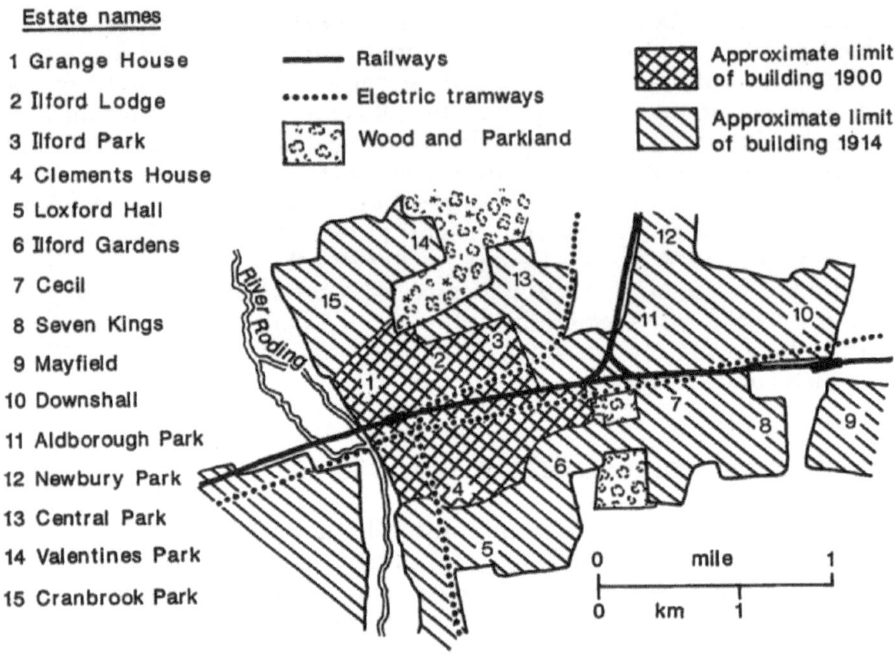

Figure 16.5 The growth of private housing estates in Ilford, 1900–14. Based on Jackson (1973) p. 60.

In England and Wales over three million private houses were constructed between the wars (69 per cent of all new building), and even in Scotland (where there was a much greater emphasis on council housing) over 110,000 houses (32 per cent of the total) were built by private builders 1920–40 (Table 16.1). As around London, most of these new houses were constructed in suburban locations and were bought, with the aid of building society mortgages, by families moving from decent nineteenth-century terraces. Costs were high in the early 1920s and, despite the provision of government subsidies for private housing, output was relatively low, but by the mid-1930s output was running at over 200,000 units per annum and the purchase price of a three-bedroom semi-detached house had almost halved. Other families, encouraged by greater affluence, stability of incomes and building society propaganda promoting homeownership, bought the houses they previously rented. Thus, although private renting remained the dominant form of housing tenure until after the Second World War, an increasing proportion of the population moved into homeownership and private renting began an inexorable decline (Table 16.2). This was particularly marked in the more affluent regions of Britain. Many northern building societies rapidly extended their branch and agency network into southern and midland England and, as most of their deposits came from people in northern industrial towns, this effectively produced a regional transfer of finance to support the private housing boom in southern England.

Although rail and tramway communications were important for much of

Table 16.1 Housebuilding in Great Britain, 1880–1940

A. Houses built in Great Britain 1880–1920 (all tenures)

Year	Houses built (all tenures)
1880	83100
1885	76700
1890	75800
1895	89800
1900	139700
1905	127400
1910	86000
1915	30800
1920	29700

B. Houses built in Great Britain 1919–1940* (divided by tenure)

Year (1st April –31 March)	Houses built for local authorities		Houses built for private owners	
	England and Wales	Scotland	England and Wales	Scotland
1919/20	576	–	–†	–†
1920/21	15585	1201	–†	–†
1921/22	80783	5796	–†	–†
1922/23	57535	9527	–†	–†
1923/24	14353	5233	71857	–†
1924/25	20624	3238	116265	3638
1925/26	44218	5290	129208	5639
1926/27	74093	9621	143536	7496
1927/28	104034	16460	134880	6137
1928/29	55723	13954	113809	5024
1929/30	60245	13023	141815	5011
1930/31	55874	8122	127933	4571
1931/32	70061	8952	130751	4766
1932/33	55991	12185	144505	6596
1933/34	55840	16503	210782	10760
1934/35	41593	15733	287513	6096
1935/36	52357	18814	272503	7086
1936/37	71740	16044	274313	7757
1937/38	77970	13341	259632	8187
1938/39	101744	19162	230616	7311
1939/40	50452	19118	145510	6411

*Various categories of housing are excluded from the figures. See source for details
†S. Merrett, *Owner-Occupation in Britain* (1982), p. 346, estimates an annual average of 25727 private completions in Britain 1919–23

Source: B. R. Mitchell, (1988), pp.390–92

suburban development, the early twentieth century also saw particular activity in road building which in turn affected the structure of towns and the nature of housing development. Rapidly increasing car ownership after 1918 (Figure 16.6) led to increased investment in road construction in several major cities, although the national government was reluctant to spend money on roads and it was not until the Trunk Roads Act of 1936 that a

Table 16.2 Housing tenure in Great Britain, 1914–1971 (percentage)

Year	Owner occupied	Public rented	Private rented	Other (including housing associations)
1914	10.0	1.0	80.0	9.0
1938	25.0	10.0	56.0	9.0
1945	26.0	12.0	54.0	8.0
1951	29.0	18.0	45.0	8.0
1961	43.0	27.0	25.0	6.0
1971	50.5	30.6	18.8	

Figures prior to 1951 are estimates. Some sources consider owner-occupation prior to 1945 was rather higher than the figures given here.
Source: M. Boddy, *The Building Societies* (London: Macmillan (1980) p.154

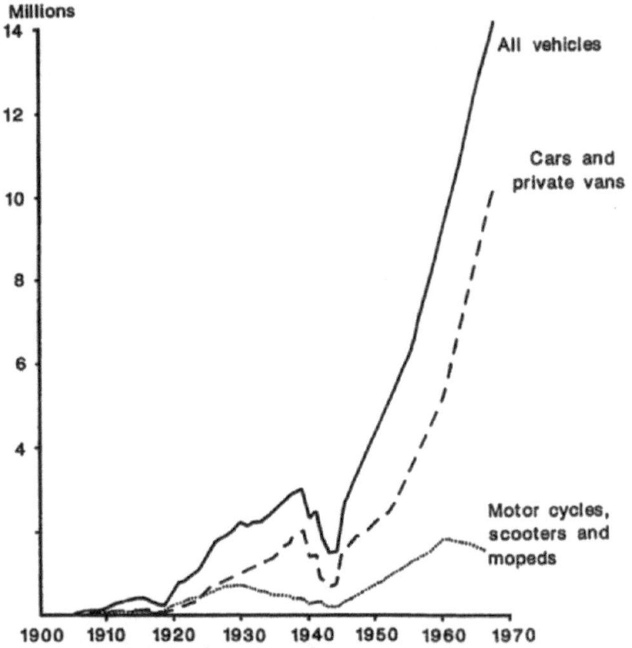

Figure 16.6 Motor vehicles licensed in Great Britain, 1904–67.
Source: Royal Commission on Local Government in England and Wales (The Redcliffe-Maud Report), vol. iii, p. 37 (Cmnd. 4040–ii, HMSO, 1969).

national road strategy began to emerge. Local government initiatives however led, among other schemes, to the early development of the inner ring road of Queen Drive in Liverpool in 1910 and the 1934 road tunnel link under the Mersey; to the new Tyne road bridge at Newcastle and the completion of the North Circular Road in London in the 1930s. Increased attention to road congestion and road traffic accidents led to the first comprehensive road safety measures (for example traffic lights, pedestrian crossings and the introduction of 30 mph speed limits in towns) and to the introduction of driving tests. Although most people could still not afford a car, increasing numbers were beginning to experience the negative effects of traffic in communities along busy arterial roads.

By the 1890s the dimensions and nature of the problems generated by Victorian urban growth were beginning to be more clearly recognized as a range of social enquiries and other evidence focused on the unacceptable conditions of housing and health experienced by the urban poor. The poor physical condition of army recruits at the time of the Boer War drew attention to health problems in towns: for example approximately 50 per cent of volunteers in Manchester in 1902 were rejected for military service on health grounds. The detailed surveys of Charles Booth in London in the 1890s also clearly indicated that urban housing conditions for the poor had not improved since the mid-nineteenth century. Booth estimated that 30.7 per cent of London's population lived in poverty and similar surveys elsewhere showed that this was not just a problem of the capital: Rowntree's survey of York (1901) showed 27.8 per cent in poverty; Glasgow, said to be the most densely populated city in Europe, had some 700,000 people living on three square miles. In 1914, 47.6 per cent of the population in all Scottish burghs lived at a density of over two per room, while in Glasgow 55.7 per cent lived at this density and over 60 per cent did so in 14 other burghs. Even in Edinburgh 32.6 per cent of the population were overcrowded, 29 per cent of houses had no separate WCs and 43 per cent of one-room flats had no sinks.

Urban living conditions were beginning to be seen as a national disgrace: C. F. Masterman summed up contemporary attitudes to big cities (in 1909) when he referred to London, Glasgow, Liverpool and Manchester as containing 'a monster clot of humanity'. The urban poor were clearly distasteful and a combination of self-preservation, humanitarian motives and rising working-class expectations led to a growing feeling that something should be done. Although an increasing proportion of the more affluent could move to the suburbs and segregate themselves from the urban poor, they found it more difficult than their Victorian counterparts to ignore conditions of life in the slums. In Glasgow, especially, working-class housing had also become a political issue – closely interlinked with Socialist politics – and the protests on Clydeside were at least partially responsible for the institution of rent control in 1915.

Growing provision of council housing in the twentieth century has had a major impact on both urban form and living conditions. For the first time the Government intervened in the free market to provide housing and, in so doing, fundamentally affected the spatial expansion and planning of towns. A few councils had begun to build houses before 1890, and the Housing Act of that year gave further impetus to such schemes. Some 24,000 council units were built in Britain before 1914, but most were concentrated in London (9,746 units), Liverpool (2,895 units) and Glasgow (2,199 units). Built with good intentions, these schemes were too few in number to make any real impact on housing needs and, in any case, rent levels and selection procedures tended to exclude the very poor. In Glasgow and Liverpool pre-1919 council housing consisted almost entirely of inner city tenement dwellings and did not compete with suburban private building. But in London the LCC used special powers under Part III of the 1890 Housing Act to acquire and build on land outside its boundaries. Low-density cottage estates were thus built in suburban areas such as Tooting and Hammersmith alongside private housing developments. Although similar schemes had been discussed in Liverpool, it was generally felt that the corporation should not intervene in a process which was being carried out effectively by private enterprise.

The Housing Act of 1919 in England and Wales (and a similar Act in Scotland) required local authorities to survey housing need and, where appropriate, provide council housing. For the first time a central government subsidy was provided and local councils were required to conform to centrally determined building standards and styles. Subsidies were mostly only available for suburban cottage estates and the Act was seen as a temporary measure to provide good quality (general-needs) housing for the relatively affluent working class at a time of acute housing need and rising housing demand. It was assumed that when private house builders had recovered from the effects of war they would be able to resume their role as principal providers of suburban housing and the Act included a clause which provided subsidies for private housebuilders. Government intervention in the housing market under the Addison Act of 1919 was the culmination of a number of processes which had developed over previous decades and growing social concern about urban housing conditions. Those concerned with political self-preservation saw it as a practical and expedient measure given postwar social and political circumstances. Whatever the various reasons, the Act initiated legislation which has had a fundamental impact on the social geography of present-day towns (Table 16.3, Figure 16.7).

Although some 214,000 houses were built under the Addison Act, it was not wholly successful in either political or social terms. Building costs were high and the national subsidy was thought to be too expensive. In 1921 the scheme was halted (although houses continued to be built until 1923) and most of the 'general-needs' houses built by local authorities were let at rents which were too high even for many skilled working-class families, let alone for those in most need. The 1923 Conservative Government introduced a Housing Act which reinstated central government subsidies for council housing, but placed more emphasis on slum clearance and the provision of low-cost housing for the poor. In 1924, however, the Wheatley Act saw a Labour Government return partially to the principles of 1919. Subsidies were lower, but local authorities were encouraged to build low-density suburban housing to meet general housing needs. This legislation resulted in the construction of over 600,000 houses in Britain by 1934 and local authorities were responsible for some 31 per cent of house building in England and Wales and some 68 per cent in Scotland between the wars. This housing was never intended for those in most housing need – rents were too high and the peripheral locations of estates were inconvenient for those who worked in central industrial areas – but it was assumed that as the more affluent working class moved to the suburbs the poor could filter into vacated housing. However, although the Wheatley Act provided good quality homes for many it never resolved the problem of housing the very poor: by the late 1920s many councils were having difficulty letting expensive suburban houses since it was as cheap to move into new owner-occupied accommodation. Moreover, high suburban rents were an acute

Table 16.3 Principal housing legislation in Great Britain*, 1890–1949

Year	Act	Main intended effects
1890	Housing of the Working Classes Act	Extended previous legislation allowing LAs to erect 'lodging houses' which included labourers' dwellings.
1909	Housing and Town Planning Act	Removed previous requirement on LAs to sell dwellings within 10 years.
1915	Rent and Mortgage Restriction Act	Introduced rent control in the private sector.
1919	Housing Act (Addison Act)	LAs required to survey housing need and provide housing. Losses beyond one penny rate to be borne by the Exchequer. Lump sum grant payable to builders of new private sector houses up to 1400 sq.ft.
1921	Housing Act	Grants to private builders extended to completions up to June 1922.
1923	Housing Act (Chamberlain Act)	Reduced subsidies available to Local Authorities who could build only if they could convince the Minister this was preferable to private enterprise. Subsidies given to builders of private dwellings for sale or rent up to £6 per year per house for 20 years or a lump sum of £75 for completions up to October 1825. Local Authorities could also make loans to private builders for house construction as well as to purchasers. Local Authorities could also give guarantees to building societies.
1924	Housing Act (Wheatley Act)	Increased subsidies for Local Authority building. LA no longer had to prove that their building did not compete with private enterprise. Subsidies to private builders extended to completions up to October 1939.
1930	Housing Act (Greenwood Act)	LAs obliged to produce a plan for dealing with slum clearance and LAs obliged to rehouse those displaced by slum clearance schemes. Subsidies available for slum clearance. Permitted local authorities to make loans for repair of private housing.
1933	Housing Act	Wheatley subsidies repealed.
1935	Housing Act	LAs required to survey overcrowding. Additional subsidies available for schemes in which it was necessary to build flats. Reduced value of loans by LAs for building and purchase of private housing.
1936	Housing Act	LAs empowered to sell council houses.
1938	Housing Act	Uniform subsidies established for dealing with slum clearance and overcrowding.
1939	Building Societies Act	Defined acceptable forms of collateral for building societies.
1944	Housing Act	£150m made available for manufacture and erection of temporary housing by LAs.
1945	Housing Act	Increased value of LA loans for house purchase or construction by private builders.
1946	Housing Act	Standard annual subsidy of £16 10s per house provided over 60 years with extra subsidies for flats and other high-cost construction by LAs.
1947	Town and Country Planning Act	Provided for development plans and development control. All development was subject to planning permission.

Table 16.3 *Continued*

Year	Act	Main intended effects
1949	Housing Act	Empowered LAs to provide housing for any member of the community not just the working classes. LA grants for private housing improvement. LA could guarantee excess advances made by building societies to private borrowers.

*Most legislation applied to Scotland as well as to England and Wales, though sometimes under a separate Act. See R. Rodger (ed) *Scottish housing in the twentieth century* (1989)

Principal sources: E. Gauldie, *Cruel Habitations* (1974); Merrett, S. *State Housing in Britain* (1979) pp.310–14; S. Merrett, *Owner Occupation in Britain* (1982) pp.338–40

problem during the depression when many councils experienced serious problems with rent arrears.

In 1930 a further change in housing policy under the Greenwood Act saw a retreat from the principle of councils providing general-needs accommodation and a switch to slum clearance and rebuilding for those dispossessed. Some subsidized general-needs building continued until 1934, and councils could build unsubsidized housing thereafter, but the assumption was that general needs could be met by the private sector and that public funds should be directed to an attack on the slums. During the 1930s most large cities developed plans for large-scale slum clearance and these were given further impetus by an overcrowding survey of 1936. Such new housing was provided both in the suburbs and the city centre. Thus suburban council housing in the 1930s was mostly built at lower cost and to lower standards than in the 1920s and the Greenwood estates rapidly developed a reputation for their poor quality and social and environmental problems.

In England and Wales these were mostly (but not exclusively) cottage estates; in Scotland tenement blocks were built in the suburbs. City centre rehousing was mostly in flats with authorities such as London, Leeds, Liverpool and Manchester building substantial blocks: for instance, the Quarry Hill scheme in Leeds consisted of 930 flats in eight-storey blocks. Rents for inner city flats were mostly low and they were much more accessible to working-class areas than remote suburban estates. Originally heralded as the answer to Britain's housing problems, in the post-war years they were rapidly caught up in the downward spiral of inner city decay and became one of the main housing problems of the late twentieth century. The Second World War caused a halt to most building schemes and air raid damage had a devastating effect on housing in many towns. But rebuilding and slum clearance were restarted by the late 1940s together with an energetic programme of general-needs housing provision by both public and private sectors. However, in 1951, 45 per cent of households still lived in privately-rented housing, much of it old, poor quality, and located in decaying inner urban areas. Not until the 1960s did the felling of vast areas of their terrace housing transform these areas, not necessarily for the better.

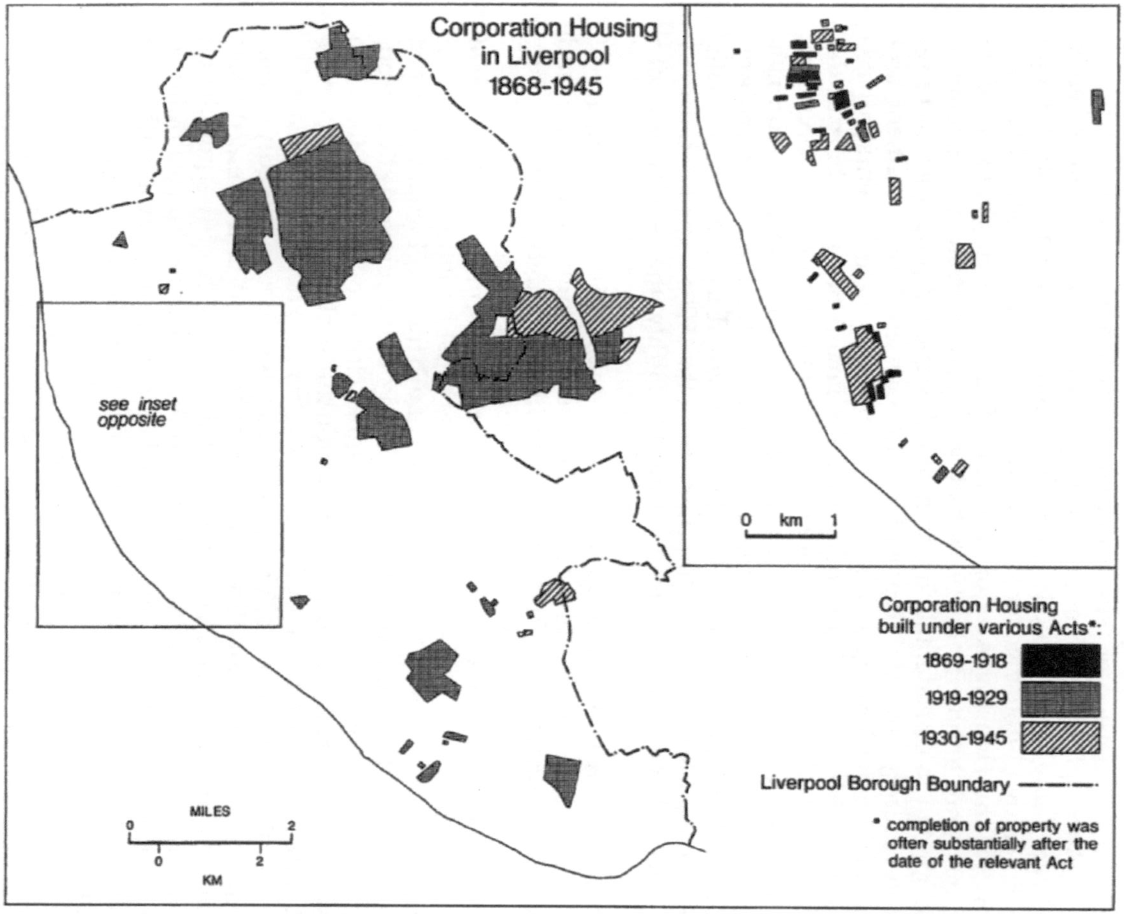

Figure 16.7 The growth of corporation housing in Liverpool, 1868–1945.
Source: Pooley and Irish (1984) p. 7.

Health and welfare

Improvements in the health of the urban population can be explained by a number of changes that occurred after 1890 (Chapter 13). First the Public Health Act of 1890, although introducing few new principles, was more effective than previous legislation in ensuring that towns took responsibility for the basic provision of pure water supply and proper sanitary arrangements. Although many slum properties never had proper facilities before they were demolished in the 1930s, an increasing proportion of the population did enjoy in-house water supply and the efficient removal of sewage. Secondly, the Housing Acts of 1890 and 1930 placed emphasis on slum clearance and, although the 1920s had seen a concentration on suburban development, a considerable number of sub-standard properties were removed. Thirdly, the development of the Garden City movement and associated town planning legislation began to stress environmental considerations which undoubtedly influenced the layout of new suburban developments and created a healthier environment on new estates. Fourthly, advances in medical knowledge and technology began to make real inroads into diseases that had been barely understood in the nineteenth century. Thus the development of immunization and new drug treatments of children against diseases such as diphtheria from the 1930s helped to cause the rapid decline in infant mortality (Chapter 13). Fifthly, the development of a state welfare policy towards health, although imperfect and selective, undoubtedly created a buffer which prevented some of the worst impacts of disease on family life. Sixthly, general increases in standards of living, and especially improvements in diet and nutrition throughout most of the population, led to greater resistance to disease and lower mortality.

The relative importance of these factors for improvements in the health of the population varied depending upon residential location and income. If you could afford to move to a new garden suburb the effect would clearly be beneficial – but then you were probably already well-nourished and reasonably healthy. If you were trapped in an inner city slum, perhaps forced to move to even more crowded accommodation due to slum clearance schemes that were not followed by rebuilding, and suffering a decline in income due to unemployment in the 1930s, your health would undoubtedly suffer. Infant and maternal mortality in particular increased in depressed areas during the recession and although overall the health of the population improved greatly between 1890 and 1940 the effects of changes outlined above were not spread evenly between classes and areas. Even within the same city there remained large variations in chances of death from common diseases (Figure 16.8).

The impact of the embryonic welfare state was also patchy. In 1911 Lloyd George introduced the first national medical insurance scheme which was intended to partially replace schemes previously run by individual friendly societies. In particular it was designed to include those who could not normally afford to subscribe to private health insurance. Weekly contributions of 4d from the employee, 3d from the employer and 2d from the state provided insured workers with sick pay of 10s per week from the approved societies administering the scheme, and also gave free medical treatment

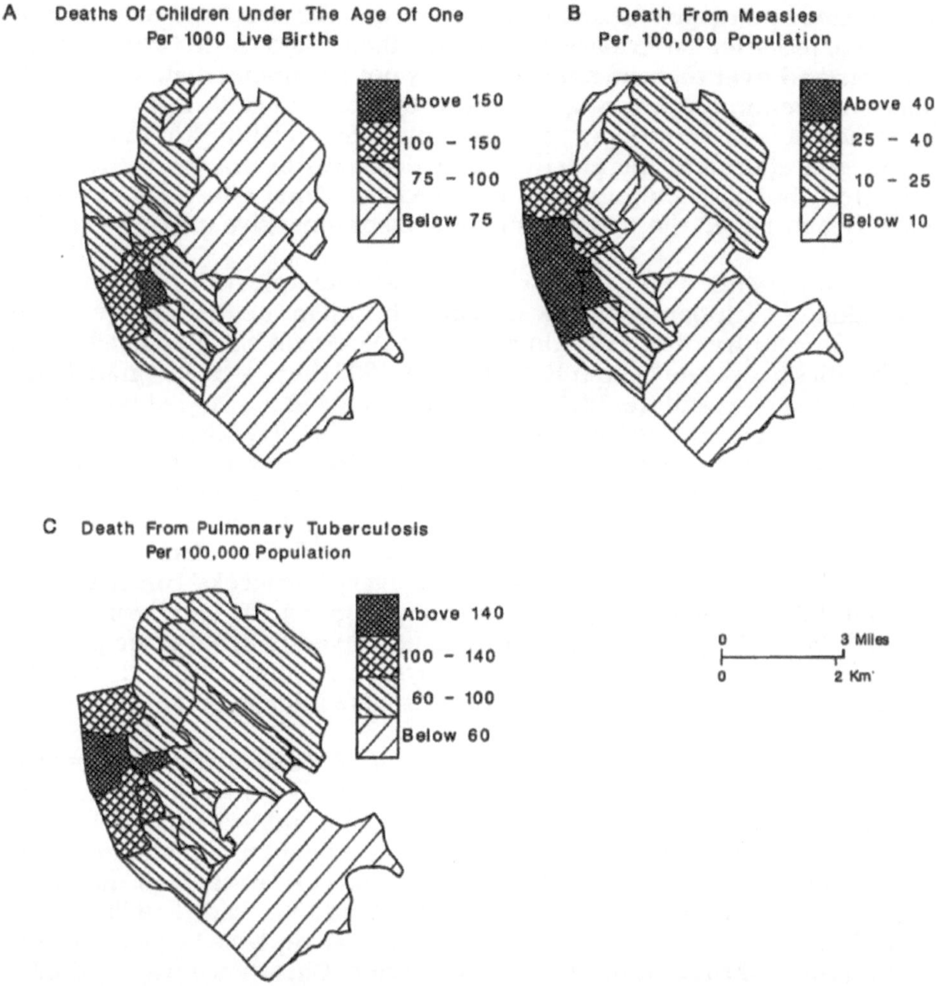

Figure 16.8 Mortality variations in Liverpool, 1935. (A) Infant mortality; (B) mortality from measles; (C) mortality from pulmonary tuberculosis.
Source: Annual Report of the Medical Officer of Health for Liverpool, 1935.

from a panel of General Practitioners. This scheme applied to all manual and non-manual workers earning under £160 per annum (raised to £250 in 1919), covering some 13 million workers by 1914 and almost 20 million in 1939. The scheme's main deficiencies were that it covered neither hospital treatment nor dependants for whom health insurance still had to be provided privately. Many were too poor to afford such cover. Hospital treatment also was paid for through private insurance schemes although the very poor could receive free treatment. With supplementary unemployment insurance benefit schemes the 1911 system of medical insurance remained largely unchanged during the interwar years and, despite its deficiencies, paved the way for the National Health Service in 1946.

In the twentieth century the extent and effect of poverty began to be recognized by a range of influential opinion, and the structural and regional effects of recession particularly highlighted the limited control which many workers had over their own incomes. It is not surprising that, along with state intervention in housing and the provision of health insurance, the government also developed welfare measures relating to unemployment, education and the elderly. As with health insurance, these were imperfect, but they cumulatively removed the worst effects of Victorian poverty and paved the way for the more comprehensive measures of the 1940s.

State intervention to relieve poverty and unemployment beyond the operation of Victorian Poor Law measures began in the first decade of the twentieth century. The 1905 Unemployed Workman's Act enabled Local Authorities to operate Labour Exchanges: in 1909 these were integrated into a national system and the 1911 National Insurance Act included provisions for unemployment as well as health insurance. The 1911 scheme was aimed principally at workers employed in industries severely affected by fluctuations in trade. It was compulsory in such industries as construction, shipbuilding, mechanical engineering, iron founding and vehicle construction – all of which tended to lay off workers when trade was slack – and was designed to provide one week's benefit for every five weeks' contributions. Thus it did not attempt to provide a living wage for the long-term unemployed, but to help those temporarily unemployed to avoid acute poverty and dependence on the Poor Law. Employers and employees each contributed 2d and the state 1d per week: benefits were initially set at 7s per week for a maximum of 15 weeks.

Although the 1920 Unemployment Insurance Act extended this scheme to most manual workers, raised the benefit levels to 15s per week and, in 1921, introduced dependants' allowances of 5s for a wife and 1s for a child, the real problem faced in the 1920s was due to the rise in long-term unemployment. Many workers had paid insufficient contributions to claim benefit and many were unemployed for longer than benefits could be paid. Families were forced to rely on Poor Law relief, placing an unacceptable burden on local rates. From 1921 the principle of unemployment insurance was gradually eroded by economic pressures and progressively replaced by the idea of indefinite maintenance. Initially unemployed workers were allowed extended benefit, beyond that to which they were entitled, on the assumption that they would eventually make good the insurance contributions. But the 1927 Unemployment Insurance Act formally introduced 'Transitional Benefits' whilst strengthening the 1921 principle that claimants had to prove that they were genuinely seeking work.

The economic crisis of the 1930s introduced much more fundamental changes. From 1931 the rising cost of benefits paid to those unemployed who had paid few insurance contributions and who seemed to have little prospect of gaining work, led to a reduction in weekly benefit (from 17s in 1928 to 15s 3d in 1931), the transference of responsibility for transitional payments to Public Assistance Committees of local authorities and the introduction of stringent means tests before the 'dole' was paid. The overall effect was to remove some 180,000 people from the receipt of unemployment benefit and to produce a substantial cut in benefit for approximately half of all claimants, although the precise impact varied from area to area because the means test was administered locally. The hatred of the punitive means test is demonstrated both in the enquiries of the Pilgrim Trust (1938) and in popular fiction. Its effects were felt most keenly in the depressed areas where the majority of the working population was often on the dole. Since it was based on family income it caused increased tension within households and even split up families: a few shillings earned by a 15-year old daughter could lead to the loss of benefit by the father and large families were particularly hard hit. The Pilgrim Trust estimated that unemployment led to an average drop in family income of 45–66 per cent. However, although the level of means-tested benefits was low and the inquisitions which accompanied them were detested, the availability of some income must have reduced absolute levels of poverty compared to the nineteenth century. Indeed, in low-wage areas such as South Wales many men may have received more in unemployment benefit than they had in work.

Administration of unemployment benefits was changed again in 1934 when the national Unemployment Assistance Board (UAB) took over transitional payments from the local Public Assistance Committees, compulsory insurance was extended to agricultural workers and benefits were increased to the levels of 1928. One problem in changing to a national system of unemployment benefit was that rates established in 1934 were in some cases lower than those operated in some areas by local Public Assistance Committees. Only in 1937 was a manageable scheme of national, means-tested unemployment benefit (as opposed to insurance) established: by then declining unemployment had begun to relieve pressure on the system. During the war the stringent means test was gradually eased and increased supplementary allowances given to dependants. In 1948 the UAB became the National Assistance Board with a much wider remit than just unemployment relief.

Significant advances in other areas of social welfare in the twentieth century saw a means-tested non-contributory pension scheme of 5s per week maximum introduced in 1908 for those over 70 years, although receipt of a pension excluded the recipient from poor relief. In 1919 pensions were raised and the means test relaxed, and in 1925 a State Pension Insurance Scheme (modelled on the Health Insurance plan) was introduced. It was not means-tested, and provided benefits for widows and orphans as well as the elderly, although like previous insurance schemes it excluded those who did not pay the premiums.

Perhaps in the long run the reforms which had the most impact on living standards were those in education. The 1902 Education Act reorganised the compulsory National Board School system of primary education established under the 1870 Act and extended it to a system of secondary education. The responsibility for this was placed on the shoulders of local government committees in Counties and Boroughs. It also provided uniform as well as free education up to the age of 12. Consolidated in the 1918 Education Act, which raised the school leaving age to 14, it enshrined the principle that children should not be barred from education through poverty. During the inter-war years the majority of children had access to a good basic education, and increasing numbers entered higher education, thus producing a labour force which was capable of seizing opportunities offered by expanding sectors of the economy.

The period from the 1890s to the 1940s produced many paradoxes and contrasts. There was no such thing as a typical British family or region: whilst some prospered others experienced depression and poverty. Even those families which apparently gained from changes after 1890 experienced some disadvantages. Jackson's (1973) vivid picture of life in a new private suburban development around London describes gains in housing standards, a healthy environment and such material possessions as new furniture, a radio and regular holidays. However, there were also drawbacks: women and children might be isolated from friends when they moved to the suburbs; wives could experience boredom and frequent worries about money and their ability to maintain their new suburban status. Men travelling daily by rail to London spent long hours away from home with consequent fatigue. Such stresses could place a strain on family life and individual health. Nevertheless, the benefits of suburban living outweighed the problems and suburban dwellers – whether in London, Manchester or Glasgow – principally shared in the initial prosperity of rising real wages brought by the 1920s and '30s. Such prosperity is perhaps best epitomized through the growth of leisure activities and the spread of the media. People travelled further for holidays, could afford a wider range of the abundance of new magazines and newspapers, and began to regularly attend the theatre or cinema and listen to the radio. One effect of these changes was to gradually produce a more uniform national culture despite the continuation of a regional press and, to a limited extent, the provision of regional broadcasting. Although initially the poor may have been excluded from many of these new opportunities and country areas took up the cinema, for example, later than the large towns, by the 1920s most places and most people had some access to national news media and a nationally-targeted leisure industry.

Elizabeth Roberts' survey of working-class women in North West England shows that by the 1920s a visit to the cinema was a regular event for working-class families, even those with little surplus income for leisure activities. Adults could get a cinema ticket for as little as 3d (children 1d) and the cinema has been associated with the parallel decline in working-class drinking and a mainly male pub society. It gradually became more acceptable for respectable women to accompany their husbands to public houses. Those who missed out on the benefits of the twentieth century, as Robert Roberts' description of Salford life at the turn of the century vividly reveals, were the 'no class' at the bottom of the working-class hierarchy. Numerous descriptions and enquiries emphasize the very real hardship endured by those on means-tested benefits in the 1930s, for whom local community- and neighbourhood-support and help in times of hardship was vital. Although in newer suburban communities unemployment could lead to relative isolation, in most working-class neighbourhoods people continued to help one another and rely on local resourcefulness to survive. The neighbourhood community and local support systems of the Victorian city continued to survive well into the twentieth century in many areas and were only slowly replaced by the privatized nuclear family encouraged by some aspects of suburban living.

Bibliography and further reading for Chapter 16

On planning and the changing structure of towns see:

ABERCROMBIE, L. P. *Greater London Plan 1944* (London: Standing Conference on London Regional Planning, 1945).

ADAMS, I. *The making of urban Scotland* (London: Croom Helm, 1978).

CHECKLAND, S. G. *The Upas Tree: Glasgow 1875–1975: a study in growth and contraction* (Glasgow: Glasgow University Press, 1976).

CHERRY, C. *Cities and plans. The shaping of urban Britain in the nineteenth and twentieth centuries* (London: Arnold, 1988).

FREEMAN, T. W. *The conurbations of Great Britain* (Manchester: Manchester University Press, 1959).

GEDDES, P. *Cities in Evolution* (London: Williams and Norgate, 1915).

GORDON, G. (ed.) *British regional cities in the UK 1890–1980* (London: Harper and Row, 1986).

JACKSON, A. *Semi-detached London: suburban development, life and transport 1900–39* (London: Allen and Unwin, 1973).

THOMAS, D. *London's green belt* (London: Faber, 1970).

YOUNG, K. and GARSIDE, P. *Metropolitan London. Politics and urban change 1837–1981* (London: Arnold 1982).

On housing and living conditions see:

BURNETT, J. *A Social history of housing 1818–1970* (Newton Abbot: David and Charles, 1978).

DAUNTON, M. (ed.) *Councillors and tenants. Local authority housing in English cities* (Leicester: Leicester University Press, 1984).

DAUNTON, M. *A property-owning democracy: housing in Britain* (London: Faber, 1987).

GORDON, G. and DICKS, B. *Scottish urban history* (Aberdeen: Aberdeen University Press, 1983) Chapters 7, 8 and 10.

MELLING, J. (ed.) *Housing, social policy and the state* (London: Croom Helm, 1980).

POOLEY, C. G. 'Housing for the poorest poor: slum clearance and rehousing in Liverpool 1890–1918' *Journal of Historical Geography* **11** (1985), pp. 70–88.

POOLEY, C. G. and IRISH, S. *The development of corporation housing in Liverpool 1869–1945* (Lancaster: Lancaster University Centre for North West Regional Studies, 1984).

POOLEY, C. G. and IRISH, S. 'Access to housing on Merseyside, 1919–39.' *Transactions of the Institute of British Geographers NS* **12** (1987), pp. 177–90.

ROBERTS, R. *The classic slum* (Manchester: Manchester University Press, 1971).

RODGER, R. (ed.) *Scottish housing in the twentieth century* (Leicester: Leicester University Press, 1989).

SWENARTON, M. *Homes fit for heroes* (London: Heinemann, 1981).

SWENARTON, M. and TAYLOR, S. 'The scale and nature of the growth of owner occupation in Britain between the wars.' *Economic History Review, 2nd Series* **38** (1985), pp. 373–92.

WHITE, J. *Rothschilds Buildings: Life in an East End tenement block 1887–1920* (London: Routledge, 1980).

WHITE, J. *The worst street in London: Campbell Bunk, Islington, between the wars* (London: Routledge, 1986).

YELLING, J. 'The metropolitan slum: London 1918–51' in Gaskell, S. M. (ed.) *Slums* (Leicester: Leicester University Press, 1990), pp. 186–223.

On health, welfare and the development of state intervention see:

CONSTANTINE, S. *Unemployment in Britain between the wars* (London: Longman, 1980).
FRASER, D. *The evolution of the welfare state* (London, Macmillan, 1973).
M'GONIGLE, G. C. M. and KIRBY, J. *Poverty and public health* (London: Gollancz, 1936).
PILGRIM TRUST, *Men without work* (Cambridge: Cambridge University Press, 1938).
ROBERTS, E. *A woman's place: an oral history of working-class women 1890–1940* (Oxford: Blackwell, 1984).
ROWNTREE, B. S. *Poverty and progress* (London: Longman, 1941).
WEBSTER, C. 'Healthy or hungry 'thirties' *History Workshop Journal* **13** (1982), pp. 110–29.
WHITESIDE, N. 'Counting the cost: sickness and disability among working people in an era of industrial recession 1920–39' *Economic History Review, 2nd Series* **40** (1987), pp. 228–46.
WINTER, J. 'Infant mortality, maternal mortality and Public Health in Britain in the 1930s' *Journal of European Social History* **8** (1979), pp. 439–62.

17

Conclusion: Continuity and change in the regional geography of Britain

Previous chapters have outlined the main dimensions of regional variation in demographic, agricultural and industrial change, and urbanization. The emphasis has been primarily on places rather than people and on economic rather than social structures. Less attention has been paid to those whose activities and cultures helped create regional diversity and distinctiveness; and to the ways in which regional and local change was perceived and experienced by people living in different parts of Britain. Yet these too are vital aspects of a changing regional geography.

Regional cultures and regional definition

From the 1740s to the 1940s the regional geography of Britain had been moulded and restructured in a variety of ways and had been perceived differently by various individuals. We argue that although the precise configuration and characteristics of regions changed over these 200 years, a distinctive regionalism persisted. Already, in the eighteenth century the nation was, in one sense, a single economic and political region with London as its undisputed head. Although the manifestation of national consciousness changed and, especially in the second half of the nineteenth century, a close and well-fostered identification of all British citizens with nation and Empire was an important force in national unity, to some extent counteracting regional differences, the view from Scotland, Wales or North East England was quite different from that in rural South East England. Likewise, the outlook of businessmen and workers in Liverpool, Birmingham or Glasgow was different from that in London.

Within the national framework strong regional distinctiveness was based on persistent cultural and economic differences: Scotland and Wales had strong cultural and linguistic identities while the English regions had cultural characteristics which transcended successive economic and social changes. To be born and bred in, for instance, Cornwall, Yorkshire or Kent was important for people of these counties; even increasing internal migration did not destroy such loyalties since many people retained contacts with

 DOI: 10.4324/9781003174783-17

and interest in their home areas. Despite an increasingly national perspective to the activities within and the perceptions of regions of Britain through the spread of mass communications and media from the late nineteenth century, regional and local newspapers and, latterly, radio and TV, have provided a counterbalance: a focus for regional issues and local responses to national and international situations.

Physical characteristics are basic to regional distinctiveness: the important highland-lowland division persisted in Scotland where, in both agricultural and industrial activities, contrasts in landscape and lifestyles demarcated two nations almost as much as their different economies and cultures. Here, as elsewhere, economic, political and administrative forces were the basis of further regional contrasts mediated through particular activities, skills and cultural attitudes. Such functional regions, defined mainly in economic or administrative terms, were not static: boundaries overlapped and changed according to their functions. Yet the nation remained a nested hierarchy of regions, ranging from the national to the local level, both for people who experienced and shaped their regional geographies as well as for those who delivered services or otherwise utilized a particular regional administrative framework.

For many people in the past the most important region was that of their own locality: in a spatially and culturally circumscribed life they identified with the neighbourhood, locality or village in which they were born, worked, raised their families, had their friends and lived out their lives. These home areas, which existed as much in the minds of ordinary folk as in measurable physical, economic or social criteria, were perceived differently by people of different age, gender, class, and race. Moreover, differing attitudes to environment and locality – reflecting such characteristics – could fundamentally affect regional perceptions.

Irrespective of where they lived an active adult travelled more widely around a town or through the countryside than a child or an elderly person whose sense of place was constrained and who identified mainly with the home and street rather than a larger region. Despite increasing female emancipation in the twentieth century, most women lived more circumscribed lives than men. Even when they worked outside the home, extra burdens of childcare and household duties meant that their time was more home-centred: the region or locality with which they identified was often smaller than that of their male counterparts. Lack of income constrained mobility for most people, regardless of age or sex. The cognitive regional geographies of the rich, who could travel widely, were different from those of the poor for whom immediate surroundings assumed much greater importance. Particular minority groups – the Irish, Jews or Blacks – discriminated against in the labour market and in housing were also constrained in their action spaces. They too had different and distinctive cognitive geographies of the regions in which they lived.

Such differences emerge from contemporary novels, accounts of local life, folk songs and poetry – for instance the rich inheritance of song and verse which comes from Lancashire cotton towns – but it is difficult (and probably fruitless) to try to place firm boundaries around such regions. It is easier to identify regions based on economic criteria, administrative boundaries or specific cultural characteristics (such as language) than to define and interpret regional values constructed through geographies of the mind. At all times, different regional geographies coexist: some are easily measured; others are speculative; all changed their boundaries and modified their character. Yet the basic regional framework of Britain changed remarkably gradually, as the following examples show, at different spatial scales and from different standpoints.

Economic regions

In late twentieth-century Britain the north-south divide between 'two nations' is marked by a wide range of indices of unemployment, job creation, income levels, share ownership, dependence on supplementary benefits, health, housing, and political allegiance. Southern England performs consistently better on all indicators than the north, Wales and Scotland. However, most commentators acknowledge that this division is spatially over-simplified: at a finer scale most inner urban areas throughout Britain perform relatively poorly whilst such smaller northern towns as Chester, York or Kendal have levels of prosperity fully equal to those in the south. Divisions are not only geographical but also social, reflecting the varying extent to which people of different gender, class and race benefit from opportunities in their localities. Some groups within society are consistently disadvantaged in all regions. Such structural forces intersect spatial variations in prosperity, although disadvantaged groups are found disproportionately in the worst areas in economic and social terms.

This is not a recent phenomenon. The forces that have produced structural and regional imbalances in contemporary Britain were apparent by the late nineteenth century: some have their origins in economic and technological change during the industrial revolution. As the economy faltered in the late nineteenth century (Chapter 12), and from the 1920s when the full blast of Britain's declining role in the world economy had begun to take effect, southern and midland England continued to grow economically and demographically at the expense of other regions (Chapter 15). As migration shifted the younger and more skilled workforce to areas of economic growth (Chapter 13), so regions of economic decline, particularly the old industrial districts of northern and western Britain, were increasingly marginalized.

During the nineteenth century structural imbalances not only produced variations in regional prosperity, but equally marginalized certain sectors of the population. Contemporary commentators were well aware of the disparities between rich and poor as epitomized in Disraeli's 'two nations' of 1845 – one rich and one poor, one privileged and one underprivileged – and Mrs Gaskell's *North and South* of 1855 – the one industrial the other rural.

Spatial expressions of disadvantage in Victorian Britain were at least as complex as those of the twentieth century. Mid eighteenth-century agricultural wages were highest in southern England, but by the early nineteenth century commercial and industrial growth in northern and midland England, South Wales and central Scotland led to generally higher wage rates than those in southern England. However, while northern industrial towns were important centres of investment and wealth creation, London remained the heart of the nation's commercial and financial activity – and of much of its industry – and it was in London that most of the really large fortunes were created in the eighteenth and nineteenth centuries: much of the vast wealth of the Grosvenors (Dukes of Westminster) came from their West End estates. Some of the greatest inequalities in Scotland separated an increasingly marginalized and dependent highland economy from the wealth generated by businesses and industries of Edinburgh, Glasgow and Central Scotland and from aristocratic landowners: in Sutherlandshire the Duke of Sutherland owned 91 per cent of the county in the 1870s and Glasgow rivalled most provincial English industrial towns for non-landed wealth by the end of the century, while crofters in the increasingly depopulated Highlands and Islands struggled to retain both livelihood and culture. Economic inequalities also reflected cyclical fluctuations in the industrial economy. Acute poverty among the urban population of northern England, London and, especially, Scotland was manifest in poor health and dire housing for the labouring poor (Chapter 11). These were for many (including Engels and Marx) the social and geographical expressions of the capitalist system. Indeed, those regions with the greatest poverty were those in which the largest numbers were exploited and marginalized by the capitalist economic system, whether among the urban or the rural proletariat.

Linguistic regions

Regions are particularly difficult to define in terms of cultural characteristics: the extent to which individuals subscribe to a particular culture and the ways in which that culture is perceived are subjective, not easily measured statistically, and vary from individual to individual and from time to time. Where cultural identity is associated with the protection and promotion of a minority language, distinctive linguistic and cultural regions may be identified. The most distinctive minority languages of eighteenth-century Britain were Welsh and Scottish Gaelic, though Manx was still quite widely spoken in the Isle of Man. The minority languages of Cornish and Norn (spoken in Shetland, Orkney and Caithness) had all but died out, but the language of southern, central and eastern Scotland was more distinct from English than other dialect variations within Britain. Dialects also had distinct cultural associations. In such regions as South Wales, the Black Country, Lancashire, Yorkshire, North East England and London, they were reflected in contemporary social and political comment as shown in the work of the many dialect poets of the industrial regions. Even within individual regions nuances in intonation and usage often distinguished between local cultures and related to the origins, activities and social background of their speakers: the true Geordie from Tyneside speaks quite differently from the Durham miner; Liverpool's clipped vowels and distinctive slang reflects a polyglot port population, especially Irish migrants, and is a world away from the open and measured speech of east Lancashire.

Of the major linguistic regions, Gaelic, widely spoken throughout the Highlands in the eighteenth and nineteenth centuries, suffered substan-

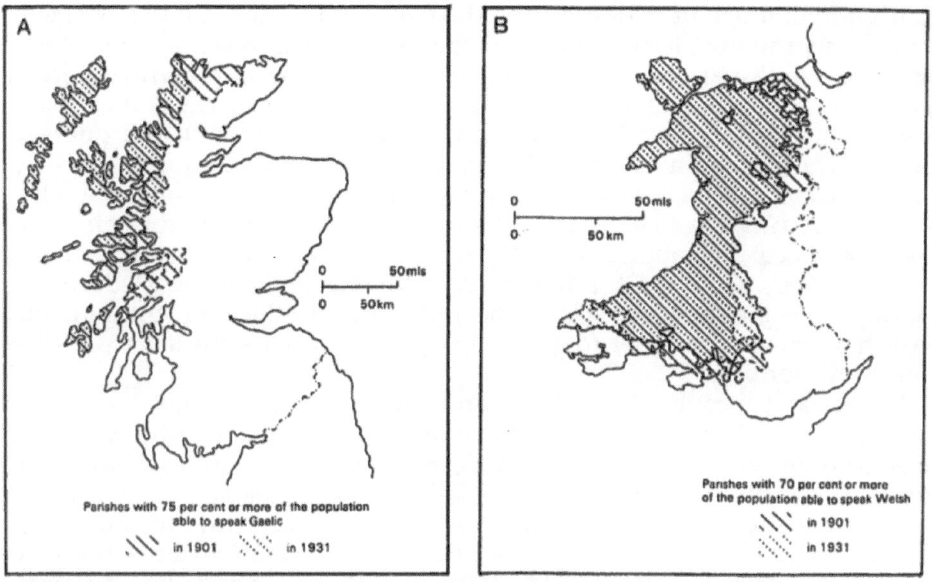

Figure 17.1 Major linguistic regions in Scotland and Wales. (A) Gaelic speaking parishes in 1901 and 1931; (B) Welsh speaking parishes in 1901 and 1931. Based on Langton and Morris (1986) p. 203.

tially in the economic and social upheavals that affected the area. In the Gaidhealtachd the language was persecuted for political reasons, and heavy migration following the 1745 Jacobite rebellion and again in the nineteenth-century clearances of population from the larger estates led to the disintegration of both the Highland economy and the role of Gaelic, as Gaelic speakers moved to lowland cities. Indeed, Glasgow became a centre for the support of Gaelic through publishing activities and the establishment of the Gaelic Schools Society in 1818, but, despite such activity, Gaelic-speaking diminished during the twentieth century: by 1931 Gaelic was a majority language only in the outer Isles and north western Scotland (Figure 17.1).

The decline in Welsh was less dramatic, partly because Welsh was recognized as a national language in a way that Gaelic never had been and partly because the transformation of rural Wales was less cataclysmic than that of the Scottish Highlands. Migrants from Wales also continued to keep their language alive in the chapels and cultural societies of Welsh communities in English towns. Nevertheless, as the bilingual zone along the Welsh border grew during the nineteenth century the area in which Welsh was the majority language progressively shrunk. By 1901 only five counties of 'inner Wales' had 90 per cent or more Welsh speakers, whereas the eight counties of 'outer Wales' mostly had fewer than 50 per cent Welsh speakers and Monmouth and Radnor had only 13 and 6.3 per cent respectively. The monoglot Welsh population declined particularly sharply from 30 per cent of the population in 1891 to only 6.2 per cent in 1921. Likewise, the extent of Welsh-speaking in migrant communities in English towns declined substantially by the early twentieth century.

Although English is now the dominant language of the Scottish Highlands and of much of Wales, during the twentieth century there have been significant attempts to halt the language decline and create bilingual communities. Cultural regions not only continue to be important for local populations but language activists have promoted their cause and, both legally and illegally, sought to extend it to social, economic and political affairs. The wish of a substantial part of the Welsh population to have its language recognized has to some extent been met by official policies. Welsh was reintroduced as a language of instruction in schools with the establishment of a Local Education Authority Welsh School in Llanelli in 1947, and by the late 1960s some 6,478 pupils were receiving their primary education through the medium of Welsh. The provision of education through the Welsh language continues to grow, though there are significant variations between counties.

Universal educational provision and the impact of a London-focused media have had conflicting impacts on regional cultures. On the one hand they have tended to destroy local dialects, and have sought to downgrade regional accents in favour of a so-called 'standard English'; on the other hand linguistic and cultural aspirations have also been stimulated by regional newspapers and broadcasting. Despite the minority status of Gaelic in Scotland compared to Welsh in Wales, the BBC began radio broadcasts in Gaelic in 1926 and established a Gaelic Department in 1935. However, the amount of time committed to Gaelic language programmes on radio, and latterly on television, has been small compared to the exposure that Welsh gets. Thus in the 1940s Gaelic broadcasts were allocated only 75 minutes of radio time per week. These responses, however inadequate, demonstrate that linguistically-based cultural regions have been recognized by successive British governments, and the expansion of regionally-based television and radio has also reflected the continued importance of cultural variations between English regions.

Community, locality and region

It is often assumed that local communities, the areas of everyday interaction for most people, became progressively less important as mobility increased, the media spread information about more distant places, slum-clearance broke up inner city communities and welfare provision made the reciprocal relationships of close-knit kin and friendship contacts less essential. Yet most community studies of the 1930s, '40s and '50s underline the extent to which neighbourhood and kin ties persisted in working-class areas, and the desire of those moving to the suburbs to recreate community life in a new environment was an aspiration not always achieved. Such studies suggest the overwhelming importance of the immediate locality in the cognitive geographies of most ordinary people.

Ruth Durant's (Glass's) (1939) study of Watling, a suburban housing estate to the north west of London, illustrates some of these themes. Durant defined community as a 'territorial group of people with a common mode of living striving for common objectives' and she based her analysis on a survey of what people actually did in their everyday activities, rather than on what they said. The estate was developed from 1927 by London County Council and by the mid-1930s there were over 19,000 residents. Early residents had difficulty adjusting to a new environment and faced hostility from people who had lived in the area before the estate was developed. This fostered community spirit and class consciousness. A Residents' Association was formed in 1928, and a local newspaper was produced later the same year. Most initial community links were through women who met in the streets, outside schools or in shops. Most men were less integrated into the new community, though some became involved in political activities. Despite these initial efforts Watling never developed a true sense of community due probably to its size, high population turnover and the economic effects of the depression.

Young and Willmott's (1957) study of the working-class, inner-city area of Bethnal Green (London) showed that not only did many families have kin living with them or close by, but most social interaction also took place within the community. Gender divisions came through strongly: mothers and daughters had particularly close contact, helping one another with everyday tasks from shopping to child-minding; men were more likely to have contacts outside the area though many were based on their place of work, and others through class-based or sporting activities such as the pub or club. Similar features characterized *The People of Ship Street* (Liverpool) and the Gorbals tenements of Glasgow. The extent to which such local communities played a part in wider regional influences – for instance sport, the arts and popular culture – is difficult to assess. Yet together these forces, in their distinctive regional settings, influenced popular perceptions of regional distinctiveness.

Administrative regions

The growth of industrial society and the gradual development of local and central government intervention in areas such as health, housing, service provision and social welfare, and in due course economic management in both transport and individual industries as well as regional planning, led to the creation of new functional regional geographies. Delivery of the increasing range of urban services that, from Victorian times, governments came to require for its citizens demanded effective spatial frameworks and was reflected both in the increasing number of clerical workers and bureaucrats who administered the system and the fact that its administrative frameworks gradually came to have meaning. However, there has never been a single, coherent spatial system, nor even a single set of coherently framed principles of regional division.

The administrative geography of eighteenth-century England familiar to most people was still the parish and the shire. The parish provided the spatial and social focus of the church; the framework within which locally-raised poor relief was dispensed; the body through which roads were maintained; and via vestry rules the vehicle through which most aspects of rural life were regulated. The shire was the link to national frameworks of civil and, in times of emergency, military organization of the region. The county sessions reflected their place in the administration of justice; the offices of Sheriff and Lord Lieutenant provided links with central government and the Crown; the assemblies in shire towns were a social focus for those with power and authority.

By the 1830s those older administrative geographies were beginning to change in response to the new demands of an industrializing society. The 1834 Poor Law Act amalgamated parishes to create Poor Law Unions for the administration of the Poor Law and the provision of workhouses, though the spatial framework, focused on historic towns and market centres, reflected a pre-railway urban hierarchy and communication system that were already fast becoming archaic. Reform of urban administration, begun in 1835, progressively replaced ancient town and borough councils by municipal corporations. Yet that process was not completed – and then ineffectively and at the expense of separation of increasingly interrelated urban and rural areas – until the Local Government Acts of 1888 and 1894 (Chapter 11). Indeed, effective regional planning for an urban-focused society evolved but slowly and in an *ad hoc*, fragmentary fashion (Chapters 11 and 16). The extent to which the multiplicity of administrative boundaries created for different purposes – health, education, housing etc. – actually had meaning for the people that lived within them can be debated. Although by the twentieth century the urban region had clearly emerged as a key planning concept, the machinery by which regional planning sought to reshape the realities of a modern economy and society lagged yet, at the same time, was in conflict with a society still bound by local and regional loyalties of an earlier age.

As shown in Chapter 16, the growth of urban regions also brought conflict between countryside and town and between county and city administrations. In official circles at least there was much debate about the amount and nature of power that should be given to urban regions at the expense of more traditional regional administrations. For instance the 1926 Royal Commission Report and the 1926 and 1929 Local Government Acts in effect protected county areas against encroachment from cities and, likewise, the Local Government Boundary Commission set up in 1945 avoided a fundamental reshaping of administrative boundaries to accommodate the reality of an economy and society dominated by large conurbations. Although in 1915 Patrick Geddes had suggested a new integrated administrative framework based on city regions, and many planners and geographers had promoted the same argument in the 1930s, there was little genuine attempt to create regional administrative units based on metropolitan areas until local government reorganization in 1974.

The continuing lack of unitary administrative frameworks for all services and related activities has caused confusion in the minds of the population and has failed to attract their loyalties. Since the 1940s health care, education, emergency services, gas, electricity and water have all been delivered within different spatial frameworks. Administrative regions continue to fail to reflect the functional reality of Britain: they have also had little impact on the cognitive geographies of their inhabitants.

Britain in a wider world

Britain's increasing involvement in a world economy from the late eighteenth century led to greater awareness and knowledge of foreign parts. The rich travelled widely for cultural and scientific enlightenment in Europe, for example the 'Grand Tour', and elsewhere scientific and commercial exploration and exploitation became worldwide as Britain accumulated an Empire that was unrivalled by the late nineteenth century when British tentacles were wrapped around the world. In British settlements and communities from the Americas and Australasia to India, South East Asia and many parts of Africa a particular view of British society was recreated. Empire migration was officially encouraged, both to ensure British territoriality and to relieve economic and social pressures at home. For some people of all social classes, awareness of Britain extended well beyond the shores of the British Isles to a vast worldwide Empire. British imperialism and economic exploitation overseas both helped to increase living standards at home and, through widespread propaganda supported by the stories of return migrants, produced stereotyped views of other cultures and of Britain's role overseas which were progressively reinforced by the educational system.

Britain's cultural and political relationships with the rest of the world have been less comfortable for most of the twentieth century and the gradual disintegration of Empire has proved difficult to manage. In part this reflects Britain's wider decline as a world power: economic decline relative to other nations, combined with the loss of cultural and political dominance, has tended to make many British people more insular. Britain's initial scepticism over and reluctance to join the European Community, and continued reservations of government about full participation, seem to reflect and to be reflected by the views of the British people, relatively few of whom view themselves as Europeans. The cognitive geographies of many stop at the British coast. Moreover, continuing emphasis on and greater identification with areas of former British culture in the white Commonwealth and the USA are reflected in trade patterns, cultural links, family contacts and the special relationships of both formal and informal political linkages.

The complex regional geography of Britain reflects a variety of perspectives and must be accommodated within a number of spatial scales. Many of the themes dominant in the past continue to be significant today: others have resulted from the transformation of British economy, society and politics since the nineteenth century. Yet the historical geography of Britain was, and still is, made by the people living in the different and distinctive regions. To understand present-day regional diversity one must have a knowledge of change in both region and nation that can only emerge from studies of the past. The systematic context of that geography has been the focus of the preceding chapters. The full story of its impact on regions throughout Great Britain requires fuller and complementary regional geographies focused on the totality of people's awareness and experience of their changing environments and activities over the past three centuries.

Bibliography and further reading for Chapter 17

Many references from previous chapters are relevant to the theme of this chapter, but see especially:

CHAMPION, A. G. et al. *Changing places: Britain's demographic, economic and social complexion* (London: Arnold, 1987).
DURANT, R. *Watling: a survey of social life on a new housing estate* (London: P. S. King, 1939).
EVERETT, A. 'Country, county and town: patterns of regional evolution in England' *Transactions of the Royal Historic Society, 5th series* **29** (1979), pp. 79–108.
FRANKENBERG, R. *Communities in Britain* (Harmondsworth: Penguin, 1966).
GILBERT, E. W. 'Practical regionalism in England and Wales' *Geographical Journal* **94** (1939), pp. 29–44.
GILBERT, E. W. 'The boundaries of local government areas' *Geographical Journal* **111** (1948), pp. 172–206.
GREGORY, D. 'The production of regions in England's industrial revolution' *Journal of Historical Geography* **14** (1988), pp. 50–58.
HUDSON, R. and WILLIAMS, A. *Divided Britain* (London: Belhaven, 1989).
KERR, M. *The people of Ship Street* (London: Routledge, 1958).
LANGTON, J. 'The industrial revolution and the regional geography of England' *Transactions of the Institute of British Geographers NS***9** (1984), pp. 145–67.
LANGTON, J. and MORRIS, R. *Atlas of industrializing Britain* (London: Methuen, 1986).
MARTIN, R. 'The political economy of Britain's North-South divide.' *Transactions of the Institute of British Geographers NS***13** (1988), pp. 389–418.
POPE, R. (ed.) *Atlas of British social and economic history since circa 1700* (London: Routledge, 1989).
PRYCE, W. T. R. 'Migration and the evolution of culture areas: cultural and linguistic frontiers in North East Wales 1750–1851' *Transactions of the Institute of British Geographers* **65** (1975), pp. 79–108.
PRYCE, W. T. R. 'The British census and the Welsh language' *Cambria* **13** (1986), pp. 79–100.
ROBBINS, K. *Nineteenth-century Britain: integration and diversity* (Oxford: Oxford University Press, 1988).
TAYLOR, E. G. R. et al. 'Geographical aspects of regional planning.' *Geographical Journal* **99** (1942), pp. 61–80.
WARD, S. *The geography of inter-war Britain: the state and uneven development* (London: Routledge, 1988).
WITHERS, C. *Gaelic in Scotland 1698–1981* (Edinburgh: John Donald, 1984).
WITHERS, C. *Gaelic Scotland: The transformation of a culture region* (London: Routledge, 1988).
YOUNG, M. and WILLMOTT, P. *Family and kinship in East London* (London: Routledge, 1957).

Traditional geographies of Britain with a strong regional component include:

MACKINDER, H. J. *Britain and the British seas* (Oxford: Clarendon Press, 2nd edition, 1907).
MITCHELL, J. (ed.) *Great Britain: Geographical essays* (Cambridge: Cambridge University Press, 1960).
OGILVIE, D. G. *Great Britain: Essays in regional geography* (Cambridge: Cambridge University Press, 1928).
SMITH, W. *An economic geography of Great Britain* (London: Methuen, 1949).

STAMP, L. D. and BEAVER, S. *The British Isles: a geographical and economic survey* London: Longman (6th edition, 1971).

See also the various *British Association Handbooks* which provide detailed accounts of British regions.

Appendix

Throughout the book statistical data are given in the measures traditionally used. This appendix provides a brief conversion to present-day metric units for principal measures of money, distance, weight and area.

Money:
one penny (d) = 0.416 new pence (p)
one shilling (s) = 12d = 5 new pence (p)
one pound (£) = 20s = 100 new pence (p)

Distance:
one inch (in) = 2.54 centimetres (cm)
one foot (ft) = 12in = 0.305 metres (m)
one yard (yd) = 3ft = 0.914 metres (m)
one mile = 1760yds =1.609 kilometres (km)

Weight:
one ounce (oz) = 28.35 grams (g)
one pound (lb) = 16oz = 0.454 kilograms (kg)
one stone = 14lb = 6.35 kilograms (kg)
one hundredweight (cwt) = 112lb = 50.8 kilograms (kg)
one ton = 20cwt = 1016 kilograms (kg)

Area:
one square foot (ft^2) = 0.093 square metres (m^2)
one square yard (yd^2) = 0.836 square metres (m^2)
one acre = 4046.9 square metres (m^2) = 0.405 hectares.

Index